KANT'S MATHEMATICAL WORLD

Kant's Mathematical World aims to transform our understanding of Kant's philosophy of mathematics and his account of the mathematical character of the world. Daniel Sutherland reconstructs Kant's project of explaining both mathematical cognition and our cognition of the world in terms of our most basic cognitive capacities. He situates Kant in a long mathematical tradition with roots in Euclid's *Elements*, and thereby recovers the very different way of thinking about mathematics which existed prior to its "arithmetization" in the nineteenth century. He shows that Kant thought of mathematics as a science of magnitudes and their measurement, and all objects of experience as extensive magnitudes whose real properties have intensive magnitudes, thus tying mathematics directly to the world. His book will appeal to anyone interested in Kant's critical philosophy – his account of the world of experience, his philosophy of mathematics, and how the two inform each other.

DANIEL SUTHERLAND is Associate Professor of Philosophy at the University of Illinois at Chicago. He has published numerous articles on Kant's philosophy of mathematics and philosophy of science, including their relation to Euclid, Newton, Leibniz, Frege, and others.

KANT'S MATHEMATICAL WORLD

Mathematics, Cognition, and Experience

DANIEL SUTHERLAND
University of Illinois, Chicago

CAMBRIDGE
UNIVERSITY PRESS

University Printing House, Cambridge CB2 8BS, United Kingdom

One Liberty Plaza, 20th Floor, New York, NY 10006, USA

477 Williamstown Road, Port Melbourne, VIC 3207, Australia

314–321, 3rd Floor, Plot 3, Splendor Forum, Jasola District Centre, New Delhi – 110025, India

103 Penang Road, #05–06/07, Visioncrest Commercial, Singapore 238467

Cambridge University Press is part of the University of Cambridge.

It furthers the University's mission by disseminating knowledge in the pursuit of education, learning, and research at the highest international levels of excellence.

www.cambridge.org
Information on this title: www.cambridge.org/9781108429962
DOI: 10.1017/9781108555746

First published 2022

A catalogue record for this publication is available from the British Library.

Library of Congress Cataloging-in-Publication Data
Names: Sutherland, Daniel, 1961– author.
Title: Kant's mathematical world : mathematics, cognition, and experience / Daniel Sutherland, University of Illinois, Chicago.
Description: Cambridge, UK ; New York, NY : Cambridge University Press, [2022] | Includes bibliographical references and index.
Identifiers: LCCN 2021024849 (print) | LCCN 2021024850 (ebook) | ISBN 9781108429962 (hardback) | ISBN 9781108455107 (paperback) | ISBN 9781108555746 (epub)
Subjects: LCSH: Kant, Immanuel, 1724–1804. | Mathematics–Philosophy. | BISAC: PHILOSOPHY / History & Surveys / Modern
Classification: LCC QA8.4 .S945 2021 (print) | LCC QA8.4 (ebook) | DDC 510.1–dc23
LC record available at https://lccn.loc.gov/2021024849
LC ebook record available at https://lccn.loc.gov/2021024850

ISBN 978-1-108-42996-2 Hardback

To Irmgard Sutherland and to the memory of
J. Paul Sutherland

CONTENTS

PREFACE AND ACKNOWLEDGMENTS

I began this work far longer ago than I would like to admit, but at a fairly specific moment. After a restorative year away from graduate school, I moved to Santa Monica into a rent-controlled apartment the size of a postage stamp with a partial view of the Pacific. With fresh mind and heart, I embarked on a project to better understand Kant's philosophy of mathematics and science, and in particular his distinction between constitutive and regulative principles. I began with the constitutive mathematical principles of experience, and hence the Axioms of Intuition in the *Critique of Pure Reason*, but I fairly quickly realized that previous commentators had not properly understood this part of the *Critique*, nor appreciated its full significance. Kant's references to magnitude and to a homogeneous manifold in intuition seemed to me to indicate an unrecognized depth to his views with implications not just for the applicability of mathematics, but also for pure mathematical cognition and the mathematical character of the world of experience. Most importantly, it almost immediately suggested to me a new understanding of the role of intuition in mathematics in representing magnitudes, a role made necessary by Kant's understanding of the limits of conceptual representation. That was in 1992. These many years later, this book is an attempt to describe Kant's theory of magnitudes and its foundational role in Kant's account of mathematical cognition and our cognition of the world; I hope it will appeal to those interested in Kant's critical philosophy and its development as well as to scholars of Kant's philosophy of mathematics and to philosophers of mathematics interested in the history of their field. I aim to publish another work in the near future that is more narrowly focused on the implications of Kant's theory of magnitudes for his philosophy of geometry, arithmetic, algebra, and analysis and his philosophy of natural science.

My investigation of Kant's views of magnitude did not blaze an entirely new path, but significantly broadens it. Reading of "Kant's Theory of Geometry" by Michael Friedman and "Kant's Philosophy of Arithmetic" by Charles Parsons first inspired me to work on Kant's philosophy of mathematics and science, and Friedman's *Kant and the Exact Sciences* provided a crucial springboard. To my knowledge, Friedman was the first to draw a connection between Kant's theory of magnitudes and the Eudoxian theory of magnitudes found in Euclid, and Friedman's interpretation of *quanta* and *quantitas* and Parsons' "Arithmetic and the Categories" were the starting point for my own investigations. This book is deeply indebted to Michael's and Charles' work, and to their help, support, and advice over the years, and especially to Michael for a postdoc in the History and Philosophy of Science program at Indiana University, and to Charles for regular conversations on two visits to Harvard, one as a graduate student and one as a visiting professor. Sometime later, Bill Tait's work and many pleasant conversations with him in Chicago, often marked by stimulating disagreement, have also been immensely valuable. But the generous help I received from these philosophers would have been for naught without the patient guidance and support of those at UCLA, especially Robert W. Adams, Tyler Burge, John Carriero, and Calvin Normore. Bob provided continual constructive feedback on drafts and had faith in me when I was struggling to find my way and lacked faith in myself; John was a terrific critical advisor, and pushed me to share my work with Friedman and Parsons; memorable conversations with Tyler helped me sharpen my core arguments; and Calvin proved a remarkable resource, imparting crucial historical insights even while racing to the airport to catch a flight.

I can't possibly list all those who have helped me in the decades since this project began; moreover, the assistance of some will be apparent only in a subsequent book focused on some of the details of Kant's philosophy of mathematics. While on a postdoc at the History and Philosophy of Science Department at Indiana University in 1998–9, regular talks with Michael Friedman, Andrew Janiak, and Konstantin Pollock helped me further work out my views. In the early 2000s, Emily Carson, Lisa Shabel, and I, sometimes joined by Ofra Rechter, gathered to share and discuss work; I hope the valuable feedback and moral support I received was in some measure returned. Bill Hart was always willing to discuss the philosophy of mathematics with me, and also generously helped me learn to better communicate my ideas. I have also benefited from the work of, and in-depth conversations with, Lanier Anderson, Vincenzo De Risi, Katherine Dunlop, and Jeremy Heis on Kant's philosophy of mathematics. My debts include the innumerable insightful comments I received during many presentations, all of which I cannot list here, but I would like to single out those who provided me the opportunity to present my views over multiple talks, which proved particularly helpful: Matt Boyle at Harvard and at the University of Chicago, Jim Conant at the University of Chicago, Vincenzo De Risi at the Max Planck Institute in Berlin, the sorely missed Mic Detlefsen at Notre Dame and in France, Michael Friedman at Stanford, Jeremy Heis at UC Irvine, and above all Ofra Rechter and Carl Posy, whose conferences in Tel Aviv and Jerusalem were unique opportunities to advance our collective understanding of Kant's philosophy of mathematics. To them and the participants at these and the many other talks, I am grateful for helpful comments.

It is impossible for me to now recollect all those who gave me particularly helpful one-on-one feedback over the years, but they include all those mentioned above as well as Mahrad Almotahari, Andrew Chignell, Kevin Davey, Lisa Downing, Katherine Dunlop, Walter Edelberg, Stephen Engstrom, Sam Fleischacker, Marcus Giaquinto, Aidan Gray, Sean Greenberg, W. D. Hart, Richard Heck, Robert Howell, Peter Hylton, Vèronique Izard, Anja Jauernig, Michael Kremer, Tony Laden, Allison Laywine, Ian Mueller, Tyke Nunez, Marco Panza, Robert Pippin, Andrew Pitel, Erich Reck, Tom Ricketts, Vincenzo di Risi, Sally Sedgwick, Lisa Shabel, Karl Schafer, Houston Smit, Daniel Smyth, Clinton Tolley, Daniel Warren, and Eric Watkins. I'd also like to thank the graduate students at University of Chicago, Harvard, Stanford, and especially UIC for aid in clarifying and correcting my views over the last twenty years.

I owe special thanks to several for their help with specific chapters in this book: Matt Boyle and Tyler Burge for helpful discussions and correspondence and Thomas Land for detailed comments on Chapter 4; Matt Boyle for prompting me to write Chapter 5; Tyke Nunez, Thomas Land, and Daniel Smyth for very helpful responses on a penultimate version; Chen Liang for conversations that helped me develop my thoughts; and Matt Boyle and Tyler Burge for crucial comments that changed my views. I hope I have been able to incorporate and respond to their criticisms; they bear no responsibility for anything that might be problematic. Marco Panza and Ken Saito provided useful comments and correspondence on the foundations of Euclidean geometry for Chapter 6; Ian Mueller and Vincenzo De Risi had an outsized influence on my understanding of Euclid and the Euclidean tradition in Chapters 6 and 9; while Kevin Davey kindly read the same chapters and helped me avoid a few mistakes and unclear statements. Andrew Pitel provided helpful feedback on Chapter 8.

This long project was munificently supported by various institutions. I would like to thank Michael Friedman and the History and Philosophy Department at Indiana University for a postdoctoral fellowship 1998–9. UIC OVCR-Institute for the Humanities grants in 2001 and 2013 supported archival research in Berlin and UIC Institute for the Humanities Fellowships in 2001–2 and 2013–14 gave me the opportunity to develop my views; thanks to Mary Beth Rose, Susan Levine, and Linda Vavra for providing an intellectual environment that was both stimulating and productive. I also received generous support from an NSF Grant in Science and Technology Studies in 2006–7 (Grant No. 0452527), an American Philosophical Society Sabbatical Fellowship in 2010–11, and an American Council of Learned Societies Fellowship in 2016–17. All this sponsorship played a crucial role in completing this book, and I am most grateful.

Although most of this book thoroughly reworks and supersedes my earlier published views, some of it still draws from those articles. An earlier version of

Chapter 2 appeared as "The Point of the Axioms of Intuition," *Pacific Philosophical Quarterly* 86 (2005): 135–59. I am grateful to the editors for permission to borrow from this article. Chapter 3 and parts of Chapter 4 are based on "The Role of Magnitudes in Kant's Critical Philosophy," *Canadian Journal of Philosophy* 34(4) (2004): 411–42. Thank you to the editors of the *Canadian Journal of Philosophy* for permission to draw from this article. The core of Chapter 7 is found in "Kant's Philosophy of Mathematics and the Greek Mathematical Tradition," *Philosophical Review* 113 (2) (2004): 157–201; I am grateful to the editors of *Philosophical Review* for permission to publish the present version. Finally, Chapter 8 is based on part of "Philosophy and Geometrical Practice in Leibniz, Wolff, and the Early Kant," in *Discourse on a New Method: Reinvigorating the Marriage of History and Philosophy of Science*, edited by Michael Dickson and Mary Domski (Chicago: Open Court, 2010). I would like to thank Open Court for permission to use it.

I also owe thanks to Hilary Gaskin at Cambridge University Press for years of patient encouragement and accommodation while this book evolved and to Hal Churchman for his assistance in the home stretch, and to Stephanie Sakson for her expert editing. Hannah Martens carefully read through the manuscript, flagging infelicities of expression and checking references, for which I am very grateful.

Finally, and most of all, I would like to thank Susan and Isaac for their patience, support, and encouragement during a project that sometimes demanded too much of my time and attention. They have been both my mainstays and the wind in my sails as I navigated this long journey; without them I would have succumbed to storms or drifted in the doldrums rather than completed this voyage.

1

Introduction

Mathematics and the World of Experience

> Philosophy is written in this grand book, the universe, which stands continually open to our gaze. But it cannot be understood unless one first learns to comprehend the language and read the letters in which it is written. It is written in the language of mathematics, and its characters are triangles, circles, and other geometric figures without which it is humanly impossible to understand a single word of it; without these, one is wandering about in a dark labyrinth. . . .
>
> —Galileo, *Il Saggiatore*, 1623[1]

1.1 Kant and the Theory of Magnitudes

My aim in writing this book is to transform our current understanding of Kant's philosophy of mathematics, and in doing so, our understanding of Kant's account of the world of experience. Mathematics and the world are more intimately intertwined in Kant's philosophy than many have appreciated. I will argue that in Kant's account, mathematics is a science of magnitudes, and the world of experience is a world of magnitudes. That is, Kant's philosophy of mathematics, pure as well as applied, is grounded in a theory of magnitudes; at the same time, all objects of experience are and all their real properties have magnitudes, so that the world we experience is a world of magnitudes. The world is fundamentally mathematical in character, and in taking magnitudes as its object of study, pure mathematics is about the world. This is particularly true of geometry – the science of continuous spatial magnitudes – which in Kant's time still enjoyed a certain pride of place in thinking about mathematics and in an understanding of the mathematical character of the world.

This reorientation in Kant's account of mathematics and his account of experience has important consequences for our understanding of both mathematical and theoretical cognition. The role of intuition in both is a major

[1] Translated in Popkin (1966, p. 65).

theme of this book. According to Kant, a magnitude is a homogeneous manifold in intuition. I will argue that in Kant's view representing magnitudes as magnitudes at all depends on intuition, because intuition allows us to represent a homogeneous manifold. The fact that intuition allows us to represent a homogeneous manifold has not been appreciated, yet it has important implications for Kant's claim that mathematical cognition and all human cognition depend on sensible intuition. Moreover, I will argue that the singularity of intuition is best understood as a mode of representing singularly, one that is compatible with representing a homogeneous manifold in intuition.

Another closely related theme is that the role of intuition in Kant's account of mathematical cognition makes both our mathematical cognition and our cognition of the mathematical features of experience more concrete than we are apt to think today. To shift for the nonce to our contemporary parlance of particulars, the role of intuition allows us to represent relatively concrete particular magnitudes in space and time. By concrete, I mean that intuition allows us to represent spatial and temporal particulars; by *relatively* concrete, I here mean that the role of intuition in representing the mathematical features of those particulars does not include the representation of causal relations, a common feature attributed to concreta in our contemporary metaphysical theorizing about them.[2] The fact that intuition represents singularly is what allows us to represent objects *in concreto*, and hence to represent particular objects. Both pure mathematical cognition and our representation of objects of experience rests on the singular representation of concrete continuous and discrete homogeneous manifolds in intuition.

Kant's account, however, is both complex and nuanced. First, Kant allows for multiple roles for intuition in mathematical cognition, not just the representation of concrete homogeneous manifolds. For example, intuition plays an important role in allowing us to represent succession, which is required for arithmetical cognition. Second, mathematical cognition does not merely rest on the intuitive representation of particular concrete magnitudes; it also essentially depends on concepts, in particular, the categories of quantity, as well as rules for the representation of magnitudes, that is, schemata. Third, Kant's primary notion of magnitude is concrete, but he also makes room for a more abstract notion of magnitude. The contrast between concrete and more abstract representations of magnitude correspond to a distinction Kant draws between two sorts of magnitude, *quanta* and *quantitas*. Furthermore, this distinction is tied to Kant's obscure and complicated understanding of number.

[2] This is a provisional characterization of "relative concreteness." I will examine Kant's notion of concreteness in more depth in Chapter 5.

A full account of Kant's philosophy of mathematics would require sorting out all the nuances and complexities of Kant's theory of magnitudes and bringing them to bear on what he says about geometry, arithmetic, algebra, and analysis. I cannot hope to do all of that between the covers of one book. Instead, this work will focus on the foundations of Kant's theory of magnitudes and its relation to both Kant's account of mathematical cognition and our cognition of objects of experience. I plan to follow with another book that will delve more deeply into the implications of the theory of magnitudes in Kant's account of geometrical, arithmetic, and algebraic cognition, as well as in the foundations of analysis. The present book, I hope, will speak to those readers interested more broadly in the foundations of Kant's theoretical philosophy, with an eye to his philosophy of mathematics and mathematical cognition and its implications for Kant's account of experience.

There are, however, a few features of the foundations of Kant's theory of magnitudes that cannot be fully and satisfactorily addressed until the details of his account of mathematical cognition have been explained. Those include aspects of the distinction between two notions of magnitude, *quanta* and *quantitas*, and their relation to number. I cannot give a complete account of number in this book, but I devote several sections to the distinction between *quanta* and *quantitas*, and discuss Kant's understanding of number and its relation to the Greek mathematical tradition.[3] We will see that in Euclid and the Euclidean tradition, the understanding of continuous magnitude is entangled with that of number in several ways. Those entanglements are also found in Kant, but unraveling them requires an in-depth focus on Kant's arithmetic. There are therefore a few claims about *quanta* and *quantitas* and their relation to number whose full defense depends on promissory notes to be redeemed in the second book.

I will argue for an interpretation of the foundation of Kant's theory of magnitudes and its relation to his understanding of experience based both on a close reading of the texts and on placing those texts in historical context. My aim is to determine Kant's views as accurately as I can without attempting to evaluate Kant's views from our contemporary perspective. It will be a sufficient accomplishment to get Kant's views right. I will press, however, to the limits of Kant's theorizing about magnitudes, that is, to the limits of how much he was able to develop and articulate his views given the time he had to devote to the topic. I will also move beyond what Kant explicitly says in order to reconstruct the key assumptions underlying his theory and to determine his views with regard to those assumptions.

[3] For an argument that Kant's conception of number includes both cardinal and ordinal elements, see Sutherland (2017). Future work will more fully address the relationship of Kant's conception of number to *quanta* and *quantitas* and also explain Kant's understanding of irrational numbers.

This work is meant to be generally accessible to philosophers interested in Kant's account of experience, and will not presuppose anything but a rudimentary understanding of mathematics nor a familiarity with the history of mathematics. From those steeped in philosophy of mathematics or its history, I beg patience, and hope that there is ample material of interest to hold their attention. In the remainder of this introductory chapter, I say a bit more about the themes mentioned above to orient the reader, before closing with an overview of the book.

1.2 Mathematics Then and Now

Kant's view of mathematics as a science of magnitudes was common in the eighteenth century, but it is strikingly different from our contemporary way of thinking. This is not the place to recount the history and philosophy of mathematics from Kant to the present, but there are two ways in which Kant's views are different that are important to highlight. The first is that mathematics has become a science of number rather than magnitudes. The "arithmetization" of mathematics over the course of the nineteenth century placed natural number firmly at the foundation of mathematics, encouraged a more abstract understanding of number, and introduced a separation between mathematics and the world. Pure mathematics is no longer about the world insofar as it is constituted by magnitudes. Instead, natural numbers are used to construct the rationals, the reals, and complex numbers, and once these foundations are complete, the relation between mathematics and the world can be taken up as an issue of applied mathematics. Further work at the end of the nineteenth century on the foundations of arithmetic itself attempted to provide a foundation of number in terms of notions more basic than natural numbers and their arithmetic. These notions were supplied by logic and the emerging theory of sets, both of which were thought of in abstract terms, which reinforced a more abstract understanding of number. At the same time, the development of axiomatics solidified the growing primacy of arithmetic over geometry, as well as a separation of pure mathematics from the world to which it was applied. "Pure" geometry came to be viewed as the study of the consequences of various sets of axioms apart from whether those axioms describe physical space. As a result of all these developments, pure mathematics shifted in the nineteenth century from being a science of magnitude to being first and foremost a science of number.[4]

The fact that the arithmetization of mathematics led to a more abstract understanding of mathematics and a separation between pure mathematics

[4] According to Petri and Schappacher (2007), the view of mathematics as a science of magnitudes was not extinguished until 1872.

and the world is closely related to a further issue: the drive to emancipate mathematics from intuition. The eighteenth century saw remarkable advances in mathematics, especially in analysis, which included what we now call calculus. Nevertheless, when it came to foundational questions, mathematicians and philosophers still reverted to thinking of mathematics as a science of magnitudes, and many of those mathematicians and philosophers particularly concerned with foundations thought that intuition, in particular, geometrical representations, played an important role in securing the meaning and certainty of the most basic concepts and propositions of mathematics. Kant was among them.[5]

Kant radically departs from previous philosophers in elevating the status and role of intuition in all human cognition. Previous philosophers distinguished between what we receive through the senses from what we represent through the intellect, and addressed how they are related. Kant argues for a deeper difference, arguing that that intuitions are a fundamentally distinct kind of representation from concepts and belong to their own faculty. Moreover, intuitions are representations in the pure forms of space and time, which allow us to represent spatial and temporal features of the world, a role which had traditionally been assigned to empirical perception. Kant argued that space and time were forms of intuition, and hence that intuitions could be not just empirical but pure and *a priori*. Kant also departs from his predecessors in holding that intuition is required for all theoretical cognition, and in particular that pure intuition is required for all mathematical cognition. Kant relies on his distinction between analytic and synthetic propositions, that is, those propositions whose truth is grounded in the content of concepts and the containment relations among them, and those propositions whose truth is not. Kant claims that mathematical propositions are synthetic, and hence require intuition, and in particular pure intuition, to ground them. Kant met almost immediate resistance from philosophers in the continental rationalist tradition following Leibniz who rejected the claim that theoretical cognition, including mathematical cognition, depends on a nonconceptual form of representation. Even some of his allies were troubled and challenged him. They thought that geometry might plausibly depend on a pure intuition of space, but it is less obvious that arithmetic and algebra depend in any way on intuition. Nevertheless, Kant's critical philosophy and his claim about the role of intuition in mathematical cognition gained wide influence.

The nineteenth-century arithmetization of mathematics and foundations of arithmetic arose against this backdrop. Mathematicians answered the call for rigor to address problems in the foundations of analysis by rejecting any

[5] This is a rather rough summary of a complex history. See Sutherland (2020b) for a more detailed account of Kant's relation to the history of analysis.

appeal to intuition, geometrical or otherwise. Far from helping establish certainty, intuition came be seen as not just unreliable but potentially misleading. The arithmetization of mathematics meant that more was at stake in Kant's claim that intuition is required for arithmetic in particular. At the same time, the rejection of intuition in arithmetic put great pressure on that claim. Frege's development of logic extended what could be expressed and derived within it, allowing Frege to expand and shift the notion of analyticity and claim that arithmetic is in fact analytic and does not depend for its justification on intuition. Russell stated that "formal logic was, in Kant's day, in a very much more backward state than at present," and that properly understood, mathematical reasoning "requires no extra-logical element" (Russell (1903), p. 457). He held that advances in both logic and mathematics itself eliminate the need for intuition. The greater logical resources that could be brought to bear and the drive to eliminate intuition led Russell to be quite dismissive of Kant's philosophy of mathematics.[6]

Not all agreed with the banishment of intuition, however. Hilbert, Poincare, and Brouwer each at some point and in their own way defended the idea that intuition has more than a heuristic role to play in our knowledge of mathematics. Some of these defenders looked to Kant for inspiration, even when their understanding of the role of intuition differed from his. What is striking from a historical perspective is that, whether one agreed or disagreed with Kant, the terms of the debate were set by him. But if we are to truly understand Kant's philosophy of mathematics, we will have to reconstruct a view of mathematics prior to its arithmetization, and that requires comprehending as best we can the idea that mathematics is a science of magnitudes, as well as Kant's account of the role of intuition in representing magnitudes both in mathematics and in experience.

One of the primary aims of this book is to bring to life this older way of thinking about mathematics. But because our modern way of thinking is deeply embedded in higher mathematics and even shapes basic mathematics education, it is difficult to shed our presumptions when reading Kant's claims concerning mathematics. The best way to recover the earlier way of thinking is to return to its roots in Euclid and the Euclidean tradition following him. The influence of Euclid's *Elements* can hardly be overstated; it was the model of mathematical reasoning and a paradigm of scientific knowledge for more than two millennia and was responsible for the dominance of geometry over arithmetic during that time. As De Risi notes, there were hundreds of translations of and commentaries on the *Elements*, and its dissemination and

[6] See Friedman (1992), especially pp. 55–6, as well as Friedman (2013) for a sustained argument that we can still learn a great deal from understanding Kant's views of mathematics and natural science, despite – in fact with the aid of – advances in our understanding of logic, mathematics, and physics.

influence throughout Europe was "only matched by the Bible and by a few other writings of the Fathers of the Church."[7] Even those who aspired to replace rather than modify the *Elements* began their studies with it and reacted against it. But Euclid's *Elements* contains more than mere geometry. An essential component is a theory of ratios and proportions among magnitudes, a theory attributed to Eudoxus.[8] This crucial part of the *Elements* set the framework for thinking about magnitudes in the Euclidean tradition and persisted into the nineteenth century. The *Elements* also contains books on number and the basic properties of numbers, including propositions governing the ratios and proportions among them. The conception of number expounded there influenced the understanding of number for nearly two millennia. The long Euclidean tradition included important challenges and modifications to the *Elements*, and there were of course remarkable advances in mathematics, particularly from the beginning of the Renaissance and through the eighteenth century. Nevertheless, the framework for thinking about mathematics, and in particular for thinking about the foundations of mathematics and about mathematical cognition, was strongly influenced by the Euclidean tradition and the Euclidean theory of magnitudes. That framework was still dominant in the eighteenth century.

We will keep Euclid's *Elements* close at hand throughout this book in order to understand Kant's very different way of thinking about mathematics. I will point out ways in which the long Euclidean tradition diverged from Euclid and describe developments during and after the Renaissance when they are important for understanding Kant. Obviously, a history of mathematics from Euclid to the eighteenth century is well beyond the scope of this work and what I highlight is quite selective. After discussing Kant's views of mathematics and magnitudes and their relation to experience in Part I, I will give a relatively brief and focused presentation of key features of Euclid's *Elements* that shaped the understanding of mathematics into the eighteenth century. That will put us in a position to dive more deeply into Kant's understanding of mathematics and its relation to the world in Part II.

Recovering Kant's understanding of mathematics requires a shift not just in an understanding of foundations, but in their aim. During and after the arithmetization of mathematics, the goal of foundations was to resolve various problems in analysis and to explain the nature of numbers by giving an account of certain mathematical notions (real, rational, natural numbers), in terms of more basic notions (rational numbers, natural numbers, logical and set-theoretic notions, respectively), and to do so in a rigorous way that would

[7] De Risi (2016, p. 592).

[8] Euclid compiled previous works of mathematics in writing the *Elements*, and the basic content of parts of it was attributed to various authors, including Eudoxus, as will be discussed in more detail in Chapter 6.

ground inferences. The primary focus was on providing a foundation for mathematics itself. Kant's aims were quite different. First and foremost, Kant wished to provide an explanation of the possibility of mathematical cognition, which includes both basic judgments, such as the judgment that there are seven apples in the bowl on the table, as well as what is required for higher mathematics, pure and applied, such as the derivation of Newton's law of universal gravitation. Kant's aim is to provide an explanation of mathematical cognition in terms of our most basic cognitive capacities. Those elements are the categories and the pure forms of intuition, so that Kant's explanation of the possibility of mathematical cognition is grounded in them.

This is not to say, however, that Kant had first settled on his theory of the categories and pure intuition, even its general shape, before addressing the foundations of mathematical cognition. Indeed, Kant's reflections on mathematical cognition, particularly in the *Prize Essay* period in the years 1762–4, was a driving force in the development of his critical philosophy, including his conviction that there is a class of truths that cannot be reduced to logical relations among concepts and that we have pure forms of *a priori* intuition. The development of Kant's critical understanding of the categories and the pure forms of intuition was strongly influenced by his philosophy of mathematics, and it offers more insights than have been generally appreciated. This is a story worthy of its own monograph, but it is one we will have to largely set aside here.[9]

What is important for our present purposes is that Kant's primary aim with respect to mathematics was to provide a foundation of mathematical cognition rather than a foundation of mathematics in our modern sense. The two sorts of foundations are inextricably linked; nineteenth-century foundations were often motivated by epistemological concerns, and Kant's understanding of mathematical cognition is conditioned by his understanding of the nature of mathematics. There is no easy division between the two. Nevertheless, the difference in focus and emphasis between Kant and post–eighteenth-century approaches is significant. In Kant's account, we attain mathematical knowledge through our cognition of magnitudes, and hence the focus of his foundations is on explaining our ability to cognize magnitudes in both pure mathematics and in experience. This is not to say that one cannot learn a great deal about Kant's philosophy of mathematics and philosophy of science by foregrounding Kant's interaction with the mathematics and science of his day, while leaving Kant's account of our cognition of magnitudes in the background; indeed, a good deal of very good work in recent decades has done just

[9] For relatively recent work focusing specifically on Kant's philosophy of mathematics in and after the Prize Essay period, see especially Carson (1992), Rechter (2006), and R. L. Anderson (2015).

that. But the focus of the present work will be on Kant's account of the foundations of mathematical cognition as a cognition of magnitudes.

1.3 Mathematics in Kant's Theoretical Philosophy

As I have already indicated, understanding Kant's theory of magnitudes is not merely crucial for his philosophy of mathematics; it is important for his entire critical philosophy. Because it is more important than is often recognized, this claim is worth defending here at the outset, though it is the book as a whole that makes the case.

Kant's mature philosophy only emerged with the *Critique of Pure Reason* (henceforth *Critique*),[10] but Kant's early reflections on mathematical cognition in 1762–4 were a key factor in moving Kant toward his view of the role of intuition in human cognition. Kant's reflections on the possibility of demonstrating God's existence during this period were certainly important to his critical assessment of and emancipation from Leibnizian and Wolffian metaphysics, as is clear in his essay *The Only Possible Argument for the Existence of God*. It was, however, his investigation of mathematical cognition in *Inquiry Concerning the Distinctness of the Principles of Natural Theology of Morals* (henceforth either *Inquiry* or *Prize Essay*) and *Attempt to Introduce the Concept of Negative Magnitudes into Philosophy* (henceforth *Negative Magnitudes*) that convinced him that Leibnizian and Wolffian rationalism based solely on conceptual representation and the relations among concepts could not account for mathematical cognition, and moved him toward his understanding of pure intuition and its role in human cognition.[11] I primarily focus on Kant's views in the critical period, in which the influence of his philosophy of mathematics on his theoretical philosophy is readily apparent, at several different levels.

The first level of that influence is well known, but bears review. Kant states in the *Prolegomena to Any Future Metaphysics* (henceforth *Prolegomena*) that metaphysicians must answer the question, "How are synthetic *a priori* cognitions possible?" and that the whole of transcendental philosophy is the answer (4:278–9). Kant claims that if metaphysics were possible, it would rest on synthetic *a priori* cognitions, but that it is disputable whether there are any

[10] Despite the deep importance of the two other critiques to Kant's philosophy as a whole, I will usually refer to *The Critique of Pure Reason* simply as the *Critique*. Our focus will be primarily on the role of magnitude in Kant's account of theoretical cognition in the *Critique of Pure Reason*, save one relatively short excursion into the *Critique of Judgment*.

[11] See R. L. Anderson (2015) for a recent particularly lucid and helpful account of Kant's reaction to Leibnizian and Wolffian rationalism, with a focus on the development of Kant's understanding of the analytic-synthetic distinction starting in the pre-critical period. For a broader account of Kant's reaction to rationalism, see Hogan (2009).

such cognitions to support metaphysics at all. Nevertheless, Kant says, to motivate and justify the question of how synthetic *a priori* knowledge is possible, we need not first establish that it is possible, because it is actual: we have clear examples of synthetic *a priori* cognitions in pure mathematics and pure natural science (4:275). Kant claims that we can "confidently say" that pure mathematics and pure natural science contain *a priori* cognitions; he adds that their status is "uncontested," and that these examples are "plenty and indeed with indisputable certainty actually given" (4:276). This is what justifies the analytic method he says he employs in the *Prolegomena*, that is, starting from the fact that we have synthetic *a priori* knowledge and seeking an explanation for its possibility (4:279).

Although Kant claims to employ the synthetic method in the *Critique*, and so presumably does not start with the assumption that we have a specific sort of cognition and then seek its explanation, he makes claims similar to the *Prolegomena* about the cognitions of pure mathematics and pure natural science. In the B-Introduction, for example, he states that, in contrast to the status of the propositions of metaphysics, pure mathematics "certainly contains synthetic *a priori* propositions" (B20). He also adds that since mathematics and pure natural science "are actually given, it can appropriately be asked **how** they are possible; for that they must be possible is proved through their actuality" (B20).

It is important, of course, to distinguish between the claim that the propositions of pure mathematics and pure natural science are indisputably certain, and the claim that their status as synthetic *a priori* cognitions is indisputably certain. The certainty of $2 + 2 = 4$ is not the same as the certainty that this proposition is synthetic *a priori*, and although Kant states that their status as synthetic *a priori* cognitions is uncontested, he gives arguments to support his claim. Even the *Prolegomena*, which he says employs the analytic method and hence assumes that we have synthetic *a priori* cognition, provides considerations in favor of this claim in the Preamble. His tone in both the *Prolegomena* and the *Critique*, however, suggests that little real argument is needed, only careful reflection in light of the proper characterizations of the *a priori*/*a posteriori* and analytic/synthetic distinctions. The considerations Kant brings to bear in the B-Introduction of the *Critique* borrow almost verbatim from the *Prolegomena* Preamble. Concerning the apriority of mathematical judgments, Kant treats it as sufficient to simply attend to marks of apriority. Kant argues that necessity and universality are sure criteria of *a priori* cognitions and that this makes it is easy to show that there are *a priori* judgments in human cognition: "one need only look at all the propositions of mathematics," and he seems to think that no more argument is required. He also states that in the proposition "every event has a cause," the concept of a cause "obviously" contains the concept of necessity, which he cites in support of his claim that the proposition is necessary (B4–5).

The same tone is present in his claim that these judgments are synthetic. Kant states that the fact that mathematical judgments are all synthetic "seems to have escaped the notice of the analysts of human reason," yet is "incontrovertibly certain." He also claims that the syntheticity of propositions of pure natural science, such as "in all alterations of the corporeal world the quantity of matter remains unaltered," is clear (A10/B14). Kant's stance on the existence of synthetic *a priori* cognitions is especially apparent in his *Prolegomena* discussion of Hume, when he suggests that if Hume had only been armed with the right distinctions and asked the right questions, it would have been clear to him that mathematics consists of synthetic *a priori* cognitions (4:272–3).

As Lanier Anderson as rightly emphasized, these passages can sound like mere table-thumping. In particular, Kant needs further argument for his claim that the propositions of pure mathematics and pure natural science are synthetic if he is going to do more than simply beg the question against the Leibnizian and Wolffian rationalists, who would argue that all *a priori* truths, indeed all properly articulated and grounded truths, are analytic.[12]

The *a priori* status of the cognitions of pure mathematics and pure natural science would have required less argument – at least for pure mathematics. The intended target of Kant's polemic included both Leibnizian rationalists and that version of a Humean empiricism that concedes the necessity of mathematical propositions. (Kant does not seem to have taken seriously the possibility of an empiricism that denied the necessity of mathematics.) Philosophers as opposed in outlook and approach as Leibniz and Hume accepted both that particular sense experiences cannot provide knowledge of necessity and that mathematical propositions are necessary, and hence *a priori*, while disagreeing on the significance of those propositions. By focusing on Leibnizian rationalism and Humean empiricism, Kant could count on the claim that pure mathematics is *a priori*.

On the other hand, the nature of the dependence of natural science on empirical observation would have led empiricists to balk at Kant's claim that there is a pure natural science containing *a priori* propositions. Kant himself holds that a great part of natural science is grounded on empirical principles. In the B-Preface, for example, Kant cites the work of Galileo, Torricelli, and Stahl as examples of natural science grounded on empirical principles (Bxii–xiii), and to this list he would have added the contributions of many others, including Copernicus and Kepler. And Kant acknowledges that natural science relies on experience in a way that mathematics does not:

> [pure mathematics] is supported by its own evidence; whereas [pure natural science], though arising from pure sources of the understanding,

[12] L. Anderson (2005) is a sustained and illuminating investigation of what that non-question-begging argument looks like.

> is nonetheless supported from experience and thoroughgoing confirmation by it – experience being a witness that natural science cannot fully renounce and dispense with, because, as philosophy, despite all its certainty it can never rival mathematics. (4:327)

For all this, Kant insists that there is a pure part of natural science that contains synthetic *a priori* principles, such as "substance remains and persists" and "everything that happens is always previously determined by a cause according to constant laws" (4:295). These are claims that empiricists would strongly resist.

How does Kant support his claim that pure natural science contains synthetic *a priori* principles? At B17–18, Kant states that he will "adduce only a couple of propositions as examples" of pure natural science and simply asserts that it is clear that they are synthetic and *a priori*, which sounds again like mere table-thumping. But it is noteworthy that in the *Prolegomena* passage in which Kant suggests that Hume could have recognized that we have synthetic *a priori* cognitions, Kant refers only to pure mathematics, and not to natural science. And in the Introduction of the *Critique*, Kant acknowledges that the *a priori* status of natural science would be contested. In a footnote following his claim that the possibility of pure mathematics and pure natural science are given by their actuality, he states:

> Some may still doubt this last point in the case of pure natural science. Yet one need merely consider the various propositions that come forth at the outset of proper (empirical) physics, such as those of the persistence of the same quantity of matter, of inertia, of the equality of effect and counter-effect, etc., and one will quickly be convinced that they constitute a *physica pura* (or *rationalis*).... (B21n)

While acknowledging the challenge, he simply reiterates that if one merely considers these propositions, one will be quickly convinced. This would hardly have moved an empiricist such as Hume, who explicitly argued against the necessity of natural laws and against the legitimacy of a concept of causality that attributed a necessary connection between events. As a consequence, Kant's arguments that were intended to target both his rationalist and empiricist predecessors are not nearly as strongly supported by his appeal to pure natural science as by his appeal to pure mathematics. Kant must still make good on his claim that the *a priori* propositions of mathematics are also synthetic; nevertheless, it is pure mathematics that carries the water. Since Kant would stake his entire philosophy on the answer to the question, "How is synthetic *a priori* knowledge possible?," pure mathematics plays a particularly important role in Kant's argument for his critical philosophy. This level of influence of Kant's philosophy of mathematics on his critical philosophy is clear to every reader of Kant.

The importance of mathematics in Kant's critical philosophy is not, however, limited to establishing the fact *that* there are synthetic *a priori* cognitions.

The answer to the question of *how* they are possible also informs us about all synthetic *a priori* cognition, and hence informs Kant's critical philosophy at a deeper level. In the *Prolegomena*, Kant states that we only need to ask how pure mathematics and pure natural science is possible "in order to be able to derive, from the principle of the possibility of the given cognition, the possibility of all other synthetic cognition *a priori*" (4:275). They do so because they "bring to light a higher question concerning their common origin" (4:280), that is, the common origin of pure mathematics, pure natural science, and metaphysics as a science.

The *Prolegomena* provides a concise statement of what he has in mind. After explaining how pure mathematics is possible, he states in §11 that the problem has been solved, and summarizes the solution:

> Pure mathematics, as synthetic cognition *a priori*, is possible only because it refers to no other objects than mere objects of the senses, the empirical intuition of which is based on a pure and indeed *a priori* intuition (of space and time), and can be so based because this pure intuition is nothing but the mere form of sensibility, which precedes the actual appearance of objects, since it in fact first makes this appearance possible. (4:284).

Kant takes himself to have established not just how mathematical cognition is possible, but important features of our empirical intuition of objects of the senses. Empirical intuition of such objects is based on a pure *a priori* intuition of space and time that is the mere form of sensibility, that precedes the actual appearance of objects, and which makes the empirical intuition of objects of the senses possible. Kant has arrived at these conclusions by reflection on the nature of mathematical cognition: "all mathematical cognition has this distinguishing feature, that it must present its concept beforehand in intuition and indeed *a priori*, consequently in an intuition that is not empirical but pure, without which means it cannot take a single step" (4:281). He then elaborates: mathematics "must be grounded in some pure intuition or other, in which it can present, or as one calls it, construct all of its concepts *in concreto* yet *a priori*." He does not argue further in the *Prolegomena* for these claims about the nature of mathematical cognition; he takes them for granted. He then argues from them for the conclusions noted above: that intuition is nothing but the form of sensibility; synthetic *a priori* propositions concerning pure intuition are valid for, and only valid for, objects of the senses; and empirical intuition is based on a pure *a priori* intuition of space and time.

Kant states that he employs the synthetic method in the *Critique of Pure Reason*, and reaches the same conclusions concerning our empirical intuition of objects of the senses in a different way. The Transcendental Aesthetic begins with reflections on our faculties and its representations, claiming that the form of sensible intuitions, and hence the form of empirical intuitions and appearances, must be *a priori*. The metaphysical expositions of the concepts of space and time then argue that space and time are pure *a priori* intuitions, and draws similar conclusions to the passage just cited from the *Prolegomena*. But in the B-edition, he only draws these conclusions after adding two additional sections, the transcendental expositions of the concept of space and of the concept of time, each expanding on a corresponding paragraph in the A-edition. Both the A-edition paragraphs and the B-edition transcendental expositions have the general form of the analytic method purportedly followed in the *Prolegomena*; they assume that we have synthetic *a priori* cognition in mathematics, and ask after the conditions of their possibility. In the case of the B-edition transcendental exposition of the concept space, Kant backs up the assumption that geometrical propositions are synthetic and *a priori* by referring to the B-Introduction discussion of apriority and syntheticity – sections that Kant borrowed from the Preamble of the *Prolegomena* in rewriting the *Critique*.[13]

Why does Kant include the transcendental exposition of space in the B-edition *Critique*, despite the fact that the transcendental exposition apparently follows the analytic method of the *Prolegomena*? The answer lies in a not overly strict application of the notions of analytic and synthetic method. It also likely lies in part in the fact that Kant finds the argument compelling, so that if the B-Introduction is successful at getting his readers to reflect on mathematical knowledge in light of the distinctions between *a posteriori* and *a priori* as well as analytic and synthetic, and to concede that at least geometry is synthetic *a priori*, they will be convinced of the role pure forms of intuition play in human cognition. The answer is also likely in part because it was Kant's early reflections on the nature of intuitive certainty in mathematics and on the limitations of Leibnizian and Wolffian rationalism that eventually led him to his view of the role of the pure forms of intuition in mathematical cognition and from that to their role in all theoretical cognition. In the elucidation of the Transcendental Aesthetic, Kant concludes:

> Time and space are accordingly two sources of cognition, from which different synthetic cognitions can be drawn *a priori*, of which especially pure mathematics in regard to the cognitions of space and its relations provides a splendid example. Both taken together are, namely, the pure forms of all sensible intuition, and thereby make possible synthetic *a priori* propositions. (A38–9/B5–6)

[13] Kant gives shorter shrift to the transcendental exposition of the concept of time. The synthetic *a priori* principles in that case are not described as mathematical. In the A-edition paragraph, the synthetic *a priori* principles include the "axioms of time in general," for example, the principles that time has only one dimension, and that different times are not simultaneous, but successive. In the B-edition, Kant adds the synthetic *a priori* cognitions of "the general theory of motion."

Kant's early reflections on pure mathematical cognition led him to develop his account of the pure forms of intuition, which he subsequently held to play a role in all synthetic *a priori* propositions and indeed even in all empirical intuitions. Kant's understanding of mathematical cognition thus had a profound influence on the doctrine of the Transcendental Aesthetic concerning human cognition: that all human cognition, empirical as well as pure, depends on two distinct kinds of representation, concepts and intuitions, and that each belongs to its own faculty, the understanding and sensibility.

This level of influence of Kant's philosophy of mathematics on his account of theoretical cognition is also familiar to careful readers of Kant. Yet the influence of Kant's understanding of mathematical cognition on his critical philosophy extends to an even deeper level that may be less familiar, and that is the topic of this book. The passages from the *Prolegomena* quoted above connect the possibility of the synthetic *a priori* cognitions of mathematics to objects of the senses, and assert that the empirical intuition of those objects is in turn based on pure *a priori* intuition. This suggests, but does not explicitly say, that the role of pure *a priori* intuition in our cognition of empirical objects is one and the same as in mathematical cognition. That is in fact Kant's view, which becomes apparent once one appreciates the role of magnitudes in both, which we will more carefully consider in Chapters 2 and 3. In brief, appearances are, with respect to their intuition in space and time, magnitudes, and mathematical cognition is cognition of magnitudes. Kant argues in the Axioms of Intuition that appearances are taken up into empirical consciousness through a synthesis that generates the representation of the determinate space or time contained in the appearance. This determinate space or time is a magnitude, and the synthesis that generates these representations is the same synthesis of space and time by means of which the concept of magnitude is constructed in intuition. That is, it is the same synthesis that underlies mathematical cognition.

As a result, our cognition of appearances and mathematical cognition are interwoven; they are both based on a common synthesis generating representations of determinate spaces and times. The Axioms of Intuition is a further elaboration of a position Kant already describes in §26 of the B-Deduction with respect to spatial magnitudes.[14] I will only briefly summarize the argument he makes there. Kant argues that all apprehension must agree with the synthetic unity of space and time represented as intuitions, and argues that this synthetic unity is the synthetic unity of an original consciousness in agreement with the categories. He concludes that the synthesis that makes perception possible stands under the categories. Kant then illustrates his

[14] I will sometimes call the Axioms of Intuition simply the "Axioms," with capitalization and in the singular, to refer to the section of the Critique with that title.

conclusion with two examples, the first of which is the perception of a house. He states that I perceive it through the apprehension of its manifold and "I as it were draw its shape [*Ich zeichne gleichsam seine Gestalt*] in agreement with this synthetic unity of the manifold in space" (B162). Furthermore, if I abstract from the form of space, the synthetic unity is the category of magnitude.

There are three points to make about his example. Kant's reference to the category of magnitude is a reference to the categories of quantity, and in virtue of standing under the categories of quantity, the synthesis of the manifold of space and time required for perception generates a representation of a magnitude. Second, the representation generated in this case is the shape of the perceived object. Third, the representation is generated by an "as it were" drawing of the shape. Geometrical cognition is based on the same synthesis in the drawing of shapes in spatial intuition. In geometry, this drawing is an act made explicit in the construction of geometrical figures, while in perception it is only implicit and, because it is an "as it were" drawing of its shape, perhaps only corresponds in some way to a temporally extended act of drawing a figure. Our perception of the house is nevertheless based on the very same synthesis that underlies the representation of spatial magnitudes in geometry. The world acquires its mathematical character, and mathematics applies to it, in virtue of the synthesis that generates representations of magnitude.

Not only the spatial and temporal features of appearances have a mathematical character. The real of appearances corresponding to sensations has an intensive magnitude, and thereby also acquires a mathematical character. It is important to appreciate that every real property of an object corresponding to a sensation has an intensive magnitude. A light source is a paradigm example of something real that has an intensive magnitude, which is also true of the light reflected by an object. It is particularly important that Kant also singles out motion and forces as having intensive magnitudes, both of which are at the heart of his account of the physical world and the laws of nature.

As a consequence, Kant's account of mathematical cognition does not simply provide a clear example of synthetic *a priori* knowledge, nor does the role of intuition in that account merely point to pure intuition as the ground of synthetic *a priori* cognitions and intuition as a necessary condition of cognition more generally. Kant incorporates mathematical cognition, and the role of pure intuition in mathematical cognition, into his account of our cognition of all features of the objects of experience. That is one reason we cannot fully appreciate Kant's account of experience without understanding his philosophy of mathematics and the theory of magnitudes on which it rests.

One might take a quite different view, attempting to separate the conditions of the possibility of experience from the conditions of the possibility of the exact sciences, drawing a line separating, on the one hand, Kant's account of ordinary everyday experience of trees and tables and chairs and, on the other, his account of mathematics and mathematical physics. This might be encouraged by the strong distinction Kant draws between philosophical and mathematical cognition, and the fact that the conditions of possible experience articulated in the system of principles of the *Critique* are established through philosophical cognition and belong to philosophy. This approach to understanding Kant would acknowledge that the principles of the Axioms of Intuition and the Anticipations of Perception establish the applicability of mathematics to objects of experience in virtue of their being magnitudes. Nevertheless, being a magnitude is a feature of appearances, and mathematical cognition has some other relation to intuition than through the representation of magnitudes. I will argue that this understanding of Kant is fundamentally mistaken. Kant's account of the conditions of the possibility of experience incorporate the conditions of the possibility of mathematical cognition, and it does so through the representation of magnitudes in intuition.

1.4 Kantian Transformations

As I remarked above, Kant's account of the role of magnitudes in mathematics and experience is nuanced and complex. While mathematical cognition and the cognition of objects of experience are grounded in the representation of magnitudes, that alone is not sufficient to explain mathematical cognition or the mathematical character of experience. The Euclidean theory of magnitudes that shaped the understanding of mathematics into the eighteenth century was based on the Eudoxian theory of the ratios and proportions that governs the mathematical relations among magnitudes. Euclid does not define the notion of magnitude, so that the meaning of magnitude is implicitly characterized by its role in the Eudoxian theory, which attributes mathematical relations to them; as a consequence, the Euclidean notion of magnitude is mathematical. But as we shall see, Kant does not simply adopt the theory of magnitudes found in the Euclidean tradition. He presses deeper, and reworks it in light of his project to explain the possibility of mathematical cognition. As a consequence, Kant defines magnitude in a way that departs from the implicit definition of magnitude found in the Euclidean tradition; his account is mereological. Kant thinks that the mereological properties of intuition and the special sort of synthesis involved in the representation of the part–whole relations of intuition are at the foundation of all mathematics. But in order to achieve properly mathematical cognition, the relation of equality and the possibility of a pure form of measurement are also required. A goal of this book is to describe these additional requirements that bridge the gap between Kant's mereological characterization of magnitude and the Euclidean mathematical conception of magnitude.

Kant's reworking of the Euclidean tradition concerning magnitudes brings out another theme of this book. Kant's philosophy is repeatedly revolutionary. He brings a new level of reflection and sophistication to old philosophical debates, quite often transforming the terms of debate in the process. Three examples we've touched on in his theoretical philosophy include his articulation of the analytic/synthetic distinction; his separation of it from the *a priori*/*a posteriori* distinction; and his introduction of a unique faculty of sensibility, to which a kind of representation distinct from concepts belongs, a representation that can be *a priori*. But what is also noteworthy is that Kant does not simply reject previous theorizing and attempt to start from a clean slate. Kant is not one who believes in throwing the baby out with the bathwater. Kant's reactions to his empiricist and rationalist predecessors is a good example. His critical philosophy is deeply influenced by and includes many elements of both: a critical stance rooted in a demand for a justification of the claims of philosophy and the project of sharply curtailing the pretensions of metaphysics, coupled with an insistence that we can have *a priori* knowledge.

Kant has a characteristic manner of not throwing out the baby with the bathwater: he alters theories at their foundations in a way that preserves many of the key claims of those theories that he thinks are, if properly understood and formulated, correct. We can even see this in Kant's relation to his earlier philosophy. The *Inquiry* of 1763, for example, provides a catalog of ways in which philosophical cognition differs from mathematical cognition. Many of the insights he gained in that essay are found in the Discipline of Pure Reason in the *Critique*, despite the fact that at the time of writing the *Inquiry*, Kant had not yet distinguished understanding and sensibility, nor articulated the properties of intuitive representations and developed a theory of pure intuition, nor developed his doctrine of the construction of concepts, nor developed his transcendental idealism. Despite tectonic shifts in the development of his views, he retains many of the same contrasts between mathematical and philosophical cognition.

Kant employs a similar strategy in reforming the rationalist metaphysics of quantity that he inherited from Leibniz, Wolff, and Baumgarten. As we shall see in Chapter 8, these philosophers provided metaphysical definitions of quality and quantity that served their larger purpose of incorporating mathematics into metaphysics. It was a criterion of success that they be able to define similarity as identity of quality and equality as identity of quantity in a way that corresponds to the geometrical notions of similarity and equality. Kant alters the metaphysics of quantity at its foundation in order to incorporate it into his theory of magnitudes and in particular the role of intuition in it; at the same time, he preserves the distinction between quality and quantity and the definitions of similarity and equality that correspond to the geometrical notions of similarity and equality.[15]

[15] Michael Friedman emphasizes the same transformative approach in Kant's philosophy of nature. In the *Physical Monadology*, written in the pre-critical period, Kant attempted to reconcile Newtonian natural philosophy with Leibniz's metaphysics by altering both. In

Most importantly for this book, the same point applies to Kant's interaction with the Euclidean tradition. Kant's fundamental reworking of the Euclidean theory of magnitudes will be the topic of Chapter 7. Kant introduces a definition of magnitude suited to explain our cognition of magnitudes in terms of our more basic cognitive capacities, in particular, the categories of quantity and the pure forms of space and time. But he does so in a way that lays the foundation for and retains what he sees as valuable in the traditional theory of magnitudes that set the basic framework for thinking about mathematics into the eighteenth century. Thus, Kant reforms both the Euclidean mathematical tradition and Leibnizian metaphysics of quantity in order to explain the possibility of mathematical cognition as a cognition of magnitudes, and he does so in a way that preserves what he thinks is correct in each.

1.5 Kant's Theory of Magnitudes and Kant Interpretation

There has been a tremendous amount of scholarship on Kant's critical philosophy and his philosophy of mathematics since the appearance of the *Critique*, far too much to survey. If a theory of magnitude is at the foundation of Kant's philosophy of mathematics and also plays a prominent role in his account of experience, one might well wonder how it could be that it has not been sufficiently appreciated until now and take this as a *prima facie* reason against the interpretation given here.

I have already mentioned the most important reason undermining an understanding of Kant's views today: conceiving of mathematics as a science of magnitudes is quite foreign to our modern understanding after the arithmetization of mathematics. If one is not attuned to the theory of magnitude, the significance of Kant's many references to magnitude are easy to pass over as merely an antiquated manner of expression.

But there are other reasons as well, which would have obscured Kant's views even in his own time. Writing in the eighteenth century, Kant could count on his readers to be familiar in a general way with the theory of magnitudes in the Euclidean tradition. It was common to think of mathematics as the science of magnitudes and their measurement. Even Euler, whom one might think unlikely to do so, invokes magnitudes and their measurement to characterize the science of mathematics. For example, at the very beginning of the *Elements of Algebra*, in "Chapter 1: Of Mathematics in General," Euler states that "Mathematics in

the *Metaphysical Foundations of Natural Science* during the critical period, Kant provides a foundation of Newtonian natural science that preserves it, while at the same time rejecting an appeal to Newtonian absolute space and time and providing an alternative understanding of them. This radically reconceives the foundations of Newton's natural science, while preserving what follows from those foundations. See Friedman (1992) and Friedman (2013).

general is nothing but a science of magnitudes, and seeks the means of measuring them," explaining that "Whatever is capable of increase or decrease, or to which something can be added or from which something can be taken away, is called magnitude."[16] This is not to say that Euler had a fully developed view of the foundations of mathematics based on the theory of magnitudes. His remarks come at the beginning of introductory texts, and he quickly moves on. The theory of magnitudes was more of a shared common background than an established and worked-out theory, and even the basic vocabulary was not fixed; the term "magnitude" was often used in ambiguous ways, sometimes by the same author, as was the term "quantity." Kant made progress by distinguishing between two senses of magnitude, *quanta* and *quantitas*, but then sometimes reverts to simply using "magnitude" without specifying which he has in mind. Kant is not renowned for the clear exposition of his views, and he would have needed to do more to make them perspicuous even to his contemporaries.

Kant would have helped matters by devoting a monograph or at least an essay to the philosophy of mathematics, but he did not.[17] One can only wish that Kant had written a work that does for his philosophy of mathematics what the *Metaphysical Foundations of Natural Science* does for his account of Newtonian physics. In fact, Kant may have had such a work in mind. He refers to the topic of metaphysical foundations of mathematics in several lecture notes and reflections, indicating that it would concern the cognition of magnitudes. In one of his reflections, for example, Kant refers to the "Metaphysics of the doctrine of magnitude or the metaphysical foundations of mathematics" (Reflexionen 14:195–6, 1764–1804), and in lectures given between the appearance of the two editions of the *Critique*, Kant is reported as saying:

> Even mathematics presents a metaphysics: it concerns objects only insofar as they have a magnitude, and reason's general application of principles to all objects lies at the foundation of all mathematics and is its metaphysics. (*Metaphysik Mongrovius* 1782–3, 29:755)

> Even mathematics requires a metaphysics, since for all mathematical cognition the principles of metaphysics [*derselben*] must lie at their foundation, which one can represent as a metaphysics of mathematics. (*Metaphysik Volckmann* 1784–5, 28:636)

[16] Euler (1802, pp. 3–4).

[17] The closest we have is his *Attempt to Introduce the Concept of Negative Magnitudes into Philosophy* of 1764. As noted above, it and the *Inquiry* provided an important impetus toward the critical philosophy, but it was written well before Kant developed some of the most important tenets of his critical philosophy. The present book focuses on Kant's understanding of magnitude in the critical period. Since there is sufficient evidence for the argument I make based on texts in the critical period, I will only incidentally refer to *Negative Magnitudes*; a work devoted to the development of Kant's critical philosophy would provide a thorough treatment.

While it is possible that Kant is simply referring to the account of mathematical cognition in the *Critique*, Kant's distinguishing and denoting a metaphysics of mathematics suggests something more systematic that would constitute its own treatise.[18]

Unfortunately for us, Kant did not write such a work, so we are forced to rely on other texts, and above all on Kant's most extensive account of theoretical cognition, the *Critique*. But the structure of the *Critique* itself is a further obstacle to understanding his account of mathematical cognition and its role in experience. The transcendental deduction of the categories aims to establish the objective validity of the categories by showing that they are conditions of the possibility of experience. The system of principles expands on the transcendental deduction to explain principles that follow from the employment of the categories under the conditions of sensibility and demonstrates that the principles are conditions for the possibility of experience. The Axioms of Intuition and Anticipations of Perception do this for the categories of quantity and quality, respectively, and in accomplishing this task, they also establish the applicability of mathematics to experience. Throughout, the focus is on the conditions of the possibility of experience, and there is no natural place in the organization of the *Critique* for a thorough account of mathematical cognition itself. Kant does give a partial account in the Discipline of Pure Reason, but his primary aim there is to distinguish it from philosophical cognition, and it is hardly an adequate exposition of his views. Moreover, the focus of the system of principles on the conditions of experience seems to relegate Kant's discussion of magnitudes in the Axioms and Anticipations to only an issue of the applicability of mathematics to objects of experience, which can be taken to suggest that pure mathematics is something separate that is then applied to them.

As a consequence, we must glean what we can from his surprisingly few and disparate discussions of mathematics. Most of these occur in the context of arguing for a particular claim: that mathematical cognitions are synthetic, for example, or that there are no axioms of *quantitas* and hence no axioms of arithmetic. The arguments are often brief even for the point he is making, and are not intended to be a full explication of mathematical cognition. Sometimes,

[18] On the other hand, Kant's reflections in the late 1790s include the claim that a philosophical foundation of mathematics is unthinkable (21:240; see also 21:242, 21:555, 22:544). He gives an example, which reveals in what sense it is impossible: philosophy cannot prove mathematical propositions. This is something on which Kant insisted throughout his career, from the *Inquiry* in the period of 1763–4, through 1797–9. The contrast with the *Metaphysical Foundations of Natural Science* derives in part because the latter work presupposes mathematics, and is able to establish *a priori* results based on mathematical constructions. In contrast, a *Philosophical Foundations of Mathematics* could at best provide an account of the cognitive foundations of mathematics. It is such a work we might have wished for.

his comments are in such opaque contexts, such as his discussion of number in the Schematism, that they bring more confusion than clarity. Any investigation of Kant's philosophy of mathematics must therefore struggle to put these pieces together.

Moving forward again to the twentieth century, there were additional factors that led to relative neglect of Kant's philosophy of mathematics by many mainstream Kant scholars, and hence a failure to uncover his theory of magnitudes. I noted above that not everyone fell in line with the Russellian view that Kant's philosophy of mathematics was deeply mistaken and had no value, but the view was widely influential. Similar judgments were passed on Kant's views on natural science after Einstein's overthrow of Newtonian physics. These assessments encouraged many who worked on Kant to jettison his views on mathematics and natural science and focus instead on Kant's metaphysics and epistemology and his broader philosophical contributions.[19] This reinforced an understanding of Kant's notion of experience as an ordinary, everyday kind of shared human experience that need not take into account Kant's philosophy of mathematics or natural science. As mentioned above, the strong distinction Kant draws between philosophical cognition and mathematical cognition may have reinforced this approach to Kant's philosophy. The result was that Kant's philosophy of mathematics mostly remained of interest to a minority of philosophers of mathematics, while it was for the most part passed over by those interested in Kant's philosophy more generally, especially in the Anglo-American tradition.

There was yet a further reason for the neglect of Kant's theory of magnitudes in much Kant scholarship. Many rightly saw Kant's philosophy as an important reaction to Hume's views, especially with regard to causation, and they paid particular attention to Kant's response to Hume in the Analogies of Experience, above all the Second Analogy, which drew attention away from the role of the Axioms of Intuition and the Anticipations of Perception and their treatment of magnitudes. This is not to say that mathematical cognition was deemed irrelevant to Kant's account of human experience. There was appreciation of the importance of Kant's claim that mathematics is synthetic *a priori* in Kant's argument for the critical philosophy, and of the fact that Kant aims to establish the applicability of mathematics to experience in the Axioms of Intuition. But there was relatively little attention paid to the role of Kant's theory of magnitudes, and his account of experience was thought to be largely detachable from the details of his philosophy of mathematics.

Despite all these obstacles, however, there was continued interest in Kant's philosophy of mathematics among some philosophers, particularly some

[19] There were notable exceptions of course; this does not do justice to the neo-Kantian movement nor to logical positivists, who attempted to learn from Kant's interaction with mathematics and science even while disagreeing with him on fundamental issues.

philosophers of mathematics. In the 1960s, the work of Jaakko Hintikka and Charles Parsons led to a resurgence of interest in Kant's philosophy of mathematics in Anglo-American philosophy, and generated a debate on the roles of the singularity and immediacy of intuition. A great amount of excellent work on various aspects of Kant's philosophy of mathematics has been published since – too much to list here – and a great deal of it does not require an appreciation of Kant's theory of magnitudes. Nevertheless, a new appreciation of the importance of Kant's understanding of magnitude emerged in Parson's 1984 "Arithmetic and the Categories," while Michael Friedman's 1992 *Kant and the Exact Sciences*, for the first time to my knowledge, explicitly connected Kant's views of magnitude to the Eudoxian theory of proportions in Euclid and gave an account of Kant's distinction between two sorts of magnitude, *quanta* and *quantitas*. The present book was inspired by and is an extension of this line of research. It provides a deeper analysis of Kant's views and their relation to the Euclidean tradition and a places greater emphasis on the systematic influence of the theory of magnitudes on Kant's philosophy of mathematics and his account of experience. It also argues that there is a role for intuition in representing magnitudes, and consequently in both mathematical cognition and the cognition of appearances, that has not been recognized before.

1.6 Overview of the Work

The book divides into three parts. The first establishes that Kant's theory of magnitudes is fundamental to his thinking about mathematical cognition and his account of experience in the *Critique of Pure Reason*. Chapter 2 gives breathing space to Kant's theory of magnitudes by addressing potential confusions about the structure of the *Critique* and Kant's treatment of space, time, and mathematics that can obscure the significance of that theory. Kant's *Critique* makes important claims about space, time, and mathematics in both the Transcendental Aesthetic and the Axioms of Intuition, claims that appear to overlap in some ways and contradict in others. Both discuss mathematics and lay claim to establishing the applicability of mathematics to experience, yet each discusses space and time in quite different ways, particularly with respect to the relative priority of parts to whole. Against this background, most interpretations of the Axioms of Intuition only attribute to it the role of establishing the applicability of mathematics to experience or introducing a metric, and have in the process overlooked the broader significance of Kant's theory of magnitudes. Chapter 2 argues for an interpretation that accords the Axioms of Intuition more importance than it is usually given for all mathematical cognition, including pure mathematical cognition.

Chapter 3 focuses on Kant's Axioms of Intuition in order to clarify his notion of magnitude and reconstruct his arguments for the principle of the

Axioms, which have not been properly understood. It shows that in Kant's view, mathematical cognition depends on the representation of magnitudes and that our apprehension of appearances is directly tied to pure mathematical cognition. It also reveals a role for intuition in representing magnitudes as *quanta*, a role that makes pure mathematical cognition dependent on the representation of concrete singulars in space and time and directly connects mathematical cognition to the representation of concrete singular objects of experience. Chapters 2 and 3 together establish that Kant's philosophy of mathematics is grounded in a theory of magnitude, a way of thinking of mathematics radically different from our own. I next take up Kant's distinction between extensive and intensive magnitudes in Chapter 4, and address an important problem for the former whose resolution points to a role for a continuous synthesis underlying our representation of continuous magnitudes. Chapter 5 then considers more closely the nature of intuitive representation, and what it means to say that intuition is singular. It reconciles the singularity of intuition with its role in the representation of a continuous homogeneous manifold. It then considers Kant's understanding of the distinction between concrete and abstract in order to explain the relation between the singularity of intuition and *in concreto* representation, and explains the sense in which *quanta* and *quantitas* are relatively abstract concepts whose objects may nevertheless be concrete.

The chapters of Part I proceed as far as possible through a close reading of Kant's texts. Before looking more deeply into Kant's views, Chapter 6 provides some historical background to the theory of magnitudes in the Euclidean mathematical tradition. This interlude begins with an examination of the *Elements* without presupposing prior acquaintance with it, briefly discusses the long Euclidean tradition that followed, and touches on a few of Kant's immediate predecessors. While necessarily incomplete and selective, it provides the backdrop for Kant's theory of magnitudes.

Part II of the book then explains the foundation of Kant's theory of magnitudes and the role of intuition in it. Kant does not simply adopt the theory of magnitudes inherited from the Euclidean tradition; he deepens and reworks it in order to explain the possibility of mathematical cognition. Chapter 7 shows that Kant defines magnitude by appeal to the notion of what I call "strict" homogeneity, a notion that reflects the limits of conceptual representation and reveals the need for intuition to represent magnitudes. It also shows that according to Kant, the categories of quantity allow us to have cognition of the part–whole relations of magnitudes. In Kant's view, the part–whole composition relations of a homogeneous manifold in intuition are a distinctive and essential feature of mathematical cognition and the mathematical properties of appearances.

Chapter 8 shows how Kant fundamentally reworks the metaphysics of quantity found in Leibniz, Wolff, and his followers by introducing a distinction between *quanta* and *quantitas* and a role for intuition in representing *quanta*. Kant thereby transforms both the theory of magnitudes derived from the Euclidean tradition and the rationalist metaphysics of quantity to ground his theory of mathematical cognition.

The work of the previous chapters reveals that Kant's account of magnitudes is mereological. This leaves a gap between Kant's mereological account of magnitudes and the rich mathematical notion of magnitudes found in Euclid and the Euclidean tradition. Chapter 9 addresses this gap. An analysis and reconstruction of the assumptions of Euclid's and Kant's theories of magnitude reveals that both presuppose a general theory of pure concrete measurement. Further analysis reveals that equality plays a pivotal role in bridging the gap between the mereology and mathematics of magnitudes. It also provides evidence that Kant was aware of that role. Kant almost certainly had more to say about the foundations of the general theory of magnitude and its relation to mathematics and the world, but it is at this point that we reach the limits of Kant's explicit theorizing in either published or unpublished texts. The concluding chapter briefly surveys the results of the inquiry while indicating further work to be done.

PART I

Mathematics, Magnitudes, and the Conditions of Experience

2

Space, Time, and Mathematics in the *Critique of Pure Reason*

2.1 Introduction

Kant makes important claims about space, time, and mathematics in both the Transcendental Aesthetic and the Axioms of Intuition, claims that appear to overlap in some ways and contradict in others. Commentators have offered various interpretations to resolve these tensions, but most of them obscure the role of the Axioms of Intuition in Kant's account of mathematical cognition and the nature of experience. Those who have considered the Axioms of Intuition agree that it is at least intended to justify the application of mathematics to the objects of experience. Some have held that the Axioms of Intuition also concerns a specific part of pure mathematics. Even these latter interpretations, however, underestimate the role of the Axioms in our cognition of both mathematics and experience.[1] I argue in what follows that the outcome of the Axioms is twofold, concerning not only the applicability of mathematics but the possibility of any mathematical cognition whatsoever, whether pure or applied, general or specific. The interpretation for which I argue clears up some potential confusions concerning the treatment of space, time, and mathematics in the Transcendental Aesthetic and the Axioms. It also allows us to see that the Axioms of Intuition contains a substantial contribution to Kant's theory of mathematical cognition that is at the heart of his account of our cognition of experience.

There are, I think, various reasons why the Axioms of Intuition and the theory of magnitudes appearing in it have not earned more attention.[2] First, many have held that Kant's primary goal in the *Critique of Pure Reason* is to respond to Hume's skepticism about causation, a response that culminates in the Second Analogy. Since the Axioms of Intuition makes little or no direct contribution to the argument leading up to the Second Analogy, many have

[1] As noted in Chapter 1, I will sometimes call the Axioms of Intuition simply the "Axioms," capitalized, to signify the section of the *Critique* with that title, and in either case refer to it in the singular.

[2] Some of these coincide with the reasons adumbrated in the Introduction for the neglect of Kant's theory of magnitudes more generally. See Section 1.5 above.

passed over them without serious investigation.[3] Second, and more importantly, the point of the Axioms is obscured by misunderstandings concerning, on the one hand, the Transcendental Aesthetic treatment of mathematics and of space and time, and on the other, the Axioms of Intuition treatment of them. Third, there is a tension between the primary aim and the results of the Axioms. On the one hand, the Axioms of Intuition is part of the System of Principles, whose role in the *Critique* is to articulate the conditions for any possible experience. On the other hand, the argument establishing those conditions grounds a principle valid for all mathematical cognition whatsoever. This result goes beyond the aims of the System of Principles and therefore might be overlooked. Nevertheless, it is fundamental to Kant's views on mathematical cognition and the mathematical character of experience.[4]

I will sort through these issues by focusing on two contrasts between the Transcendental Aesthetic and the Axioms. Section 2.2 examines the apparently conflicting claims Kant makes concerning the possibility and applicability of mathematics in the Aesthetic and the Axioms. I consider two accepted strategies for resolving the conflict and show that they are both wanting. Section 2.3 shifts to Kant's treatment of space and time in the Aesthetic and the Axioms. Here, too, there is an apparent conflict, which is resolved by appealing to a distinction between space and time and determinate spaces and times. This resolution is a clue to resolving the conflict over the Aesthetic and Axioms' treatments of mathematics. Section 2.4 examines Kant's understanding of determination in general, the determination of spaces and times, and construction. Section 2.5 focuses on the Axioms and the role of determination, construction, and magnitudes in Kant's account of mathematical cognition. The interpretation that emerges resolves a lingering issue concerning the

[3] See Peter Strawson (1989), p. 31, and Robert Paul Wolff (1963, pp. 228–231, 238), for explicit articulations of this view. More recently, Sebastian Gardner has stated that "the section which is crucial for the *Critique's* legitimation of the metaphysics of experience is the Analogies of Experience (supplemented by the Refutation of Idealism) . . . the Axioms and Anticipations make a relatively slight (and uncontentious) contribution to Kant's objectivity argument" (1999, p. 166). Gardner accordingly gives the Axioms relatively little attention.

[4] While neglect of the Axioms of Intuition and magnitudes has been the norm, it is not universal. Michael Friedman, for example, devotes attention to Kant's notion of magnitude and its two species, *quanta* and *quantitas* (see Friedman (1992), chapter 2), work that inspired my investigations. My approach differs from his, however, in focusing above all on the Axioms, and in explaining the point of the Axioms in order to highlight the importance of Kant's theory of magnitudes for Kant's views on any mathematical cognition whatsoever. Beatrice Longuenesse's careful and penetrating treatment of the Axioms also does not fit the norm (see Longuenesse (1998)). My interpretation of the role of the Aesthetic and the Axioms is, I think, in broad agreement with hers, although we differ on the nature of the determination underlying determinate spaces and times. I will return to this point in Section 2.3 below.

relation between the possibility and applicability of mathematics, which is addressed in Section 2.6. That puts us in a position in Section 2.7 to resolve any apparent conflict between the Aesthetic and Axioms claims concerning mathematics. I respond to two objects in Section 2.8 and clarify the way in which Kant formulates the principle of the Axioms. Section 2.9 concludes that the Axioms of Intuition contains substantive claims about any mathematical cognition whatsoever and that the concept of magnitude is at the heart of Kant's account.

2.2 Mathematics in the Transcendental Aesthetic and the Axioms

I begin with a few clarifications. The title "Axioms of Intuition" is misleading, because it does not argue for any axioms. The section immediately following the Axioms of Intuition is the Anticipations of Perception, and Kant calls the principles of the Axioms and Anticipations mathematical. He explains, however, that these principles are not themselves mathematical; they are called mathematical because they explain the possibility of mathematical principles (A162/B201–2). The mathematical principles whose possibility is explained presumably include geometrical axioms, such as Euclid's postulates; nevertheless, the Axioms of Intuition argues for just one principle that is not itself an axiom.

Kant altered the formulation of the principle from the first to the second edition of the *Critique*. The principle in the A-edition is: "All appearances are, in accordance with their intuition, extensive magnitudes" (A162). In the B-edition it is "All intuitions are extensive magnitudes" (B202). We will return to the significance of Kant's reformulation of the principle in Section 2.8; for now, I will defer to the B-edition formulation.

Kant states that the Transcendental Aesthetic is a science of all principles of *a priori* sensibility, and for humans, those are the *a priori* forms of space and time (A21/B35). Kant's primary goal is to argue that our concepts of space and time originate in pure *a priori* intuitions of space and time. In the process, however, Kant establishes further important features of space and time. For example, he argues that space and time can only be represented as single; that a plurality of spaces and times can be viewed only as parts of the single unique space or time, and not as composing it; and that space and time are regarded as infinite.

Since Kant states that geometry is a science that determines the properties of space synthetically and *a priori* (B40), and the Transcendental Aesthetic is the science of all principles of *a priori* sensibility, one might expect Kant to establish some claims of or about mathematics as well. Indeed, geometry does figure importantly in the B-edition's transcendental exposition of the concept of space (and the corresponding third argument in the A-edition), where Kant explains the possibility of geometry as synthetic *a priori* cognition (B40–1; A24).

Kant also appeals to geometry to establish that it is "indubitably certain" and in fact "as certain and indubitable as can ever be demanded of a theory that is to serve as an organon" that space is an *a priori* form of intuition and hence a merely subjective condition of all outer objects (A46–9/B63–6).

Finally, geometry plays a role in Kant's dispute with the Leibnizian-Wolffian account of space.[5] Kant draws a contrast between his claim that space and time are ideal and the Leibnizian-Wolffian claim that space and time are mere relations of appearances abstracted from experience. On Kant's account, space and time are ideal, and all objects of the senses are mere appearances; as a consequence, space and time have (and only have) appearances as the field of their validity. But space and time are also two sources of synthetic *a priori* cognition, "of which especially pure mathematics in regard to the cognitions of space and its relations provides a splendid example"; hence, these synthetic *a priori* cognitions apply to (and only apply to) appearances (A38–9/B55–6).[6] Kant contrasts his account with the Leibnizian-Wolffian:

> they can neither offer any ground for the possibility of *a priori* mathematical cognitions (since they lack a true and objectively valid *a priori* intuition), nor can they bring the propositions of experience into necessary accord with those assertions. On our theory of the true constitution of these two original forms of sensibility both difficulties are remedied. (A40–1/B57)

Thus, Kant takes his Transcendental Aesthetic arguments that space and time are ideal and are the sources of the synthetic *a priori* cognitions of mathematics to show that experience must necessarily accord with the propositions of mathematics and, in particular, geometry.

Kant, however, seems to attribute some of the same roles to the Axioms of Intuition. He states, for example, that the principle of the Axioms makes pure mathematics applicable to objects of experience "in its complete precision" (A165/B206). Kant also asserts that the application of mathematics to experience rests on the principles of the Axioms and the Anticipations, and that the objective validity of mathematical principles (which requires their applicability to objects of experience) is grounded in these principles (A160/B199).

[5] I use the term "Leibnizian-Wolffian" to describe various aspects of the rationalism that was widespread in seventeenth- and eighteenth-century Germany and was so strongly influenced by Leibniz. Wolff was influenced at least as much by the scholastic tradition as by Leibniz, and their views differed in ways that are sometimes important; I will often elide differences, however, when what they had in common is more important.

[6] Kant's emphasis at this particular point in the text is on the *restriction* of the validity of space and time to appearances, but it is clear that he is taking it for granted that they are valid of appearances, and similarly for the synthetic *a priori* cognitions of mathematics; the rest of the paragraph, including the passage next quoted, makes this clear.

Moreover, Kant seems to hold that the Axioms and the Anticipations ground the possibility of mathematical principles themselves, stating that mathematical principles all acquire their possibility from the principles of the Axioms and Anticipations (A162/B202). Kant also refers to both the possibility of mathematical principles and their objective validity, the latter of which entails their applicability to objects of experience. He explains that mathematical principles rest on the synthetic principles of pure understanding for "their application to experience, thus their objective validity, indeed the possibility of such synthetic *a priori* cognition" (A159–60/B198–9), and then repeats the point when he states that he will include among the synthetic principles of pure understanding "those on which the possibility and the objective *a priori* validity of the latter [i.e., mathematical principles] are grounded" (A160/B199). The role for the Axioms in grounding the possibility of mathematical cognition is reinforced by Kant's claim that they "provide the principle of the possibility of axioms in general," about which he says "even the possibility of mathematics must be shown in transcendental philosophy" (A733/B761).

One might think that Kant's references to the possibility of mathematical principles are to the possibility of their applicability. In that case, the Axioms would concern only the applicability of mathematical principles after all. However, I think that the context of the passages from A159–60/ B198–9 shows that by "such synthetic *a priori* cognition" Kant means the synthetic *a priori* cognitions that belong to mathematics itself, not cognition of their applicability. And the discussion at A733/B761, which refers to axioms "in general" fairly unambiguously refers to the possibility of mathematical principles themselves.[7]

Finally, the Axioms of Intuition includes a discussion of arithmetical and geometrical propositions and the status of axioms, which concerns mathematics itself rather than merely its applicability.[8] Thus, the Axioms, as well as the Aesthetic, seems to concern both the possibility of the synthetic *a priori* propositions of mathematics and their application. Kant's assertions can give the impression that the Axioms of Intuition is a redundant reworking of what was established in the Aesthetic, and this has been the opinion of some Kant scholars.[9]

[7] There is, however, a more sophisticated version of this objection. One might hold that Kant is discussing the possibility of mathematical principles themselves but maintain that he means their *real* possibility, and that their real possibility is tantamount to their applicability. We will be in a position to respond to this objection after we have a better understanding of Kant's views and will return to it in Section 2.6 below.

[8] While both the principles of both the Axioms and the Anticipations are called mathematical, the Axioms of Intuition is often singled out for their relation to mathematics. I will address the reasons for this below.

[9] This view is most notably endorsed by Vaihinger. He thinks that Kant establishes the "apriority of pure mathematics as such," only because he confuses two notions of space articulated in the Aesthetic. Vaihinger claims that Kant's account of space as "an

To avoid attributing a redundancy that undercuts the need for the Axioms and the claims Kant makes in it, it is more charitable to look for a difference between the Aesthetic and the Axioms that accords each its own role and corresponds to some very general division in mathematics. One such strategy claims that the Aesthetic concerns the synthetic *a priori* status of pure mathematics, while the Axioms of Intuition establishes the validity of applying mathematics to experience.[10] This suggestion has the virtue of corresponding to a distinction between pure and applied mathematics that appears in Kant and, albeit with a different understanding, also appears in modern mathematics.

In fact, the issue of the applicability of mathematics was forced on Kant by Christian Wolff and his followers. They wished to avoid conflict between the conclusions of mathematics and metaphysics and, in particular, the conflict between the geometrical proof of the infinite divisibility of extension and a metaphysical doctrine that rejected the infinite divisibility of physical monads. They did so by claiming the ideality of space and denying that mathematics applies to metaphysical reality.[11] In the last paragraph of the Axioms, Kant is at pains to refute the claim that mathematics does not apply to the objects of experience, which confirms that the Axioms of Intuition concerns the applicability of mathematics.

Unfortunately, this interpretation cannot be sustained; as noted above, the Transcendental Aesthetic already asserts the applicability of mathematics to experience, and the Axioms of Intuition concerns the possibility of mathematics itself, not just its application.[12]

unconscious, potential transcendental form of intuition, which is there before all experience" is "sufficient for the validity of the application of the propositions of mathematics to all empirical objects" (Vaihinger (1922, vol. 2, p. 88)). Vaihinger's interpretation therefore leaves no need for the Axioms of Intuition in establishing the applicability of mathematics to experience. Bennett states that the Axioms should not have been located in the Analytic at all, since they are a direct consequence of the Aesthetic (Bennett (1966, p. 170)); see also Robert Paul Wolff: "According to the division of the *Critique* which Kant establishes in the Introduction, [the Axioms] should not really exist at all, for mathematics is supposed to be dealt with in the Aesthetic" (1963, p. 228).

[10] For examples of this view, see Kitcher (1982, §5), and Walsh (1975, pp. 110–11).

[11] The Wolffian position was openly attacked at the Berlin Academy in 1745–7. The Introduction of Friedman (1992) gives a helpful account of these issues.

[12] Some have thought that the Axioms of Intuition argues for substantive claims about mathematics apart from its application to the objects of experience but are nevertheless committed to the view that the Axioms *should* only concern the application of mathematics. Kemp Smith, for example, thinks that the Axioms of Intuition makes substantive claims about mathematics, but holds that these claims are only "very externally connected to the main argument." He sees little relation between these claims and the rest of the Axioms; see Kemp Smith (1979, p. 348). Walsh states that he finds Kant's "attempt to connect his general principle [of the Axioms] with the part played by mathematics baffling"; see Walsh (1975, p. 116). At least part of his bafflement can be traced to his

One might nevertheless attempt to shore up the interpretation by assigning different dates of composition to the two sections. One could then claim that the Aesthetic reflects a less mature view concerning the applicability of mathematics, a view that Kant supplements and corrects in the Axioms.[13] This solution is dubious, however. If the Axioms reflected Kant's mature position and was inconsistent with the Aesthetic, we should presume that he would have altered the latter, or at the very least given us some indication that the Axioms of Intuition supersedes the Aesthetic on this issue. Kant does neither through two editions. Thus, the strategy of assigning pure mathematics to the Aesthetic and applied mathematics to the Axioms is not as attractive as it might at first seem. This is not to say that Kant doesn't have a distinction between pure and applied mathematics, only that he does not use it to distinguish the treatment of mathematics in the Aesthetic from that in the Axioms.[14]

consistent description of the Axioms as concerning merely the applicability of mathematics to the objects of experience.

In their translation of the *Critique*, Guyer and Wood cite R5585 (18, pp. 241–2), one of Kant's reflections dated 1779–81, which suggests to them that the Axioms and the Anticipations are called mathematical principles "because they are the conditions of the possibility of *applied* mathematics" (Kant (1998, p. 284n58, my emphasis)). The passage reads:

1. Possibility of (pure) mathematics.
2. Application [struck out]. Possibility of applied. For all things (as appearances) have a magnitude: extensive and intensive. (Through it mathematics receives objective reality. It does not pertain to entia rationes.)

If the reflection is indeed about the Axioms and Anticipations, the first item actually supports the idea that they concern *both* pure and applied mathematics.

In contrast, others have recognized that it is part of the purpose of the Axioms to argue for substantive claims about mathematics apart from its application. See Brittan (1978), pp. 97ff.), Guyer (1987, pp. 190ff.), and Longuenesse (1998, p. 274).

13 Walsh, for example, endorses this view. He states that as Kant writes the Analytic, he "now realizes that the notion of a space-time world, i.e. of a set of objects in space and time, is far more difficult than was at first suggested.... It was only in the Analytic that Kant made any attempt to work out this story in detail, and it accordingly follows that the discussions of the latter must take precedence over those of the Aesthetic" (1975, pp. 112–13).

14 The modern distinction between pure and applied mathematics can also lead to anachronistic misunderstandings of Kant. In contemporary accounts of pure mathematics, geometry is axiomatized and formalized in a way that is often interpreted as abstracting from any spatial interpretation. In Kant, on the other hand, all pure mathematics is intimately related to pure space and time. The contemporary distinction between pure and applied mathematics may suggest that in Kant's view non-Euclidean geometries are logically possible, while only Euclidean geometry is really possible – a real possibility determined by the form of our intuition of space. Kant, however, has no place for non-Euclidean geometries as logical possibilities, since for him any geometry will necessarily

A more promising strategy for distinguishing between the two treatments is to claim that the Aesthetic establishes some general "topological" features of space and time, and hence of the objects that appear in them, while the Axioms of Intuition establishes that a metric can be applied to space and time, and hence to the objects that appear in them.[15] On this interpretation, the intended features of space and time only loosely correspond to the modern understanding of topology in being very general, non-metrical properties of space and time; they are general features of space and time, such as their singularity and infinity, and the feature that the parts of space and time can be viewed only as parts of a single unique space or time, not as composing them. Kant claims in the Aesthetic that space and time have these general features; moreover, Kant calls geometry the science that determines the properties of space synthetically and *a priori*; understanding this claim as concerning very general, non-metrical properties lends support to this line of thought. Furthermore, it is plausible that the introduction of a metric into space and time explains the applicability of mathematics "in all its precision." This interpretation has the virtue of corresponding, at least in rough, to a modern conception of a possible division within mathematics between more general properties of space, on the one hand, and more specific sorts of properties, on the other.

One difficulty with this interpretation concerns Kant's claim in the Aesthetic that his account of space and time both grounds the possibility of *a priori* mathematical cognitions and brings the propositions of experience into necessary accord with mathematics (A40–1/B57). I have already noted

require going beyond logic to intuition, and the only form of intuition available is Euclidean. For more on resisting an anachronistic use of the distinction between pure and applied mathematics, see Friedman (1992, pp. 55–6, 66, 98–104). See also Longuenesse (1998, pp. 274–5).

15 See, for example, Gordon Brittan (1978, pp. 92–4) and Paul Guyer (1987, p. 191). Guyer holds that Kant establishes that certain topological "features of shape" hold for all objects of empirical intuition, while the Axioms of Intuition establishes that a metric can be applied to those objects. It is important to note, however, that Guyer holds that this is a description of what Kant actually accomplishes rather than what he thought he was establishing. See Guyer (1987, p. 191, especially n. 6).

Brittan (1978, pp. 92–4) holds that the principle of the Axioms serves to rule out merely possible worlds that are not "really" possible worlds, that is, worlds in which there are objects that do not stand in relations that allow the introduction of a metric. Note that the claim that the purpose of the Axioms is to establish the possibility of introducing a metric is not incompatible with the view that the Axioms of Intuition merely concerns the applicability of mathematics to experience, if the purpose of the Axioms is restricted to establishing the ability to introduce a metric *in the field of experience*. Walsh (1975, p. 111), for example, appears to hold this view. Brittan, in contrast, holds that the metrical properties are established for space and time, which in turn explains the applicability of mathematical properties to the objects presented in them.

that the very general properties of space and time he establishes in the Aesthetic, such as their unity and the priority of the whole over the parts, do not constitute topology in the modern mathematical sense. It is also the case that Kant does not think of them as mathematical; even if they might have consequences for mathematics, he never refers to them as mathematical propositions. This undermines a topological reading of Kant's claims about mathematics in the Aesthetic.

Although the topological/metric distinction will not allow us to reconcile Kant's claims about mathematics in the Aesthetic with those in the Axioms, I think it is closer to the mark than the pure/applied interpretation.[16] We are nevertheless left without an adequate interpretation of the role of mathematics in the Aesthetic and the Axioms. There is, however, another tension between the Aesthetic and the Axioms, and a resolution of that tension will lead to a different understanding of the point of the Axioms and a better account of the difference between the Aesthetic and Axioms treatments of mathematics.

2.3 Space and Time in the Transcendental Aesthetic and the Axioms

The Aesthetic claims that space and time can be represented only as single and all-encompassing and in such a way that their parts cannot precede them. The Aesthetic argues that for this reason the original representations of space and time are intuitions (A24–5/B39, A31–2/B47). On the other hand, the Axioms of Intuition asserts that all intuitions are extensive magnitudes and that an extensive magnitude is one for which "the representation of the parts makes possible the representation of the whole (and therefore necessarily precedes the latter)" (A162/B203). But space and time are intuitions. Hence, we have an apparent contradiction between the two texts: space and time can be represented only in such a way that the whole precedes the parts, and yet the representation of the parts makes possible and precedes the representation of the whole.[17] I will call this the part–whole priority problem.

[16] Kant is concerned with the conditions of measurement, and the Axioms of Intuition establishes an important necessary, although not sufficient, condition of measurement, taken in a particular sense. We will first be a position to explain this point in Chapter 9.

[17] More carefully, Kant states in the Aesthetic that "the parts of space cannot precede the single all-encompassing space as its components [*Bestandteile*] (from which its composition would be possible)" (A24–5/B39). In the case of time, Kant does not mention composition (A31–2/B47). Nevertheless, one might attempt to dodge the contradiction by claiming that the Aesthetic only asserts that the parts of space (and time) cannot precede the whole in such a way that allows the parts to *compose* the whole. That will not solve the difficulty, however, since the Axioms' claim that the whole presupposes the parts explicitly concerns parts that compose the whole (B201n). The solution to the contradiction lies elsewhere.

One might think that the Aesthetic's and Axioms' accounts of the representation of space are indeed contradictory and that this contradiction is a sign of tensions in Kant's views or even a deep flaw in Kant's philosophy.[18] It would be more sympathetic, however, to hold that the account Kant offers in the Aesthetic suppresses the role of the understanding in order to focus on sensibility. In this reading, Kant provides an incomplete account in the Aesthetic, which he elaborates in the Axioms once he is in a position to discuss the contributions of the understanding to cognition.[19]

There is good evidence that Kant gives a provisional account of space and time in the Aesthetic. Kant asserts that in the Aesthetic he will "isolate sensibility by separating off everything that the understanding thinks through its concepts" (A22/B36). A footnote to §26 of the B-Deduction, although fraught with interpretive challenges, indicates what Kant may have in mind. Kant draws a distinction between the form of intuition and the formal intuition of space or time. The latter contains a unity in the representation of space or time, and that unity presupposes a synthesis by means of which the understanding determines sensibility. Yet that unity precedes all concepts and belongs to space and time, not to the understanding. Kant states that "in the Aesthetic I ascribed this unity merely to sensibility, only in order to note that it precedes all concepts" (B160–1n). He thereby explicitly acknowledges that a full account of our representations of space and time departs from that given in the Aesthetic. At the same time, he indicates that the unity of space and time presupposes the activity of the understanding, while claiming that this unity is prior to concepts. This suggests that the understanding provides a unity to our representations of space and time but does so without the use of concepts; this is what Kant thinks allows him to isolate sensibility in the Aesthetic and to abstract from what the understanding thinks through concepts.

One might react to this elaboration of his account by declaring that Kant is papering over what is in fact a deep tension; he ought not to assert that he can isolate sensibility from the understanding, and the only recourse is to abandon Kant's Aesthetic account of space in favor of that outlined in the Axioms.[20] I think, however, that this is to give up too easily. Kant clearly thinks that the

[18] For examples of this view, see Vaihinger (1970, vol. 2, pp. 224–8), Kemp Smith (1979, pp. 94–9), and Robert Paul Wolff (1963, pp. 228–9).

[19] Cassirer holds that many of the statements of the Aesthetic are incompatible with claims Kant makes later in the *Critique* because they reflect pre-critical views articulated more than a decade before in Kant's *Concerning the Form and Principles of the Sensible and Intelligible World* (henceforth *Inaugural Dissertation*). He holds that the arguments of the Aesthetic must therefore be taken provisionally (1954, pp. 23–4). See also Walsh (1975, pp. 112–15), Kitcher (1982, pp. 237ff.), and, in particular, Longuenesse (1998, pp. 212–13 and 274n70).

[20] As does Robert Paul Wolff (1963, pp. 229–30).

priority of the whole over parts is not only a fundamental property of intuition, but one that can support a crucial argument in the Aesthetic. The account of our representations of space in the Transcendental Analytic is no doubt an elaboration of Kant's views in the Aesthetic in light of the role of the understanding, but we should resist a repudiation of the Aesthetic's important claims if there is an alternative.

We are presented with an alternative once we see that in Kant's view, the Aesthetic concerns space and time as given intuitions, while the Axioms concerns "determinate" spaces and times – that is, geometrical figures, durations, and the spatial and temporal features of appearances. Consideration of space and time as given intuitions treats them as representations of our sensible faculty abstracted from the concepts of the understanding as well as from all empirical content (A22). According to Kant, this leaves us with *a priori* representations of space and time as single and all-encompassing, representations in which their parts cannot precede the whole. In contrast, concepts of the understanding are employed in the cognition of *determinate* spaces and times. In Kant's view, the conditions for cognizing a determinate space or time require that the representation of its parts precedes the representation of the whole. The conditions for representing space and time as given intuitions are therefore quite different from the conditions for cognizing determinate spaces and times.

This resolution of the part–whole priority problem rests upon Kant's account of the conditions of representing space and time as given intuitions and the conditions of cognizing determinate spaces and times, a distinction rooted in his theory of cognition. It is a resolution that has been endorsed by various commentators.[21] Nevertheless, I think its implications for the role of mathematics in the Aesthetic and the Axioms have not been fully appreciated, for this resolution also allows us to distinguish the role of mathematics in each without appeal to a distinction within mathematics – between either pure and applied mathematics or topological and metrical mathematical properties.[22]

[21] Melnick describes the position particularly clearly (1973, pp. 18–19), as does Brittan (1978, p. 97). See also Walsh (1975, p. 114), Allison (2004, p. 114), Guyer (1987, p. 191), and Longuenesse (1998, p. 274).

[22] Melnick, for example, distinguishes between space and time as given intuitions and determinate spaces and times to differentiate the Aesthetic's and Axioms' treatments of them. Nevertheless, he does not explain why Kant discusses geometry in the Aesthetic or why he claims to have established the applicability of mathematics to experience there. A similar lack of appeal to determinate spaces and times to explain the different treatments of mathematics holds for Walsh (1975, pp. 114–15), Brittan (1978, pp. 95–9), Allison (1983, pp. 94–8), and Guyer (1987, p. 191, especially n. 6).

Longuenesse, in contrast, highlights the importance of determinate spaces and times and their determination in Kant's account of mathematical cognition in her penetrating

In Kant's view, mathematics consists of mathematical cognitions, which are grounded in our cognition of determinate spaces and times, a determination that is effected through concepts and carried out by the understanding. Thus, the Axioms' explanation of the possibility of the synthetic *a priori* cognitions of mathematics takes into account the particular contributions of the understanding. Moreover, the Axioms of Intuition concerns the possibility of all mathematical cognition, whether pure or applied, topological or metrical. In contrast, the Aesthetic explanation of the possibility of synthetic *a priori* cognitions is more general in the sense that it abstracts from the particular contributions of the understanding by means of concepts. Furthermore, the Aesthetic concerns the possibility of synthetic *a priori* cognitions more generally, not just the synthetic *a priori* cognitions of mathematics, although the latter serve as paradigm examples. The remainder of the chapter will flesh out these claims.

2.4 Determination, Construction, and Mathematics

Kant's notion of a determinate space or time rests upon an appeal to the notion of determination, which is found in the philosophy of Christian Wolff and his followers, including Alexander Baumgarten; their views were deeply influenced by Leibniz's rationalism. Kant used Baumgarten's *Metaphysica* for many years in his lectures on metaphysics.[23] Baumgarten introduces the notion of determination in an early section entitled "Being [*Ens*]"; he begins by defining when something is determined, undetermined, determinable, or is a determination:

> What is either posited to be A, or posited not to be A, is DETERMINED. What is however only posited to be A or not-A, is UNDETERMINED.... That which can be determined is DETERMINABLE ...
>
> Those things (notes or predicates [*notae et predicate*]) that are posited in something by determining [it] are DETERMINATIONS. A determination is either positive and affirmative, and if true REALITY, or it is negative, and if true NEGATION.[24]

It might strike the reader as surprising that in a metaphysical treatise discussion of being, Baumgarten writes about positing and not positing predicates that may or not be true of the thing of which they are posited. According to Baumgarten, however, metaphysics is not simply, as one might think of it

account of the Axioms. Longuenesse goes further in calling for a radical rereading of the Aesthetic (1998, pp. 212–28).

[23] Baumgarten's text is presented with Kant's marginalia in volume 17 of *Kant's Gesammelte Schriften*. I will use Baumgarten (2014), and cite references to both.

[24] Baumgarten (2014, p. 107); AA17:33–5. I have left out some of the text, as well as section headings and cross-references.

today, the science of the fundamental nature of reality in contrast to epistemology or conditions of cognition. Baumgarten defines metaphysics as the science of the first principles of human cognition, and ontology is defined as the science of the more general predicates of a being (Baumgarten (2014, pp. 99–100)). There is no clear demarcation between metaphysics, on the one hand, and epistemology or cognition, on the other.[25] "Predicate" and "subject" refer not to linguistic items but to concepts of properties and things; they also refer to properties and things denoted by those concepts, and hence to properties and things themselves insofar as we cognize them, while still allowing that the way something appears to us may not be the way it is.

Although Baumgarten uses the word "concept" sparingly, his references to positing, predicates, and principles of human cognition make clear that what is posited of something is a predicate concept.[26] He also defines the essence of a possible thing as the collection of the essential determinations in it, glossing essence both as substance and as the first concept of a thing (Baumgarten (2014, p. 108)).[27] What is important for us presently is that according to the Leibnizian-Wolffian theory of cognition, determinations are posited by means of concepts. We will return to this point shortly, after presenting an important example of Kant adopting the notion of determination found in Wolff and Baumgarten.

Both Wolff and Baumgarten incorporate Leibniz's doctrine of the complete concept of a thing in their principle of thoroughgoing determination: for a thing to be an individual is to be determined with respect to every possible pair of contradictory predicates.[28] In Baumgarten, Wolff, and Leibniz, the

[25] Baumgarten's *Metaphysica* includes a treatment of psychology, which divides into empirical and rational; the former includes a study of the cognitive faculties "based on experience that is nearest to hand." This analysis of the cognitive faculties nevertheless belongs to metaphysics because it is defined as "the science of the general predicates of the soul," and proceeds through an analysis of such predicates (Baumgarten (2014, p. 198)).

[26] Baumgarten distinguishes between natural metaphysics and artificial metaphysics, and clarifies that metaphysics as the science of the first principles of human cognition is artificial. This sort of metaphysics includes the development of concepts (*conceptuum*) as well as the determination and conception of first propositions. Eberhard, a follower of Baumgarten famous for his attacks on Kant, reworked Maier's German translation of *Metaphysica* in 1783. He added a note to Baumgarten's definition of metaphysics, stating that the first principles of human knowledge contain the most general and abstract concepts (Baumgarten (2014, p. 99)). While the *Metaphysica* is of course attempting to increase our knowledge of metaphysics, there is no attempt to define or study that science in abstraction from the conditions of human cognition.

[27] Eberhard, a follower of Baumgarten and ardent Wolffian famous for his attacks on Kant, reworked Meier's German translation Baumgarten's *Metaphysics* in 1783. Concerning Baumgarten's definition of metaphysics as the science of the first principles of human cognition, he noted that "these first principles of knowledge contain the most general and abstract concepts." See Baumgarten (2014, p. 99).

[28] Baumgarten, *Metaphysica* §§148–54 (17:56–8).

thoroughgoing determination of an individual with respect to concepts exhausts the possible determinations of a thing.

Kant says of both a concept and a thing that they are determined with respect to contradictorily opposed predicate–concept pairs. Thus, Kant too uses determination with respect to both a thing and the concept of a thing. According to Kant's principle of the determinability of concepts, at most one of two contradictorily opposed concepts can apply to a concept, which follows from the principle of contradiction. According to Kant's version of the principle of thoroughgoing determination, at least one of every pair of contradictorily opposed concepts must apply to a thing. Kant calls this "the principle of the synthesis of all predicates which are to make up the complete concept of a thing" (A572/B600).[29] We will return to these principles in Chapter 5, when we more closely consider Kant's understanding of the representation of individuals.

Although Kant adopts the notion of determination, he alters it in light of his distinction between conceptual and intuitive representation. Kant holds that the Leibnizian-Wolffian tradition treats intuitive representations of sensibility as merely confused conceptual representations of the understanding. It therefore overlooks the fundamental difference between concepts and intuitions and the unique contribution intuitions make to cognition (A270–1/B326–7). This fundamental change in his theory of cognition alters his understanding of determination. In Kant's view, our cognition of an object requires both concepts and intuitions. Kant nevertheless retains the view that in our cognition of objects, determination belongs to conceptual representation; we determine an object of cognition through an act of the understanding employing concepts. Sensibility and intuitions play the passive role of being determinable.

Kant's understanding of the determining/determinable relation is at the heart of his understanding of the form/matter distinction. Form and matter are one of the four pairs of concepts of reflection, but are the most basic and "ground all other reflection, so inseparably are they bound up with every use of the understanding" (A266/B323). Kant says of matter and form:

> The former signifies the determinable in general, the latter its determination (both in a transcendental sense, since one abstracts from all differences in what is given and from the way in which that is determined). (A266/B323)

Kant then discusses five manifestations of the form/matter distinction corresponding to the determining/determinable distinction, in the course of which he states that "the understanding . . . demands first that something be given (at least in the concept) in order to be able to determine it in a certain way"

[29] See Wood (1979, pp. 37ff.) for a helpful discussion of these matters.

(A267/B322–3). In contrast, he says, "the form of intuition (as a subjective constitution of sensibility) precedes all matter (the sensations)," adding that the form precedes "the things and determine[s] their possibility" (A267/B323–4). What we see here is something that will be striking to every reader of the *Critique*, namely, that the form/matter relation, and *eo ipso* the determining/determinable relation, is relative to the level of analysis and to that which is being considered. The form of intuition is determining relative to sensation, which is determinable.

In the *Critique*, however, the most important context in which the determining/determinable relation appears is in regard to our cognition of objects, and in that context, it is the understanding which determines by means of concepts, while intuitions play the role of the determinable. Kant repeatedly appeals to this relation in the context of our cognition of objects. It appears, for example, in his explanation of the categories. He states that the categories "are concepts of an object in general, by means of which its intuition is regarded as **determined** with regard to one of the **logical functions** for judgments" (B128). He also says of understanding that it is "generally speaking, the faculty of **cognitions**. These consist in the determinate relation of given representations to an object" (B137). This is in contrast to the contribution of sensibility, which emerges in Kant's description of the imagination in the B-Deduction:

> Now since all of our intuition is sensible, the imagination, on account of the subjective condition under which alone it can give a corresponding intuition to the concepts of understanding, belongs to **sensibility**; but insofar as its synthesis is still an exercise of spontaneity, which is determining and not, like sense, merely determinable, and can thus determine the form of sense *a priori* in accordance with the unity of apperception, the imagination is to this extent a faculty for determining the sensibility *a priori* . . . (B151–2)

Imagination either mediates between the understanding and sensibility or is the understanding insofar as it determines sensibility, but in either case, sensibility plays the role of the determinable. Since our cognition of objects is the most important context for the determining/determinable relation, it is not surprising that when Kant discusses the contribution of the forms of space and time to our cognition of objects, he almost always refers to them as conditions, rather than as determinations, of what appears in them. In Kant's footnote to §26 of the B-Deduction, which we briefly discussed in Section 2.3 above, Kant states that the understanding determines sensibility through a synthesis that precedes all concepts. This suggests that in this special case, the understanding can determine without employment of concepts. Nevertheless, in the context of our cognition of objects, Kant's view is that the understanding determines by means of concepts.

Kant uses the notion of determination to give an account of the determination of spaces and times. Kant's discussions in the B-Deduction, the Axioms of Intuition, and the Discipline of Pure Reason show that he thinks of determinate spaces as lines and surfaces that we are capable of drawing in thought (B137–8, A162–3/B203, and A715/B743). A line, circle, and cone constructed in Euclidean geometry are paradigm examples. Determinate spaces also constitute the spatial properties of objects. In §26 of the B-Deduction, Kant states that we, as it were, draw the shape of a house when we apprehend it (*zeichne gleichsam seine Gestalt*) (B162), suggesting that a determinate space constitutes part of our representation of the house.[30] The Axioms of Intuition confirms this; it argues that the apprehension of an appearance is possible only by means of a synthesis through which the representations of a determinate space or time are generated (B202).[31] Determinate spaces therefore comprise any spatial figure that can be either "as it were" or actually "drawn in thought" as well as Euclidean figures constructible in accordance with a straight-edge and compass.

Following Baumgarten's metaphysics, Kant divides determinations into inner and outer determinations, that is, determinations that belong to the object alone, and determinations that are relations between the object and other objects. The determinate space of a house is an inner determination. A full determination of the spatial properties of an object will include its outer determinations as well and hence its spatial relations to other objects. The inner determinations of an object, however, are primary and spatial objects include a determinate space.[32]

Determination involves a combination or synthesis, and the determination that generates a representation of a determinate space or time is quite special. The act of understanding involved is called figurative synthesis and acts on the manifold of space and time through the productive imagination. It is only through this synthesis that a determinate intuition is possible:

> inner sense . . . contains the mere **form** of intuition, but without combination of the manifold in it, and thus does not contain any determinate

[30] Kant's use of the phrase "as it were (*gleichsam*)" ties mathematical cognition to our cognition of experience, while at the same time allowing that the latter does not literally require, e.g., sketching the outline of a house in cognizing it. I will say more about the connection between the two kinds of cognition below, in particular in Section 5.5.

[31] We will consider this point and Kant's argument for it in more detail in the next chapter.

[32] Kant says less in this context about determinate times, but an analogous account would construe a determinate time as the relative temporal ordering and duration of states of a single object, to the extent that those can be determined independently of other objects (if that were indeed possible). In contrast, the outer temporal determinations would be the temporal duration and ordering of an object and its states relative to other objects. The conditions for the determination of these relations are further specified in the Analogies of Experience.

> intuition at all, which is possible only through the consciousness of the determination of the manifold through the transcendental action of imagination (synthetic influence of the understanding on the inner sense), which I have named the figurative synthesis. (B154)

Kant emphasizes that inner sense, that is, time as a mere form of intuition, does not itself contain any determinate intuition. The representation of a determinate intuition requires the action of the understanding on inner sense, which runs through the parts of a space or time to be represented. Here, as elsewhere, Kant cites as examples of this figurative synthesis the drawing of a line and the outlining of a circle (B154, B203).[33] The representations of the parts themselves are generated through a successive synthesis, which begins from a point, and the representation of the whole determinate space or time presupposes a representation of these parts (B203).

Kant's doctrine of construction appeals to and further elaborates his views on the generation of the representations of determinate spaces and times. Kant defines construction as that by means of which an *a priori* intuition corresponding to a concept is "exhibited" (*dargestellt*) (A713/B741). Kant states that through construction, mathematics considers a concept *in concreto*, because exhibiting an intuition *a priori* corresponding to the concept provides it with an object (A715/B743).[34]

As already noted, Euclidean straight-edge and compass constructions are paradigm examples of determinate spaces, but Kant's notion of construction includes any figures that can be generated by the drawing of a line or the sweeping out of a line to generate a surface. At the heart of his understanding of construction is the notion of determination grounded in a special synthesis particular to mathematics. In the Postulates of Empirical Thought, Kant argues that a thing is possible only if its concept agrees with the conditions of experience, and his example explains construction as a determination of space; he states that the impossibility that two straight lines enclose a figure "rests not on the concept in itself, but on its construction in space, i.e., on the

[33] Michael Friedman provides a detailed account of figurative synthesis that emphasizes its importance not only for mathematics but for physics (see Friedman 1992, chapters 1 and 2). See also Beatrice Longuenesse for a thorough but quite different analysis of figurative synthesis, which Kant also calls "*synthesis speciosa*" (Longuenesse 1998, part III).

[34] Kant allows that we can intuit an indeterminate *quantum*, by which he means a space or time that is enclosed within boundaries, and that does not require construction (A426n/B454n). (This claim appears in a footnote to the thesis of the First Conflict of Transcendental Ideas in the Antinomy of Pure Reason, but appears to reflect Kant's own views.) Nevertheless, the cognition of a determinate space or time requires construction in intuition. I will return to indeterminate *quanta* repeatedly below, especially in Chapters 4 and 5.

conditions of space and the determination of it" (A220–1/B268).[35] In the Analogies, Kant claims that the mathematical principles of the Axioms and Anticipations teach us, concerning appearances, how

> both their intuition and the real in their perception could be generated in accordance with rules of a mathematical synthesis, hence how in both cases number-magnitudes [*Zahlgrößen*] and, with them, the determination of the appearance as magnitude, could be used. E.g., I would be able to compose and determine *a priori*, i.e., construct the degree of the sensation of sunlight out of about 200,000 illuminations from the moon. (A178–9/B221)

Kant glosses construction as *a priori* composition and determination. In the Discipline of Pure Reason, Kant states that the form of intuition (space and time) can be cognized and determined completely *a priori*. He then adds that with regard to space and time

> we can determine our concepts *a priori* in intuition, for we create the objects themselves in space and time through homogeneous synthesis, considering them merely as *quanta* ... [this] is the use of reason through construction of concepts ... (A723/B751)

The role of determination in construction is brought out especially well in §38 of *Prolegomena*, where Kant discusses the mathematical basis of the laws of nature:

> Space is something so uniform [*gleichförmiges*] and as to all particular properties so indeterminate, that we should certainly not seek a store of laws of nature in it. On the other hand, that which determines space to assume the shape of a circle, or the figures of a cone and a sphere, is the understanding, so far as it contains the ground of the unity of their constructions. The mere universal form of intuition, called space, must therefore be the substratum of all intuitions determinable to particular objects; and in it, of course, the condition of the possibility and of the variety of these intuitions lies. (Ak.4:321–2)

Space as a given intuition is an indeterminate substratum in which we can determine particular objects through the construction of a concept such as the concept of a circle, cone, or sphere.[36]

[35] Note that Kant mentions two contributions, the determination of sensibility by the understanding, and the conditions of sensibility, and that the latter are described as conditions rather than as a determination, as noted above.

[36] See also A714/B742. Kant's general definition of construction as "exhibiting an *a priori* intuition corresponding to a concept" may seem to leave open the possibility of constructions that are not effected through a mathematical determination. An example that appears to exploit this possibility appears in *The Metaphysics of Morals*: "The law of the reciprocal coercion necessarily harmonizing with everyone's freedom under the

The fact that Kant explains construction by appealing to mathematical determination has important consequences for his conception of mathematics. Since Euclidean geometry requires construction of concepts, geometry depends upon the determination of spaces by means of concepts through the activity of the understanding. Arithmetic also depends upon a construction of "number-magnitudes" in pure intuition, and algebra depends on symbolic construction, both of which require a determination of intuition through concepts by the understanding.[37]

In fact, Kant delimits the field of mathematics by appeal to construction. In the Discipline of Pure Reason, Kant makes clear that mathematics consists of mathematical cognitions and states that mathematical and philosophical cognitions are distinguished not by their objects but by the method according to which they are established. The method of mathematical cognition is the construction of concepts, while the method of philosophical cognition is rational cognition through concepts alone.[38]

Kant explicitly warns against delimiting mathematics and philosophy by their objects. He gives examples of objects that are considered by both: magnitudes (with respect to their totality and infinity), lines, spaces, and the continuity of extension (i.e., its infinite divisibility) (A714–15/B742–3). Even a motion as a description (*Beschreibung*) of space, that is, a figurative synthesis in productive imagination exemplified by the drawing of a line "belongs not only to geometry but even to philosophy" (B155n). Mathematics thus consists of those cognitions established through the construction of concepts, and only those cognitions. Hence, mathematics consists only of cognitions which

principle of universal freedom is, as it were, a **construction**, that is, an *a priori* exhibition of that concept in pure intuition *a priori*, by analogy with [exhibiting] the possibility of bodies moving freely under the law of the *equality of action and reaction*" (6:232–3).

Note that Kant states that it is "as it were [*gleichsam*]" a construction. This does not by itself set it apart from the kind of construction found in theoretical cognition, since Kant uses the same expression in §26 of the B-Deduction to describe the construction of a house (B162). On the other hand, Kant only states that there is a construction of the concept that is *analogous* to presenting the possibility of bodies in accordance with Newton's third law. Kant adds that the doctrine of right *wants* a mathematical exactitude in its determination but that this exactitude cannot be expected (6:232). Kant's tentativeness is clarified in the only other mention of construction in *The Metaphysics of Morals*: "any moral proof, as philosophical, can be drawn only by means of rational cognition *from concepts* and not, as in mathematics, by the construction of concepts" (6:403). Every other mention of construction in Kant is in the context of theoretical cognition and is mathematical, and passages in which Kant describes construction in any detail refer to determinate spaces or times or some determinate object.

[37] For arithmetic, see A178–9/B221, A240/B299, A717/B745, A720/B748, and 4:283; for algebra, which relies on a symbolic construction, see A717/B745, A734/B762. I will give a more detailed analysis of the nature of arithmetic and algebraic construction in an analysis of Kant's philosophy of these particular branches of mathematics.

[38] See Emily Carson (1999) for a very helpful discussion.

ultimately rest on a determination of intuition through a figurative synthesis guided by the understanding.

As we saw in Section 2.2, Kant calls geometry the science that determines the properties of space synthetically and *a priori*. This might suggest that any synthetic *a priori* truth about the nature of space (at least those that are not clearly epistemological in character, for example, that space is ideal) should belong to mathematics. Thus, one might think that the claims of the Aesthetic that space is single and infinite belong to mathematics in virtue of the fact that they are general truths about space. This line of thought supported the interpretation according to which these count as very general mathematical propositions, described as "topological." We rejected this interpretation on the grounds that these propositions do not seem to be mathematical (even if they have implications for mathematics), and because Kant does not refer to them as mathematical. We now see that Kant has a principled way of distinguishing philosophical and mathematical cognition. If, for example, the singularity and infinitude of space can be established through construction, then these claims will belong to geometry in virtue of the construction justifying them; otherwise, they will not. The same condition holds for the purported claims of any other field of mathematics. Note as well that there is nothing to rule out having both philosophical and mathematical cognition of the same thing, such as the infinitude of space.

This is not to say that we cannot consider whether either mathematics as a whole or subfields of mathematics concern a particular sort of object. The point is that one cannot start with the objects to demarcate mathematics. Once we have delimited which cognitions count as mathematical, we are free to consider whether there are objects particular to mathematics, or whether, for example, geometry and arithmetic have objects particular to them. As we shall see shortly, Kant does identify a kind of object particular to mathematics.

2.5 Determination, Construction, and Magnitudes in the Axioms

We have seen that in Kant's account the determination of our cognitions requires the application of concepts of the understanding. Since the Aesthetic concerns sensibility in abstraction from what understanding thinks through its concepts, a treatment of the determination of our cognitions does not belong in the Aesthetic.[39] That treatment can be given only within the Analytic, whose purpose is to explain the contributions of the understanding together with the contributions of sensibility. Thus, the Analytic would be the proper home of a positive account of mathematical cognition.

[39] This is not to say that the claims of the Aesthetic are established through a simple act of abstraction, only that its object of study is intuition apart from the activity of the understanding.

The Analytic contains a systematic presentation of all the synthetic principles that flow *a priori* from pure concepts of the understanding under the sensible conditions of the forms of space and time (A136/ B175). Kant holds that these principles are nothing other than rules for the objective use of the categories. Given Kant's classification of the categories, one would presume that a general treatment of mathematics would fall under the categories of quantity and the Axioms of Intuition correspond to the categories of quantity. Furthermore, as noted earlier, the principles of the Axioms and the Anticipations are called the mathematical principles, and they concern extensive and intensive magnitudes, respectively. Since the principle of the Axioms asserts that all intuitions are extensive magnitudes, one would expect to find a treatment of mathematical determination, and hence of mathematical cognition, that depends on the concept of magnitude.[40] In fact, the concept of magnitude does play this central role.

Kant distinguishes between two kinds of magnitude in the Axioms: *quanta* and *quantitas*. The distinction is quite important to Kant's theory of magnitude, and we will examine them more closely in the next two chapters; we can pass over the differences between them for now, however. What is important at present is that Kant states the following:

> Now of all intuition none is given *a priori* except the mere form of appearances, space and time, and a concept of these, as *quanta*, can be exhibited *a priori* in pure intuition, i.e. constructed ... (A720/B748)

Thus, in Kant's view, concepts of space and time can be exhibited in *a priori* intuition, and hence constructed, if they are constructed as *quanta*, that is, through employment of a concept of magnitude. In his discussion of algebra, it is *quantitas* that is constructed; that is, algebra is also constructed under a concept of magnitude (A717/B745). In a passage cited above concerning the difference between mathematical and philosophical cognition, Kant asserts the centrality of the concept of magnitude in mathematical construction:

> the form of intuition (space and time) ... can be cognized and determined completely *a priori* ... With regard to the [form of intuition] we can determine our concepts *a priori* in intuition, for we create the objects themselves in space and time through homogeneous synthesis, considering them merely as *quanta* ... [this] is the use of reason through construction of concepts ... (A723/B751)

[40] Although Kant calls both the Axioms and Anticipations mathematical principles, and the principle of the Anticipations also concerns magnitude, Kant thinks that the Axioms, which correspond to the categories of quantity, are more fundamental to mathematical cognition. I will discuss the reason for this in Chapter 4.

Kant holds that we can construct the concept of magnitude itself. He goes much further, however. When he discusses the demarcation between mathematics and philosophy, Kant states that mathematical cognition concerns *only* magnitudes:

> The form of mathematical cognition is the cause of its [i.e., mathematics] pertaining solely to *quanta*. For only the concept of magnitudes [*Größen*] can be constructed, i.e. exhibited *a priori* in intuition. (A714–15/B742–3)

Mathematics concerns only magnitudes for the reason that only concepts of magnitudes can be constructed.[41] In other passages we have examined, Kant refers to constructing concepts of Euclidean figures, such as the concept of a circle or cone. We see here that whatever is constructed must fall under the concept of magnitude; in other words, the concept constructed either must itself be the concept of magnitude or, if it is a concept of a specific geometrical figure, such as a circle, must fall under the concept of magnitude.[42]

Kant says almost nothing about construction in the Axioms, but he does discuss the role of the concept of magnitude and the conditions for cognizing determinate spaces and times. Kant defines a magnitude as a homogeneous manifold in intuition in general. We will look more closely at this definition and his argument for the principle of the Axioms in the next chapter; for now, we will focus on the relation between this concept and the determination of spaces and times. In the course of his argument, Kant contends that determinate spaces or times are generated through the synthesis of a homogeneous manifold in intuition, that is, a synthesis "through which space and time in general are determined" (B202–3). A determinate space or time is generated through the synthesis of the homogeneous manifold of space or time, a figurative synthesis which constructs the concept of magnitude, that is, exhibits *a priori* an object in intuition that falls under the concept of magnitude. The result is the representation of a determinate magnitude.[43]

2.6 The Possibility of Mathematics Revisited

Kant's argument of the Axioms makes two key claims that are directly relevant to understanding the role of mathematics in it. The first is the one just

[41] See also A178/B221, quoted above, which suggests that the product of mathematical synthesis is the representation of a magnitude.

[42] See also Kant's discussion of Plato in *A Recently Prominent Tone of Superiority in Philosophy*, written in 1796. Kant speaks of Plato's admiration for the fitness of geometrical figures, such as circles, to resolve a variety of problems "just as if the requirements for constructing certain magnitude-concepts [*Größenbegriffe*] were laid down in them *on purpose*" (8:391). Concepts of particular spatial figures are a kind of magnitude concept.

[43] Kant uses the expression "determinate magnitude" at B203, a point to which I will return below.

mentioned: mathematical cognition is grounded in the synthesis of a homogeneous manifold in intuition, a synthesis "through which space and time in general are determined" (B202–3), and through which representations of determinate spaces and times are generated. Kant's larger argument of the Axioms turns on the second claim: the apprehension of an appearance is grounded in a synthesis that generates a determinate space or time for the appearance, which is the very same synthesis on which mathematical cognition is grounded (B202, B203). The first claim explains the possibility of mathematical cognition itself (A159–60/B198–9, A733/B761), while the second guarantees the applicability of mathematics to experience (A166/B207, A224/B271). Thus, the Axioms of Intuition explains both the possibility of mathematical cognition itself and its applicability.

We are now in a position to respond to an objection raised in note 7 above, which denies that the Axioms of Intuition explains the possibility of mathematics itself by collapsing Kant's claims concerning the possibility of mathematical cognition into its applicability. The objection interprets the Axioms' references to demonstrating the possibility of mathematics as demonstrating its real possibility, where real possibility is tantamount to the applicability of mathematics to the objects of experience.[44] It is apparently supported by Kant's insistence that mathematical cognitions are only of the forms of objects as appearances, and that we have mathematical cognition only on the assumption that there are things that can be presented in accordance with that form; that is, that mathematics is possible only insofar as it applies to things given in intuition (B146–7 and A224/B271). We can now see, however, that what explains the possibility of mathematics is Kant's account of the role of the understanding in employing concepts to determine sensibility through a figurative synthesis that generates representations of determinate spaces and times, and this is distinct from explaining its applicability. The connection between explaining the possibility of pure mathematics and establishing its applicability is tight, because it is forged on the claim that the synthesis underlying mathematical cognition is one and the same as that underlying the apprehension of appearances.[45] Nevertheless, explaining the possibility of mathematical cognition and establishing its applicability to objects of experience is still distinct.

[44] Friedman suggests such a view of the matter when he states that "only *applied* mathematics yields a substantive body of truths" (Friedman (1992, p. 94)).

[45] This point is made very clearly by Longuenesse: "For Kant, explaining the possibility of pure mathematics and deducing the possibility of its application to appearances in a pure (mathematical) natural science are one and the same procedure … because explaining the possibility of pure mathematics is also explaining the possibility of its use in empirical science – that is, the transcendental basis of its application to appearances" (Longuenesse (1998, p. 275)).

Kant's explanation of the possibility of mathematics also sheds further light on the distinction between pure and applied mathematics. As pointed out in footnote 14 above, Kant's understanding of the difference between pure and applied mathematics is quite different from our modern understanding. In Kant's view, pure mathematics is possible only through a determination of space and time and the representation of determinate spaces and times; there is no possibility of describing geometry in a set of axioms that abstracts from all relation to space, for example, and then asking whether the space of experience satisfies those axioms.[46]

2.7 Mathematics in the Transcendental Aesthetic and the Axioms Revisited

We are now in a position to distinguish the Aesthetic treatment of mathematics from that in the Axioms. The Aesthetic is not concerned with explaining the contribution of the understanding in making mathematical cognition possible through the generation of representations of determinate spaces and times. The Aesthetic instead focuses on the contribution of sensibility in making synthetic *a priori* cognition possible. The first contribution is in explaining how it is possible to have *a priori* cognition that goes beyond what is analytically contained in concepts. Kant's answer is that space and time are *a priori* intuitions: that space and time are intuitions provides an extra-conceptual connection between concepts; that they are *a priori* makes it possible for *a priori* cognitions to arise from them (B73). It is in this sense that the Aesthetic explains the possibility of the synthetic *a priori* cognitions of mathematics. As Kant states:

> Time and space are accordingly two sources of cognition, from which different synthetic cognitions can be drawn *a priori*, of which especially pure mathematics in regard to the cognitions of space and its relations provides a splendid example. (A38–9/B55–6; see also A46–7/B64)

The Aesthetic account of the possibility of mathematical cognition nevertheless omits the more particular conditions of mathematical cognition that depend on the understanding. Similarly, the Aesthetic explanation of the applicability of mathematics considers only the contribution of sensibility. The ideality of space and time explains the possibility of *a priori* knowledge of their properties, but it thereby also explains why mathematics applies (and applies only) to those objects. Namely, those objects are given in the pure *a*

[46] This is a point Friedman makes several times, including in his discussion of real possibility: "There is only one way even to think such properties [i.e., "topological" properties of space]: in the space and time of *our* (Euclidean) intuition" (Friedman (1992, p. 93)).

priori forms of space and time and are therefore necessarily subject to those properties of space and time articulated in mathematics (A26/B42, B73, see also A149/B188).

I have appealed to the role of the understanding in generating representations of determinate spaces and times to distinguish between the treatment of mathematics in the Aesthetic and in the Axioms. One might think that the exclusion of determinate mathematical cognitions from the Aesthetic is inconsistent with the transcendental exposition of the concept of space, whose argument presupposes that we have determinate geometrical cognitions and seeks to explain its possibility (B40–1). In my account, however, the Aesthetic is precluded only from explaining or appealing to the intellectual conditions for the mathematical determination of geometrical figures. The argument of the transcendental exposition turns not on how geometry arrives at its synthetic *a priori* cognitions, only on the fact that geometry consists of them; consistent with the form of a transcendental exposition, Kant takes the fact that we have geometrical knowledge as a starting point. For support, he refers the reader to the Introduction, where Kant establishes that the synthetic character of geometrical cognitions follows from a consideration of the content of the concepts involved in a few representative mathematical propositions, while he establishes their apriority by appeal to their necessity and strict universality (B4, B14–17). At no point does Kant appeal to the nature of mathematical determination or explain how the synthetic *a priori* claims arise.

A similar worry could be raised concerning the general 'topological' properties of space and time Kant argues for in the metaphysical expositions; one might think that Kant at least tacitly appeals to the determinate cognitions of geometry when he argues that space is single and infinite. Recall, however, that mathematics and philosophy are demarcated by their method, not by their objects, and that philosophy and mathematics concern some of the same objects, including space. Thus, the same claim concerning space might, in some circumstances at least, be demonstrated both philosophically and mathematically. What is important, however, is that even if Kant tacitly appeals to mathematics to establish such claims, the Aesthetic would be relying on a result of mathematics, not on an account of the mathematical determination required to establish it. Only the latter would overstep the proper scope of the Aesthetic.[47]

[47] Friedman endorses the view that Kant appeals to mathematical truths to establish the claims of the Aesthetic. See Friedman (1992, pp. 63–5) for a discussion of the infinite extent and infinite divisibility of space. Friedman also argues that the claim in the Aesthetic for the priority of "the singular intuition *space* rests on our knowledge of geometry. Our cognitive grasp of the notion of space is manifested, above all, in our geometrical knowledge" (1992, pp. 69–70). Friedman further develops and refines his views in important ways in a later article (Friedman (2000)). I am not here taking a

It is in precisely in this light that I think we should understand an alteration in the metaphysical exposition between the A- and B-editions. In the former, Kant makes a claim about the determination of magnitudes to argue that space is an intuition:

> 5) Space is represented as a given infinite magnitude. A general concept of space (which is common to a foot as well as an ell) can determine nothing in respect to magnitude. If there were not boundlessness in the progress of intuition, no concept of relations could bring with it a principle of their infinity. (A25)

In the B-edition, Kant replaces this with an argument that appeals only to the nature of conceptual representation, and makes no mention of determination. Kant, I think, removes the first argument because it depends on a claim about the determination of magnitude, a claim that he is not in a position to either explain or defend in the Aesthetic if he is going to abstract from the contributions of the understanding.

2.8 Two Objections and Kant's Formulation of the Axioms Principle

I would like to anticipate two objections to my interpretation of the distinction between the Aesthetic's and Axioms' treatment of mathematics, an interpretation which was inspired by the resolution of the part–whole problem to distinguish between space and time as intuitions and determinate spaces and times. The Axioms of Intuition argues for the principle that all intuitions are extensive magnitudes, and it defines an extensive magnitude as one in which the representation of the parts precedes and is presupposed by the representation of the whole. Since space and time are intuitions, this principle applies to them. The Aesthetic, however, claims the contrary: space and time are represented as single and all-encompassing and are represented in such a way that their parts cannot precede them. We avoided this impasse by restricting the Axioms to a claim about determinate spaces and times, while the Aesthetic concerns space and time as given intuitions. Given this restriction, the principle of the Axioms should actually be: "All determinate intuitions are extensive magnitudes." Yet this is not how Kant formulates it, which seems to be evidence against my interpretation.

There is a two-part response to this objection. First, the context of the Axioms implicitly restricts Kant's claim in the Axioms to determinate intuitions. As noted earlier, the Transcendental Analytic takes into account the contributions of the understanding in determining cognitions, and the System of Principles articulates principles for the determination of the categories

position on the extent to which Kant relies on geometry to establish claims outside the transcendental exposition of space.

under the conditions of sensibility. This sets the context for the Axioms principle, and is sufficient to restrict it to determinate intuitions. Second, the key claims of the Axioms discussed in Section 2.5 concern the role of the understanding in employing concepts in order to generate representations of determinate spaces and times. Moreover, Kant's argument that all intuitions are extensive magnitudes, which we will examine more closely in the next two chapters, appeals to the successive synthesis of space and time required for the representation of a determinate space, such as a line, or a determinate time. Thus, the Axioms can at best establish a principle that determinate spaces and times and appearances are extensive magnitudes; it simply cannot establish that space and time as given intuitions are extensive magnitudes.

Since the results of determining space or time as a given intuition are extensive magnitudes, it is tempting to think that space and time are themselves extensive magnitudes, especially if one thinks that the distinction between extensive and intensive magnitudes is exhaustive. If that were the case, then we would still have a contradiction between the priority of the whole over the parts of space and time in the Aesthetic and the priority of the parts over the whole in Kant's account of extensive magnitudes in the Axioms. That is not, however, Kant's view, for the extensive/intensive magnitude distinction applies only to determinate magnitudes, that is, magnitudes determined by the understanding through concepts. Space and time as given intuitions are magnitudes in virtue of containing a homogeneous manifold in intuition. Nevertheless, they are not determined by the understanding through concepts, and hence are neither extensive nor intensive.

A final objection concerns the point of the Axioms. The System of Principles clearly focuses on the conditions for the cognition of objects of experience. For example, in introducing the System of Principles, Kant states that "experience . . . has principles of its form which ground it *a priori*, namely, general rules of unity in the synthesis of appearances" (A156–7/B195–6). Kant adds that the supreme principle of all synthetic *a priori* judgments is that every object stands under the necessary conditions of the synthetic unity of the manifold of intuition in a possible experience. This focus on principles of experience would seem to restrict the role of the Axioms to explaining the applicability of mathematics to the objects of experience; the possibility of mathematics itself is simply not on the table. As I have pointed out, however, Kant also describes the Analytic as containing a systematic presentation of all the synthetic principles of pure understanding, that is, those synthetic principles that flow *a priori* from pure concepts of the understanding under the sensible conditions of the forms of space and time (A136/B175). This more general description of the Analytic clearly encompasses principles of mathematical determination. Furthermore, as we saw in Section 2.5 above, the Axioms' argument rests on a subargument that *all* determinate intuitions, both pure and empirical, are extensive magnitudes. Kant has argued for and is entitled to this claim, which goes beyond establishing a principle of experience and has important consequences for the nature of mathematical cognition.

There is, however, a tension between the narrower and broader aims of the Analytic that surfaces in the Axioms, which led Kant to alter his formulation of the principle from the A- to the B-edition.[48] In keeping with the narrower aim of establishing a principle of possible experience, Kant focuses on the taking-up of appearances into empirical consciousness and on the apprehension of appearances that makes experience possible. The A-edition version, "All appearances, according to their intuition, are extensive magnitudes," does nothing to dispel the impression that the principle is only about appearances whose apprehension constitutes experience. The B-edition version, "All intuitions are extensive magnitudes," more clearly states the broader conclusion, which includes all determinate intuitions, including those that underlie pure mathematical cognition, and not merely those which underlie appearances. Most importantly, Kant explains our cognition of determinate spaces and times as the result of the determination of intuition through the employment of concepts by the understanding, a determination carried out through a figurative synthesis that constructs the concept of magnitude. Kant thereby reveals the foundation of his account of mathematical cognition.

2.9 Conclusion

The aim of the *Critique of Pure Reason* is to provide not a system of transcendental metaphysics, but only a critique that will explain the possibility of human cognition and show the way toward a system. Its primary focus is on the conditions of experience, and includes no focused treatment of mathematical cognition. Kant instead distributes his account of mathematics across the *Critique*: the Introduction provides reasons to think that mathematics consists of synthetic *a priori* cognitions; the Aesthetic articulates the sensible conditions of mathematical cognition; the Axioms of Intuition takes into account the contributions of the understanding; and the Discipline of Pure Reason presents its method. Kant's organization and presentation of the *Critique* can easily lead to misunderstandings of the role of the Axioms that obscure his views on the nature of mathematical cognition. If we think that the Axioms of Intuition is restricted to arguing for either the applicability of mathematics to objects of experience or the possibility of introducing a metric into space and

[48] Most commentators do not address the change in formulation of the principle. Guyer is an exception; he states that the A-edition formulation suggests an ontological distinction between intuitions and appearances that is not conveyed by the B-edition formulation. However, he also thinks that Kant came to emphasize this distinction between intuitions and appearances in the B-edition of the *Critique*, so the motivation for Kant's change in formulation is left unclear. See Guyer (1987, pp. 441–2n3).

time, we may miss the fact that the principle of the Axioms concerns any mathematical cognition whatsoever. Correcting these misunderstandings clarifies the treatment of mathematics and space and time in the Aesthetic and the Axioms. It also allows us to see that magnitudes are at the heart of Kant's philosophy of mathematics, and that Kant's Axioms of Intuition, and the theory of magnitudes it contains, warrants close investigation. That will be the task of the next chapter.

3

Magnitudes, Mathematics, and Experience in the Axioms of Intuition

3.1 Introduction

In the *Critique of Pure Reason*, Kant argues for two principles that concern magnitudes. The first is the principle that "All intuitions are extensive magnitudes," which appears in the Axioms of Intuition (B202); the second is the principle that "In all appearances the real, which is an object of sensation, has intensive magnitude, that is, a degree," which appears in the Anticipations of Perception (B207).[1] A circle drawn in geometry and the space occupied by an object such as a book are paradigm examples of extensive magnitudes, while the intensity of a light is a paradigm example of an intensive magnitude. These principles justify and explain the possibility of applying mathematics to objects of experience. As we saw the previous chapter, the Axioms principle also explains the possibility of any mathematical cognition at all, and hence the possibility of pure mathematics.[2]

The principles of the Axioms and the Anticipations are the first to appear in Kant's System of Principles and are central to Kant's theory of human cognition. Nevertheless, Kant's notions of magnitude, the principles in which they appear, and the arguments for them have not been well understood. Kant defines the concept of magnitude in the Axioms of Intuition, which contains claims about magnitudes in general that are central to understanding the role of magnitudes in Kant's philosophy of mathematics. The primary focus of this chapter will therefore be on the Axioms. It will argue that, in Kant's view, mathematical cognition depends on the properties of magnitudes, and that

[1] These are the principles as they appear in the B-edition; I discussed the change in formulation of the Axioms principle in Section 2.8.

[2] Invoking the distinction between pure and applied mathematics can be misleading, a point discussed in the previous chapter that bears repeating. Kant does not think of pure mathematics in the way that we might – that is, as a body of knowledge that can be developed without reference to space and time, and whose applicability to space and time is a further question. For Kant, pure mathematics is essentially related to pure space and time, while the issue of the applicability of mathematics concerns its application to objects of experience appearing in space and time, not its applicability to space and time. (See Chapter 2, notes 14 and 46.)

intuition plays an important role in mathematical cognition by allowing us to represent those properties. The role of intuition in representing magnitudes has passed unnoticed; revealing it opens new ways of understanding the role of intuition in Kant's philosophy of mathematics.

Section 3.2 reconstructs and explains Kant's argument for the Axioms principle. It argues that Kant's argument divides into two stages, the first concerning magnitudes in general and the second concerning extensive magnitudes. Section 3.3 clarifies Kant's views on definition in general before Section 3.4 examines Kant's definition of magnitude. It argues against a widely accepted modification to it suggested by Hans Vaihinger. Clarifying Kant's definition allows us to identify what are, in Kant's view, the fundamental properties of magnitudes. Section 3.5 introduces Kant's distinction between two notions of magnitude, *quanta* and *quantitas*, while Section 3.6 considers this distinction in light of the Schematism. The analysis clarifies Kant's theory of magnitudes and makes it possible to further unfold and analyze Kant's theory of magnitudes, as well as its role in mathematical cognition and experience, in subsequent chapters.

3.2 The Magnitude Argument

The A-edition version of the Axioms of Intuition begins by explaining the concept of extensive magnitude and argues that all appearances are extensive magnitudes. Kant seems to give further thought to the nature of magnitudes between the two editions of the *Critique*, or at least seems to think more explanation is required, since many of the changes he introduces concern or discuss magnitudes.[3] In the B-edition of the *Critique*, Kant inserts a paragraph at the beginning of the Axioms of Intuition. We will examine this passage in detail, so I quote it in full as well as the first sentence of the next paragraph:

> All appearances contain, in accordance with their form, an intuition in space and time, which grounds all of them *a priori*. They can thus not be apprehended, i.e., taken up into empirical consciousness, except through the synthesis of the manifold through which the representations of a determinate space or time are generated, i.e., through the composition of the homogeneous and the consciousness of the synthetic unity of this manifold (homogeneous). Now the consciousness of a homogeneous manifold in intuition in general, insofar as through it the representation of an object first becomes possible, is the concept of a magnitude (*quanti*). Thus, even the perception of an object, as appearance, is only possible through the same synthetic unity of the manifold of the given sensible intuition through which the unity of the composition of the manifold

[3] See B115, B162, and B201n for examples. Section 26 of the B-deduction (B162), in particular, gives magnitudes a more prominent role.

> homogeneous is thought in the concept of **magnitude**; i.e., the appearances are all magnitudes, and indeed **extensive magnitudes**, because as intuitions in space or time they must be represented through the same synthesis as that through which space and time in general are determined.
>
> I call an extensive magnitude that in which the representation of the parts makes possible the representation of the whole (and thus necessarily precedes it) . . . (B202– A167/B203)

Commentators who have discussed the added paragraph virtually universally hold that it either reiterates the A-edition argument or provides a new argument for the principle that all intuitions are extensive magnitudes.[4] This is mistaken and obscures both Kant's understanding of magnitude and the overall argument of the Axioms. The closing reference to "extensive magnitudes" notwithstanding, the new paragraph really concerns only magnitudes. It explains the concept of magnitude and argues that all appearances are magnitudes. The definition of extensive magnitudes and the argument that all appearances are extensive magnitudes are left to the second paragraph, that is, the first paragraph of the A-edition. Thus, the B-edition Axioms of Intuition contains a new argument that supplements the old, and together the new and old arguments constitute a larger two-stage argument.

The two-stage structure of the argument has not been appreciated. This has been, I think, because both arguments turn on the nature of determinate spaces and times, and because the new first paragraph ends with the claim

[4] Kemp Smith (1979, p. 394) is representative: he thinks the inserted paragraph needs no special comment after his account of the A-edition argument, though he points out that it gives needed prominence to the role of synthesis. According to H. J. Paton (1965, vol. II, pp. 111–16), this inserted paragraph contains an argument for the principle of the Axioms that is largely independent of the A-edition argument. Walsh (1975, p. 113) considers the new paragraph a summary proof that "simply stands alongside the old," though it cannot stand on its own. Allison (2004) does not discuss the Axioms of Intuition at any length, and does not mention the paragraph added in the B-edition; Guyer (1987, pp. 191–2) holds that the opening paragraph of the B-edition is only a slightly more elaborate argument for a two-step recapitulation of the argument found at the end of the Axioms of Intuition.

Some have pointed out that the inserted paragraph ties the Axioms more closely to the transcendental deduction. Wolff (1963, p. 229) thinks that the purpose of the added paragraph is to ground the A-edition proof in the transcendental deduction. While the paragraph does tie the Axioms more closely to the deduction, that is not its primary purpose. Brittan (1978, pp. 111–12) does not clearly distinguish the argument of the added paragraph from that of the A-edition, although he too claims that it is more closely connected to the Deduction. Longuenesse (1998, p. 274n70) claims that the paragraph added in the B-edition repeats a portion of the argument in §26 of the B-Deduction. She also states, however, that §26 is itself a repetition of an argument of the A-edition, so it is unclear to me what relationship she thinks the paragraph has to the A-edition argument.

that appearances are magnitudes, and then adds "and indeed extensive magnitudes . . ." But the argument clearly does not establish that appearances are extensive magnitudes; in fact, extensive magnitude has yet to be defined. The last sentence is intended to point to the A-edition argument, which provides a definition of extensive magnitude and then argues that determinate intuitions and appearances are not only magnitudes but extensive magnitudes in particular. Kant's transitional comment works because the A-edition argument that follows it begins with "I call an extensive magnitude that in which . . ."; that is, he is now focusing on one kind of magnitude. In the A-edition, Kant had assumed that the notion of magnitude was sufficiently clear; in the B-edition, he clarifies this notion and provides an additional argument based on it.

The thrust of the added paragraph can be summarized in three steps: (1) The apprehension of an appearance requires a synthesis that generates a determinate space or time, which constitutes a part of the appearance. (2) Every determinate space or time is a magnitude in virtue of the homogeneous manifold that it contains. (3) Therefore, every appearance is a magnitude in virtue of the determinate space or time it contains. This outlines the argument strategy, but a full understanding of Kant's theory of magnitudes turns on the details, so a close reconstruction will be helpful.

1. All appearances contain an intuition in space or time, which grounds all of them *a priori*.[5]
2. Apprehension of appearances thus requires a synthesis of the manifold whereby the representation of a determinate space or time is generated. In other words, appearances, as intuitions in space or time, must be represented through the same synthesis as that through which space and time in general are determined.[6]

[5] Kant actually states in the first line that all appearances contain an intuition in space *and* time. A paradigm appearance of, say, a table would contain an intuition in both. Kant believes, as he came to emphasize in the B-edition, that our representation of time ultimately depends on spatial representations, so that a world consisting only of appearances represented in time would not be possible. This does not, however, rule out the possibility of nonspatial temporal appearances in a world that includes objects of spatial representation. Kant's claim at the beginning of this argument would exclude the possibility of any appearances that occur in time but not space, but Kant goes on to say that appearances require a determinate space *or* time and ends the argument by referring to appearances insofar as they are intuitions in space *or* time; it seems that he does not wish to limit appearances to those in both space and time. (On the other hand, Kant did not think an appearance could be in space without also being in time – the condition of all representations.)

[6] "Appearances as intuitions in space or time" simply refers to appearances insofar as they contain an intuition in space or time, as described in the previous step of the argument.

3. The synthesis generating (representations of) a determinate space or time requires:
   (a) composition of the homogeneous and
   (b) consciousness of the synthetic unity of this (homogeneous) manifold.
4. Definition: Consciousness of the homogeneous manifold in intuition in general, insofar as through it the representation of an object first becomes possible, is the concept of a magnitude (*quanti*).
5. Therefore, even the perception of an object, as appearance, is possible only through the same synthetic unity of the manifold of the given sensible intuition as that whereby the unity of the composition of the homogeneous manifold in the concept of magnitude (*quanti*) is thought.
6. Therefore, appearances are all without exception magnitudes (*quanta*).

This argument rests on implicit premises that can be roughly divided into two classes: those concerning the nature of intuition and those concerning the conditions for the apprehension, and hence the perception, of appearances. The first class of implicit premises includes the claims that space and time are forms of sensible intuition that impart spatial and temporal properties to both pure and empirical intuitions represented through them, and that all appearances contain an intuition. Kant supports these claims in the Aesthetic.[7] Kant's argument also assumes that every intuition contains a homogeneous manifold, a property that Kant first mentions with respect to magnitudes in a B-edition discussion of the Table of Categories (B115) and with respect to intuitions in the B-deduction (B162). We will see below that homogeneity is a crucial property of both intuitions and magnitudes.

The second class of implicit premises concerns the conditions for the apprehension of an appearance. These include the claim that the apprehension of an appearance as an object requires a synthetic unity of the manifold in a given intuition. It also includes the claim that apprehending an appearance as an object requires the representation of determinate properties, and that this representation of determinate properties is accomplished through a synthesis of a manifold contained in the appearance. In the case of a spatial or temporal manifold, the resulting representation is of a determinate space or time. The clearest support for these claims about apprehension is found in the B-Deduction, the culmination of which is found in §26. Kant states there that the synthesis of apprehension requires a unity of the synthesis of the given manifold, and that this unity must be the unity of combination of the manifold in intuition in general according to the categories. Since perception itself requires this synthesis according to the categories, Kant concludes that the categories have objective validity for all objects of experience (B161), which is the aim of the deduction.

[7] See especially A19–21/B33–5, B41, A26/B42, A34–6/B50–2, and A38–41/B55–58.

In the process of establishing the objective validity of the categories, Kant also establishes important implicit premises of the Axioms argument. Section §26 of the B-deduction discusses two examples of the necessary use of the categories. I discussed the first already in Section 2.4 above, which directly concerns the Axioms. Kant states that in the apprehension of a house, the synthetic unity required in the synthesis of the spatial manifold is the category of magnitude, which he glosses as "the category of the synthesis of the homogeneous in an intuition in general" (B162). Kant does not further explain the concept of magnitude here. That further explanation, and its implication for the possibility of mathematics as well as the applicability of mathematics to experience, constitutes the unique contribution of the Axioms argument.[8]

As discussed in Sections 2.4 and 2.5, determination is at the root of Kant's account of cognition, including mathematical cognition and the cognition of magnitudes. According to Kant, lines, surfaces, durations, and the particular spatial and temporal features of an object of experience all count as determinate spaces or times, which we can "draw in thought" (B137–8, A162–3/B203, A715/B743). Thus, determinate spaces comprise any spatial figure that can be drawn in thought as well as figures constructible in geometry. In §26 of the B-Deduction, Kant states that we "as it were" draw the shape of a house when we apprehend it (*zeichne gleichsam seine Gestalt*) (B162), suggesting that our representation of the space occupied by a house rests on the synthesis of a determinate space. Step 2 of the Axioms' argument outlined above makes the connection between apprehension of appearances and the generation of determinate spaces and times explicit: the apprehension of an appearance is possible only by means of a synthesis through which the representations of a determinate space or time are generated.

Kant states in the *Prolegomena* that space is something uniform and indeterminate and is the substratum of all determinable intuitions; it is the understanding that determines space to assume forms such as circles, cones, or spheres (Ak.4:321).[9] Moreover, the understanding produces determinate spaces and times through a special synthesis called synthesis of composition.

[8] Others have noted this connection between the B-deduction and the first paragraph added to the B-edition of the Axioms. As mentioned in note 4 above, Wolff (1963, p. 229) thinks that the purpose of the added paragraph is to ground the A-edition proof in the transcendental deduction, and Brittan (1978, p. 112) notes that the added paragraph is more closely connected to the Deduction. Longuenesse (1998, pp. 243, 274n70) claims both that the passage concerning apprehension of the house is merely an example of what Kant has established and that it contains an argument that Kant repeats in the first paragraph of the Axioms. I read this part of §26 as an example of what Kant has just established, but I do not think that the argument is simply repeated in the Axioms. Rather, it is largely implicitly presupposed in the Axioms, and the Axioms argument builds on it by focusing on the concepts of magnitude and extensive magnitude.

[9] See also A714/B742.

Just before presenting his system of principles in the A-edition, Kant states that the Axioms and Anticipations principles are mathematical, while the Analogies and Postulates principles are dynamical. Kant adds a footnote in the B-edition to explain the distinction further, and the explanation appeals to a special synthesis that pertains only to magnitudes. Kant distinguishes between two sorts of combination (*Verbindung*), namely, composition (*Zusammensetzung*) and connection (*Verknüpfung*), which he also contrasts using the Latin terms *compositio* and *nexus*. The synthesis of composition is the synthesis of a manifold of what does not necessarily belong to each other, adding that "it is the same as the synthesis of the **homogeneous** in everything that can be considered **mathematically**" (B201n).

It is because the representation of determinate spaces and times rests on the composition of a homogeneous manifold that both they and appearances are magnitudes. Furthermore, that which explains the applicability of mathematics to experience also explains the possibility of mathematics itself. That is, the Axioms' argument establishes the applicability of mathematics to appearance by describing the conditions for mathematical cognition (the representation of determinate spaces and times) and showing that the perception of appearances presupposes those conditions.

The magnitudes argument crucially depends on Kant's definition of the concept of magnitude, which we will examine more closely, but before doing so I would like to make a point about Kant's understanding of definitions.

3.3 Kant on the Definition of Concepts

In the Discipline of Pure Reason, Kant distinguishes between explanations, expositions, explications, declarations, and definitions. The term "definition" (*Definition*) can be used in a broad sense to mean any explanation or clarification (*Erklärung*) of a concept,[10] but Kant would like to carefully distinguish between the explanation of two sorts of concepts: given and made (A727–32/B755–60). The given/made distinction cuts across that between *a priori* and empirical concepts, so that we have both empirical and *a priori* given concepts and empirical and *a priori* made concepts. Kant's examples of given empirical concepts are the concepts of gold and water; his examples of concepts given *a priori* are substance, cause, right, and equity (A728/B756). We face a fundamental epistemological limitation in explaining or clarifying given concepts, because we can never be sure that all the marks of the concept are clearly and

[10] Kant allows, somewhat reluctantly, this broader notion of definition in his description of philosophical definitions (A730/B758 and A731/B759n). The term "clarification" in English sounds fairly minimal: any identification of the marks of a concept might count as a clarification. By *Erklärung*, Kant seems to have in mind something a bit more substantial, so I prefer "explanation" as a translation.

sufficiently represented. In the case of given empirical concepts, we have "only some marks of a certain kind of objects of the senses," so that it is never certain that "one doesn't sometimes think more of these marks but at another time fewer of them" (A727–8/B755–6). For given *a priori* concepts, I can never know that it is "adequate to the object" since the concept of the object "can contain many obscure representations, which we pass over in our analysis." As a consequence, we cannot be certain that the analysis of a given concept has been exhaustive, in the sense that all the marks of the given concept have been clearly and sufficiently revealed. Kant is at pains to emphasize this point, stating that we can never be certain that we have exhaustively developed the "distinct representation of a (still confused) given concept," so that the exhaustiveness of the analysis is at best "**probably** but never **apodictically** certain" (A728–9/B756–7).

Kant stipulates the meaning of "definition" in a narrower and proper sense in a way that marks this limitation in our knowledge of given concepts: to "define" in the proper sense is to "exhibit originally the *exhaustive* concept of a thing within its boundaries" (A727/B755, my emphasis). Thus, no given concept, whether empirical or *a priori*, can be defined in the proper sense. Kant proposes instead the use of "exposition" (*Exposition*) for the results of analysis of given concepts, and, in the case of empirical concepts, "explication" (*explizieren*).[11] Kant states that the word "exposition" signifies an analysis of the concept "which is always cautious, and which the critic can accept as valid to a certain degree while yet retaining reservations about its exhaustiveness" (A729/B7557).

In contrast, Kant holds that the marks of made concepts can be exhaustively presented, because such concepts are deliberately made (*vorsätzlich gemacht*), so that I know what I wanted to think in it. His example of an empirical made concept is a ship's clock (*Schiffsuhr*); for *a priori* made concepts, he has in mind mathematical concepts, and as an example he gives the concept of a circle (A729/B757 to A732/B760). Made concepts are "electively thought" (*willkürlich gedacht*);[12] that is, I know what I want to think in them and

[11] The relation between the meaning of these different terms is not entirely clear in the Discipline discussion of these issues. Definition (*Definition*) can be used in a broad sense that corresponds to any clarification (*Erklärung*) of a concept, but also in the narrow sense indicated. Exposition (*Exposition*) is applied only to given concepts; it may be further restricted to *a priori* given concepts, and in many contexts, that seems to be what Kant has in mind, but what Kant says of it – that it indicates that the analysis is always cautious and warrants reservations – applies to given empirical concepts as well. On the other hand, Kant states that a given empirical concept can only be explicated (*expliziert* or *explicirt*), the Latin-derived nominalization of which is "explication [*Explication* or *Explikation*]."

[12] *Willkürlich* is normally translated as "arbitrary" or "voluntary," neither of which quite fits Kant's sense; Kant means not that one chooses any old characteristics, or that one's act of

I choose which marks to include in the concept. Kant later says, in contrasting philosophical "definitions" of given concepts with mathematical definitions of *a priori* made concepts, that the former are only expositions of given concepts and come about only analytically through analysis, while the latter come about synthetically, that is, through the combination of concepts.

Kant's account of made concepts leads to a puzzle concerning the concept of magnitude and mathematical concepts that we will set aside here.[13] What is important for us here is that Kant almost certainly regards the concept of magnitude as a given concept; hence, its definition would be definition in the broad sense and not the narrow sense, so that the clarification of this concept is not one concerning whose exhaustiveness we can be sure. As a result, there might be necessary properties of magnitudes that are neither given in the exposition of the concept nor flow from it, either analytically or synthetically. For ease of expression, I will simply use "definition" understood in the broad sense in what follows. What is important is that Kant does not assume that his characterization of the concept is exhaustive, and we should keep in mind the open-ended nature of our inquiry into magnitudes and their properties.[14]

3.4 Kant's Definition of the Concept of Magnitude

Kant's definition of the concept of magnitude states that:

> Now, the consciousness of the homogeneous manifold in intuition in general, insofar as through it the representation of an object first becomes possible, is the concept of magnitude (*quanti*). (B203)[15]

choosing is free rather than compelled, but that there are no constraints on which characteristics one may choose to synthesize. One can use "arbitrary" in this last sense, but I think it is clearer to translate *willkürlich* as "elective."

13 Briefly, the puzzle concerns the possibility of exhaustively exhibiting a made concept if the definition of the concept appeals to given concepts. I will address this puzzle in future work.

14 I would like to thank Matt Boyle for prompting me to clarify the sense in which Kant is providing a definition of the concept of magnitude.

15 In the German, the passage is "Nun ist das Bewußtsein des mannigfaltigen Gleichartigen in der Anschauung überhaupt, sofern dadurch die Vorstellung eines Objekts zuerst moglich wird, der Begriff einer Größe (*quanti*)." Kant uses "manifold" as an adjective and nominalizes "homogeneous," which emphasizes the property of homogeneity. A closer translation of the phrase that preserves this emphasis would be "manifold homogeneous," but it is difficult to hear "homogeneous" as a noun in English. On the other hand, "manifold of homogeneous elements" suggests a discrete manifold and departs further from Kant's own words. I will therefore revert to "homogeneous manifold." The emphasis Kant places on homogeneity is important, however, and should be kept in mind. I will discuss the significance of restricting the definition of magnitude to *quanta* in the next section.

I will address the nature of a homogeneous manifold in more detail below. Before doing so, however, I would like to discuss an emendation to the definition proposed by Hans Vaihinger. He suggested adding the phrase "synthetic unity" to the definition, so that the concept of magnitude is the consciousness of the synthetic unity of the homogeneous manifold in intuition. Vaihinger claims that the context compels the addition of "synthetic unity." In support, he states only that the cause of the omission is apparent to one familiar with the psychology of the typesetter, because these same words are found in the previous sentence; he does not explain which contextual factors he has in mind nor how he understands "synthetic unity."[16] Nevertheless, influential commentators and translators have followed Vaihinger. His suggestion is noted in Raymund Schmidt's edition of the *Critique*, and Norman Kemp Smith adopted the recommendation in his translation of the *Critique*, the standard English translation for most of the twentieth century.[17] Others followed Vaihinger and Kemp Smith in their analyses of the Axioms – for example, H. J. Paton, Ernst Cassirer, and Gordon Brittan.[18]

Vaihinger's addition has led to serious misunderstandings of Kant's concept of magnitude. It has confused the relationship between the concept of magnitude and the categories of quantity and obscured the distinction between two concepts of magnitude, *quanta* and *quantitas*. The distinction between *quanta* and *quantitas* and the relation between them are among the most vexing issues in Kant's account of mathematical cognition, so it is important to start out with as clear an understanding of Kant's concept of magnitude as possible. The next several sections will focus on clarifying Kant's definition of the concept of magnitude.

3.4.1 *The Concept of Magnitude and Categorial Unity*

In order to evaluate Vaihinger's suggested emendation, we must clarify what sort of unity is meant by "synthetic unity." One obvious possibility is that "synthetic unity" refers to the category of unity. In that case, the concept of magnitude subsumes the homogeneous manifold in intuition under the category of unity. While this may not have been what Vaihinger had in mind, there are considerations that speak in favor of it, and this is how many commentators who have followed Vaihinger's recommendation have interpreted it.

16 See Vaihinger (1900).

17 Schmidt's 1926 edition, p. 217n3; Kant (1965, p. 198).

18 Paton (1965, vol. II, p. 115), Cassirer (1954, p. 127), and Brittan (1978, p. 111). In Guyer and Wood's 1998 edition, however, they do not follow Vaihinger in their translation of the *Critique*.

In fact, the relation between the Axioms principle and the categories seems to give quite powerful reasons for adding "synthetic unity" to the concept of magnitude and understanding it as the category of unity. The principle of the Axioms in some way or other corresponds to the conditions of judgments about objects in accordance with all three categories of quantity (unity, plurality, and allness) under the conditions of sensibility. Immediately following the Table of Categories, Kant states that allness (totality) is the result of a distinct act of the understanding, but that it nevertheless arises from thinking of a plurality as a unity (B111). If we add consciousness of synthetic unity to the concept of magnitude, then the concept of magnitude will require consciousness of a unity of a plurality, that is, of a homogeneous manifold. Thus, it will require a consciousness of plurality, unity, and allness, and the concept of magnitude will correspond to all three categories of quantity. On the other hand, if we do not add "synthetic unity," it is difficult to see how all the categories of quantity will be manifested in a principle about magnitudes.

Kant reinforces this reading of the role of the concept of magnitude when he refers in several passages to "categories of magnitude" or the "category of magnitude" (B115, B162, B193, B201). Furthermore, his discussion of the schemata of the quantitative categories is about the schema of the concept of magnitude (A142/B182).

Moreover, the argument itself seems to give credence to this reading. This support comes in two distinguishable parts. First, the logical structure of the argument apparently requires adding "synthetic unity" to the exposition of magnitude; second, the focus of the argument seems to support interpreting this synthetic unity as categorial unity. Let's consider each of these parts in turn.

In step 5 of the reconstruction of the argument (see p. 62 above), Kant refers to thinking of "the synthetic *unity* of the combination of the homogeneous manifold in the concept of magnitude" (my emphasis). He therefore seems to think that a representation of unity is part of the content of the concept of magnitude. Step 3(b) states that generating the determinate space or time of an appearance requires "consciousness of the synthetic unity of a homogeneous manifold." Step 4 gives the definition of the concept of magnitude as "consciousness of a homogeneous manifold," and step 5 infers that the concept of magnitude is required for any appearance. If we do not add "synthetic unity" to the definition of magnitude, then step 3(b) will require something distinct from the concept of magnitude given in step 4, which is in turn distinct from Kant's apparent description of the concept of magnitude in step 5. Thus, the argument will not, in the strictest sense, be valid, unless we add "synthetic unity" to the exposition of the concept of magnitude in step 4.

If the logical structure of the argument warrants adding synthetic unity to the concept of magnitude, then the argument's primary focus seems to show that this synthetic unity is the category of unity. Kant's argument is about conditions for the apprehension of appearances, and Kant asserts that this apprehension depends on the synthesis that generates the representation of a determinate space or time, which is a determinate magnitude. And a determinate space or time is in fact represented as a categorial unity of a plurality and hence as a totality under the category of allness. Thus, in step 3, when Kant states that the representation of a determinate space or time requires both (a) composition of the homogeneous and (b) consciousness of the synthetic unity of this homogeneous manifold, it is possible that he has in mind the category of unity. And this makes it likely that the synthetic unity referred to in step 5 is also the category of unity.

It must be noted, however, that the correspondence between the quantitative categories and the principle of the Axioms is at best an imperfect one, for the Axioms principle concerns not merely magnitudes but extensive magnitudes (in contrast to the Anticipations principle, which concerns intensive magnitudes). It is easy to suppose that the imperfect correspondence is simply a result of the difficulty Kant has in fitting magnitudes into his architectonic scheme. Nevertheless, this mismatch speaks against taking the relation between the categories and the concept of magnitude too literally.

Despite this worry about the mismatch between the categories of quantity and the principle of the Axioms, I think Vaihinger's suggested emendation, the general correspondence between the categories of quantity and the Axioms of Intuition, the logical structure of Kant's argument, and the focus on determinate spaces and times, all likely influenced commentators who have thought that categorial unity should be included in Kant's definition of the concept of magnitude. Norman Kemp Smith, who follows Vaihinger's suggestion in his translation of the *Critique*, only briefly discusses the Axioms in his commentary,[19] but he is generally disposed to see the architectonic as a driving force in Kant's philosophy. H. J. Paton, who also adopts Vaihinger's suggestion, explicitly endorses a correspondence between the categories of quantity and the concept of magnitude when he describes the concept of magnitude as "the pure category of quantity (totality)."[20]

I think, however, that this understanding of Kant's characterization of the concept of magnitude at the beginning of the Axioms is mistaken, and that it leads to serious misunderstandings of Kant's theory of magnitudes. We have good reason not to include the category of unity in his definition of the concept of magnitude and good reason to reject the view that the concept of magnitude corresponds to all three categories of quantity. In Kant's view, the concept of magnitude is not the concept of a totality, that is, a plurality unified under the category of unity. I will argue that the concept of magnitude is

[19] Kemp Smith (1979, p. 347).
[20] Paton (1965, vol. II, p. 115).

simply the concept of a multiplicity or a manifold, and it is related to the category of plurality rather than all three categories. Concepts such as plurality, multiplicity, manifold, and magnitude do not require that their objects fall under the category of unity.

Reflection on the Table of Categories bears this out. Under the heading of "quantity," we find the categories of unity, plurality, and allness (or totality) (A80/B106). As mentioned above, Kant states that the last category under each heading derives from a combination of the first two through a distinct act of the understanding, and he asserts that allness is plurality considered as unity (B111). Kant does not elaborate on the nature of this distinct act of the understanding; if, however, the category of plurality already required subsuming the plurality under the category of unity, then there would be no difference between the categories of plurality and allness (as totality). Thus, the category of plurality itself does not include the category of unity.

This point is of general importance for concepts of multiplicity, manifold, and other concepts besides magnitude. In Kant's view, to think of something by means of such a concept may require a unity in representing it, but does not entail that we be conscious of a unity in the represented manifold falling under the category of unity. In other words, the articles in phrases such as "a multiplicity" or "the homogeneous manifold" may imply some sort of unity, but not one that falls under the category of unity. I will return to what this noncategorial unity might be below.

There is more evidence that the category of plurality does not require subsuming the plurality under the category of unity in §11 following the Table of Categories. Kant states that the category of allness (as totality)[21] is not employed to represent the infinite, even though the concepts of unity and multitude, that is, plurality, can be employed in its representation (B111). I will not attempt to give an account of Kant's views on infinity here; it is clear, however, that employment of the category of allness (as totality) is incompatible with the representation of infinity, while the category of plurality is not.

This point bears directly on the concept of magnitude. Kant states in the Aesthetic that space and time are represented as infinite magnitudes (A25/B39). Since application of the category of allness is incompatible with the infinite, the concept of magnitude includes neither the category of unity applied to a plurality nor the category of allness (as totality). The concept of magnitude is closely associated with the category of plurality only.

Section 21 of the *Prolegomena* (AA4:302–4) provides further evidence that the concept of magnitude corresponds to the category of plurality and does not

[21] I will discuss the category of allness and its relationship to totality in further work on Kant's understanding of *quanta*, *quantitas*, and number; we can pass over their difference here.

subsume a homogeneous manifold under the category of unity. Kant presents the Table of Categories and lists the categories of quantity as shown below:

Transcendental Table
of the Concepts of the Understanding
1.
According to Quantity [*Quantität*]
Unity (Measure [*das Maß*])
Plurality (Magnitude [*die Größe*])
Allness (Whole [*das Ganze*])

The concept of magnitude is explicitly associated with the category of plurality rather than with allness. These texts decisively undermine a reading of the concept of magnitude that includes the category of unity applied to a manifold, and entails that the concept of magnitude Kant introduces at the beginning of the Axioms does not correspond to all three categories of quantity. This compels us to look for another interpretation.

3.4.2 The Concept of Magnitude and Noncategorial Unity

I noted that Vaihinger states that the context of the argument compels adding "synthetic unity" to the concept of magnitude. We have seen that the logical structure of the argument apparently does require it, and perhaps that is the context to which he is referring. As I also noted, however, Vaihinger does not say that he had in mind the category of unity. This leaves open the possibility that he intended some sort of noncategorial synthetic unity of the sort mentioned in the previous section, a possibility we should consider.

In Kant's view, every concept requires that a manifold be brought to the synthetic unity of apperception, including the concept of magnitude. It is important to distinguish between the unity of consciousness in a concept and the consciousness of a unity. The former is a unity in the act of representing; the latter is a unity in that which is represented.

One might nevertheless hold that thinking of any multiplicity in one act of consciousness requires not merely a unity in the act of representation but also at least some minimal sort of unity on the side of the represented multiplicity. In the B-deduction, Kant argues that the analytic unity of apperception presupposes a possible synthetic unity of a manifold combined in one consciousness; in fact, this possible synthetic unity is the ground of the identity of apperception.[22] Thus one might think that the unity of consciousness in the

[22] **Begin Note** In §16 of the B-Deduction, Kant states:

> it is only because I can combine a manifold of given representations **in one consciousness** that it is possible for me to represent the **identity of the**

concept of magnitude presupposes some sort of unity in what is represented. And as I noted in the previous section, even if concepts like magnitude, manifold, multiplicity, and plurality do not require the category of unity, the use of an article in phrases such as "a manifold" or "the homogeneous manifold of intuition" might be thought to imply some minimal sort of unity.

In fact, Kant makes a point of distinguishing the synthetic unity thought in any combination of a manifold from the category of unity. He calls the former a qualitative unity, which is presupposed by every category; he calls the latter a quantitative unity and identifies it only with the category of unity (B131). Kant says that we must seek the qualitative unity someplace higher, and what he goes on to say makes it clear that he has in mind the unity of apperception as the source of the qualitative unity of any combination. Kant also describes qualitative unity as a property of any concept, in contrast to its object; it is, he says, a logical criterion of the possibility of a concept (B114–15).

Although Kant closely ties qualitative unity to the unity of apperception and to the unity of a concept, and hence to the act of representing, he also attributes qualitative unity to the represented. In §12 following the Table of Categories, Kant states:

> In every cognition of an object there is, namely, **unity** of the concept, which one can call **qualitative unity** insofar as by that only the unity of the comprehension of the manifold of cognition is thought, as, say, the unity of the theme in a play, a speech, or a fable. (B114)

This suggests that the unity of a concept corresponds to at least a minimal sort of unity of the manifold of representations unified under it.

I will not attempt to give an account of the various notions of unity in Kant, but we have seen reasons that at least suggest that the content of a concept includes the representation of a synthetic unity. We thus have two reasons for thinking that synthetic unity ought to be added to Kant's definition of the concept of magnitude: every concept represents at least a minimal synthetic unity of the manifold that falls under it, and the logical structure of the argument appears to require it. But we should ask: Why did Kant not include it? It is of course possible that it was simply a slip, but I think there was a reason for it. To begin, however, we should first reconsider the structure of Kant's argument.

> **consciousness in these representations** itself, i.e., the **analytical** unity of apperception is only possible under the presupposition of some **synthetic** one . . . Synthetic unity of the manifold of intuitions, as given *a priori*, is thus the ground of the identity of apperception itself. (B133–4)

This synthetic unity is the unity presupposed by the combination of any manifold in one consciousness.

3.4.3 Explaining the Concept of Magnitude without Synthetic Unity

First, a preliminary point. Kant might appear to include unity in his reference to the concept of magnitude in step 5, since he refers to "that whereby the unity of the composition of the homogeneous manifold in the concept of magnitude is thought." Nevertheless, the phrase "in the concept of magnitude" can be read as modifying "homogeneous manifold" and not "the unity of the composition."[23] I would not want to place too much weight on this, however, as there is a more important point.

Whether or not Kant's references to unity in steps 3 and 5 are to categorial unity or to a minimal synthetic unity, the validity of the argument does not require that the content of the concept of magnitude contain synthetic unity. It only requires that the generation of a determinate space or time includes a consciousness of whatever is thought in the concept of a magnitude. And it seems obvious, or at least overwhelmingly plausible, that consciousness of the synthetic unity of a homogeneous manifold includes consciousness of the homogeneous manifold.[24] Thus, even if minimally enthymematic, the argument is valid. Kant could have made it valid in the strictest sense if he had added the premise:

3.5 Consciousness of the synthetic unity of a homogeneous manifold includes consciousness of the homogeneous manifold.

Even if Kant's overall argument is not as perspicuous as it should be, he could be forgiven for thinking this step need not be made explicit.

The key point is that the cognition of determinate spaces and times requires consciousness of a synthetic unity of a homogeneous manifold in intuition, but that consciousness of the homogeneous manifold alone is sufficient for determinate spaces and times (and hence appearances) to fall under the concept of magnitude, which is the claim Kant wishes to establish in this argument.

One might still argue that the content of the concept of magnitude requires the representation of at least some minimal synthetic unity, and hence that it is more precise to include "synthetic unity" in the concept of magnitude. Kant, however, had good reasons not to. First, the focus of the overall argument is on appearances and the representation of determinate spaces and times on which they rest, and as noted above, these do require the category of unity applied to

[23] In German, the passage is "Also ist selbst die Wahrnehmung eines Objekts, als Erscheinung, nur durch dieselbe synthetische Einheit des Mannigfaltigen der gegebenen sinnlichen Anschauung moglich, wodurch die Einheit der Zusammensetzung des mannigfaltigen Gleichartigen im Begriffe einer Große gedacht wird."

[24] One might think it is at least possible to be conscious of the unity of some manifold without being conscious of the manifold itself. However, to be conscious of the unity as a unity *of a manifold* would seem to require at least some sort of consciousness of the manifold.

the plurality of the manifold, and it is thus likely that Kant is referring to a categorial unity in step 3b and in step 5 as well. Thus, if he had included a minimal noncategorial unity in the concept of magnitude, it would likely be mistaken for the category of unity. That is, in fact, what seems to have been the case with a large number of commentators influenced by Vaihinger's recommendation, if not Vaihinger himself. Second, Kant wishes to emphasize the fact that what is characteristic of the concept of magnitude, and sets it apart from other concepts, is that it picks out a homogeneous manifold in intuition. Had he included the synthetic unity common to the content of all concepts, it would have detracted from this point. In fact, on close reading, Kant's argument teases apart three features of the representation of a determinate space or time: the composition of the homogeneous, the consciousness of the synthetic unity of this (homogeneous) manifold, and the consciousness of the homogeneous manifold. These distinguished features appear as steps 3a, 3b, and 4 in the argument reconstruction.

Thus, adding a reference to a noncategorial unity invites confusing it with a categorial unity of a plurality, which in turn would lead to a misunderstanding of the relation of the categories to the concept of magnitude Kant defines in the Axioms. It would also needlessly complicate the exposition of the concept and draw attention away from the fact that Kant's characterization of magnitude focuses on the property of being a manifold that is homogeneous. We should therefore accept Kant's definition of the concept of magnitude as it stands, and focus on the marks of a magnitude which his definition highlights.

3.4.4 The Content of the Concept of Magnitude

In defining magnitude as a homogeneous manifold in intuition in general, Kant abstracts from our particular human forms of intuition, space, and time. Whether or not there are other forms of intuition is not something Kant thinks we can know, but his definition leaves open that he is not singling out our particular forms of intuition, that is, space and time. It is possible that he has only human intuition in mind here; at the very least, however, he is defining magnitude with respect to what they have in common as intuitions, and abstracting from any particular properties of either space or time. This abstract characterization is also present in the house example of §26 of the B-Deduction mentioned earlier. It states that if we abstract from the form of space, we arrive at the "category of the synthesis of the homogeneous manifold in an intuition in general, that is, the category of **magnitude**" (B162). This is one of the passages that describes the concept of magnitude as corresponding to the category of magnitude, an issue I will address in the next section. However it is that we are to understand this correspondence, this passage also suggests that features of any intuition in general, rather than the particular features of space and time, are important for cognition and especially for mathematical cognition. I will also return to this point below.

There is a further important point about the meaning of "manifold," one which I will discuss in more detail in Chapter 5. Kant uses the term "manifold" with a quite broad meaning: he uses it for any kind of multiplicity or manifoldness whatsoever. In particular, it includes not just "many-ness" but "muchness," that is, not just what can be said to fall under a count noun, but that which can fall under a mass noun, a "stuff."[25] It is thus a sense of manifold that also encompasses a continuous extent or amount of something, which will include the manifolds of space and time, as well as intensive magnitudes, such as the amount or intensity of a light or a force. It also encompasses a multiplicity of discrete things.

The key property of a magnitude is that it is a homogeneous manifold. Furthermore, it is a homogeneous manifold in intuition. Thus, to be a magnitude requires nothing more than being a many-ness or muchness in intuition that is homogeneous. The homogeneity of the manifold gains particular prominence once we limit Kant's definition to how he states it in the first paragraph of the Axioms. It is the sole property of a manifold in intuition to which he calls attention. In fact, a more literal translation of Kant's definition states that a magnitude is a "manifold homogeneous" (*mannigfaltigen Gleichartigen*), taking manifold as an adjective and nominalizing homogeneous, further emphasizing the notion of homogeneity.[26] As we will see, there was a very long history of regarding homogeneous magnitudes as those that stand in mathematical relations. We will consider more closely what Kant has in mind by homogeneity and its relation to that history in Chapter 7.

Kant's definition also states that the concept of magnitude is the consciousness of a homogeneous manifold in intuition in general, "insofar as through it the representation of an object first becomes possible" (B203). Kant's point here, I think, is to draw attention to the intentional content of the concept, and hence to the importance of the consciousness of the representation of a homogeneous manifold in representing objects in intuition. This does not preclude sensation itself, in addition to what corresponds to it in the object, from having an intensive magnitude. In fact, according to Kant, even consciousness has intensive magnitude. It is nevertheless noteworthy that in Kant's B-edition reformulation of the principle of the Anticipations, he omits asserting that the sensation of an appearance has intensive magnitude, and asserts only that the object of sensation has one. This reformulation also reflects Kant's desire to focus attention on the object represented, even if sensation itself also has intensive magnitude.

[25] To be clear, by "muchness," I do not mean "howmuchness" or amount, but just "stuff."
[26] See note 15 above.

Kant's definition of the concept of magnitude leaves us with a puzzle and some unanswered questions. Those commentators who included "synthetic unity" in Kant's concept of magnitude and interpreted that to correspond to the category of unity did so because the Axioms principle is supposed to correspond to the categories of quantity, as indeed it in some way should. How, on the present account, can we make sense of that correspondence? Moreover, how do we explain why Kant sometimes refers to the category or categories of magnitude? We also do not yet have an account of how the concept of magnitude contributes to the mathematical determination of spaces and times at the root of mathematical cognition. The answer to these questions turns on Kant's distinction between two notions of magnitude: *quantum* and *quantitas*.

3.5 *Quantum* and *Quantitas*

The distinction and the relation between *quantum* and *quantitas* are central features of Kant's account of mathematical cognition, but are also among the most difficult. In order to proceed, we need to have at least a preliminary sense of the meaning of these terms, which will be all that will be attempted in this section; we will sharpen our understanding of them in the coming chapters. Even then, however, we will not be in a position to give a complete interpretation of them that includes their relation to number in this work; some claims will be fully defended only in a monograph focused on the details of Kant's understanding of arithmetic in particular.

Kant's definition of magnitude at the beginning of the Axioms is, as his parenthetical insertion indicates, the definition not just of magnitude but of a *quantum* (*der Begriff einer Größe* (*Quanti*)). Hence, all that we have discussed up to this point concerns *quanta*. Later in the Axioms of Intuition, Kant draws a distinction between a *quantum* and *quantitas*. Kant uses the term magnitude (*Größe*) for both *quantum* and *quantitas*, using these latter terms or adding them in parentheses when he chooses to be careful. As mentioned in Chapter 1, however, Kant is often not careful; in many contexts, we are left to sort out for ourselves which notion of magnitude Kant has in mind, and commentators can hardly be blamed for having difficulty keeping them apart. Furthermore, as noted above, Kant sometimes refers to the categories or even the category of magnitude rather than to the categories of quantity, and he does not specify which sense of magnitude he has in mind.

The Latin *-itas* ending generally denotes a related but more abstract thing, concept, or name than that indicated by the stem word to which it is affixed. This leads one to expect that *quantum* is used to refer to something relatively concrete, something that is a magnitude, while *quantitas* refers to the magnitude of a thing in some more abstract sense that concerns the quantity of the *quantum*.[27] Mathematicians and philosophers writing in Kant's time quite often used the term "magnitude" ambiguously for a thing insofar as it can have a quantity or amount and for the quantity or amount itself, for example, a piece of paper and its length, which roughly corresponds to the distinction suggested by the terms *quantum* and *quantitas*. Although Kant also often uses "magnitude" ambiguously, his appeal to the distinction between *quantum* and *quantitas* marks a difference between a relatively concrete notion of magnitude as that which can have an amount and the amount it has.[28] The relative concreteness of *quanta* is reflected in Kant's claim in the Axioms principle that all appearances *are* extensive magnitudes, rather than that they *have* extensive magnitudes, although the Anticipations principle refers to the real in all appearances *having* an intensive magnitude, so that Kant is not entirely clear on this point. The fact that Kant draws a distinction between *quanta* and *quantitas*, where the latter refers to something more abstract, does not mean that Kant is committed to the real existence of some abstract thing corresponding to *quantitas*; rather, he is drawing a distinction within the contents of our representations, a distinction between that which the concepts *quantum* and *quantitas* represent.[29]

What is important at present is that in Kant's account, a *quantum*'s being relatively concrete does not entail that a *quantum* has properties particular to space or time. The only human forms of intuition are space and time; this fact together with the definition of a *quantum* entails that for humans, concrete *quanta* that contain intuition will be something spatial or temporal. As noted in the previous section, however, Kant defines *quantum* as a homogeneous manifold in intuition in general, a definition that abstracts from any features particular to either space or time. This point is significant because it suggests that one role for intuition in cognition, especially mathematical cognition, is simply to represent something relatively concrete insofar as it is a homogeneous manifold. This is a property common to both space and time, and hence does not depend upon the more particular properties of either.

[27] I say "relatively" concrete for now to indicate that the sense of concrete introduced here is relative to the contrast. I will say more about concreteness in Chapter 5. There, we will see another sense in which a *quantum* is relatively concrete, one that contrasts to a stronger sense of concreteness that is common in our contemporary metaphysics.

[28] Note that there is no similar distinction between the concrete and abstract form of the German term for magnitude (*Größe*); e.g., there is no term such as *Größheit*. This may have prompted Kant to revert to the Latin *quantum* and *quantitas*, though it is not uncommon for Kant to employ Latin terms when making finer distinctions. Kant does use the term "quantity" (*Quantität*) for the quantitative forms of judgment and the categories of quantity, and sometimes uses it in a way that corresponds to *quantitas*, though he also sometimes uses it as a synonym for magnitude (*Größe*); see the *Metaphysik Volckmann* (1784–5, 28:424n).

[29] I will return to this point in Chapter 5.

Nevertheless, defining a *quantum* as a "homogeneous manifold in intuition *in general*" does not suggest something abstract.[30] The important point for us presently is that a *quantum* is something relatively concrete despite its definition abstracting from features particular to space or time, or perhaps from the human forms of intuition altogether.[31]

Fully determining the sense in which Kant thinks *quanta* and *quantitas* are concrete and abstract is an important but difficult task that will take some time; we will take as our starting point that *quanta* and *quantitas* are related to each other as relatively concrete and abstract. The point worth reiterating now is that Kant's argument that all determinate intuitions and hence all appearances are *quanta* appeals to the concept of a *quantum*; appearances fall under this concept in virtue of the determinate intuitions on which their apprehension depends also falling under this concept. But the content of that concept is something that does not, at least according to the explication he gives, refer to anything but a homogeneous manifold in intuition. He does include in his definition of the concept of magnitude the qualification "insofar as it makes the representation of an object first possible," but the properties that characterize what it is to be a *quantum* concern only what is presented in intuition. This is not to say that we can have cognition of a *quantum* as a *quantum* without employing concepts and indeed the categories. But to be a *quantum* is simply to be a homogeneous manifold in intuition. This marks a contrast with *quantitas*, which is closely associated with the categories of quantity. As we shall see, this is one of the most important differences between *quanta* and *quantitas*.

Unfortunately, despite defining a *quantum*, Kant nowhere defines *quantitas*, and determining its nature is difficult. The Axioms, the Discipline of Pure Reason, and related texts help explain the nature of *quanta* and *quantitas* and how mathematics relates to them. After examining these texts, I will turn to the Schematism, which sheds further light on the relation between *quanta* and *quantitas*.

[30] This is a point that might be obscured by the expression "in general." The German term for general here is *überhaupt* rather than *allgemein*, a term Kant also sometimes uses but declines to use in the present context. Kant uses *überhaupt* quite frequently to make a general claim, but it also carries a connotation of "at all" and need not suggest that the topic of the claim, in this case a *quantum*, is itself something general or abstract. For example, reference to a "horse in general" might suggest something abstract in a way that "any horse at all" or "any horse whatsoever" does not. And like these phrases, *überhaupt* can also sound like an intensifier. In fact, Kant sometimes even uses *allgemein* and *überhaupt* in the same sentence: in the Transcendental Aesthetic, Kant refers to "the general concept of any spaces whatsoever [*der allgemeine Begriff von Räumen überhaupt*]." A more perspicuous translation of *überhaupt* might therefore be: a *quantum* is a manifold homogeneous "in any form of intuition whatsoever."

[31] I will return to general concepts of relatively concrete things in Chapter 5.

In the Axioms, Kant characterizes *quantitas* as that which answers the question, "How big is something?," for example, "How big is the notebook on my table?" Kant introduces his discussion of *quantitas* as follows:

> But concerning magnitude (*quantitas*), i.e., the answer to the question "how big is something?" ... (A163–4/B204)

The text in German reads:

> Was aber die *Größe*, (*quantitas*) d.i. die Antwort auf die Frage: wie groß etwas sei? betrifft ...

The term for "big" in German is *groß*, which has the same root as magnitude, *Größe*, a fact to which Kant seems to be calling attention. As an answer to this question, *quantitas* is about the amount of something, a point reinforced by what he next claims: that some mathematical truths concern *quanta*, but not *quantitas*. To explain, he states that, for instance, strictly speaking, the axioms "Between two points only one straight line is possible" and "Two straight lines do not enclose a space" relate only to magnitudes (*quanta*) as such (A163/B204). These axioms concern spatial magnitudes, but not their amounts, in contrast to *quantitas*, which does concern the amount of something. A paradigmatic answer to the question, "How big is something?," and hence what its *quantitas* is, requires a measurement – that is, a specification of a unit of measure and a number that specifies how many units, taken in their totality, either constitute or are equivalent to that which is measured. Nevertheless, the amount of something is a property of the object that is measured and is independent of the unit chosen: the length of a book, for example, is the same whether we describe it as 11 inches or 28 centimeters.

In what sense is *quantitas* abstract? First, let me reiterate that the Latin *-itas* ending was used to indicate a more abstract thing, concept, or name, a reflection of different philosophical positions staked out in response to the problem of universals. I am not here raising the ontological issue, but asking what sort of thing or property is picked out by the concept of *quantitas* regardless of its ontological status; that is, What are the characteristics of *quantitas*?[32] For all that has been said up to this point, a *quantitas* could be the particular amount of a particular *quantum* – this (and only this) book's length, for example. The book might have a *quantitas* that is equivalent or equal to the *quantitas* of this sheet of paper, but each would have its own *quantitas*. Alternatively, a *quantitas* might be the amount shared in common by different *quanta*: the *quantitas* of this book might be one and the same *quantitas* as this sheet of paper, and this is at least one possible sense in which *quantitas* might be considered more abstract than *quanta*.

[32] As we will see in Chapter 5, Kant uses the term "abstract" to refer to ways of regarding and ways of using concepts. He does not use it to refer abstract objects.

In other passages, however, Kant suggests a yet broader understanding of *quantitas* restricted to neither particular nor common amounts of *quanta*. He claims, for example, that statements of numerical relation such as 7 + 5 = 12 are about *quantitas* (B204–5). On the face of it, such statements are most naturally understood as about numbers rather than as covert references to amounts of *quanta*; they are not obviously about how big some concrete thing is, or how many there are of some concrete kind of thing. This at least suggests that Kant employed a more abstract notion of *quantitas*, a notion that includes numbers apart from the things numbered.

Kant further contrasts *quanta* and *quantitas* in the Discipline of Pure Reason, where he seems to push toward an understanding of *quantitas* that abstracts and detaches from *quanta*. He states:

> mathematics does not construct mere magnitudes (*quanta*) as in geometry; it also constructs mere magnitudes (*quantitatem*), as in algebra, in which it wholly abstracts from the properties of the object [*Gegenstand*] that is to be thought in accordance with such a concept of magnitude. (A717/B745)

Abstracting wholly (*gänzlich*) from the properties of an object suggests that any concrete object, and hence, any *quantum*, has been left entirely behind. This would be in keeping with an abstract conception of *quantitas* that is about numbers rather than *quanta*. One possible and relatively natural reading is that the *quantitas* of arithmetic refers to particular numbers, such as 5 or 7, while the *quantitas* of algebra abstracts from the properties of particular numbers, refers to numbers in general by means of variables, and concerns generalizations covering all numbers, such as the associative law of addition. Alternatively, one might take what Kant says in this passage as indicating a broader notion of *quantitas* that encompasses irrational as well as whole and rational numbers or quantities.[33] In either case, Kant's emphasis on the completeness of abstraction might lead one to think that algebra has as its object an object in general, or even that it has no object at all.[34] For many early modern mathematicians, algebra was viewed more as a problem-solving technique than its own discipline, which lends this latter view historical plausibility.[35]

[33] Michael Friedman rejects the first interpretation and argues for the second. See Friedman (1992, pp. 107–12).

[34] For the view that algebra has no object at all, see Friedman (1992, pp. 112–22, especially 113, 114, and 122; but see also 114n34). I will discuss these issues in more detail in work devoted to the specifics of Kant's philosophy of algebra and its relation to symbolic representation.

[35] That algebra was viewed as a problem-solving technique is a point emphasized by Friedman (1992, chapter 2, esp. pp. 113–15). Shabel (1998) also emphasizes this view of algebra, citing Christian Wolff as an example.

Nevertheless, the Discipline of Pure Reason passage describes the *quantitas* of algebra as that concept by means of which an object is thought, and the contrast to *quanta* in the same sentence suggests that the objects whose properties are abstracted from are *quanta*. Moreover, early modern mathematicians applied algebra widely to geometrical problems, and in their interpretation of this application, the objects referred to seem to be geometrical *quanta*.[36] Finally, Kant states that algebra abstracts wholly from the properties of objects, but if it truly abstracts from *every* property, it is unclear how the concept of *quantitas*, employed in algebra, relates to any object at all as its amount. So, what are the objects that fall under *quantitas*, and what, more generally, are the objects of mathematics?

Several texts outside the *Critique* shed light on what Kant had in mind. In §4 of the *Inquiry*, Kant states that "the object of mathematics is magnitude" and adds that algebra is the general doctrine of magnitudes (2:282). This pre-critical view is echoed in writings from the critical period. In a letter to Schultz on November 25, 1788, Kant describes algebra as "that general doctrine of magnitude [*jener allgemeinen Größenlehre*]" (10:555). Kant also indicates in the *Metaphysical Foundations of Natural Science* that *mathesis* is the general doctrine of magnitude (4:489).[37] And in his May 19, 1789, letter to Carl Rheinhold, Kant further clarifies his understanding of algebra. He argues vehemently against the claim of Eberhard, a Wolffian, that mathematicians do not concern themselves with the reality of the objects of mathematics, stating:

> [Mathematicians] cannot make the smallest assertion about any object whatsoever without exhibiting it in intuition (or, *if we are considering only magnitudes without qualities*, as in algebra, exhibiting in intuition the magnitude-relations [*Größenverhältnisse*] thought under the adopted symbols). (11:42, my emphasis)

Kant makes it clear here that what algebra abstracts from is not every property of an object, but rather only its qualities.[38] It is therefore likely

[36] Lisa Shabel has argued that Kant's conception of algebra should be understood as intimately connected to geometry and gives a historically contextualized account of Kant's understanding of symbolic construction. See Shabel (1998).

[37] Kant is also recorded as having called algebra a *Mathesis universalis* in his mathematical lectures recorded by Herder in 1763–4 (29:49).

[38] Kant is not thereby contradicting the claims he makes in the *Critique* concerning the objective reality of concepts. Kant states there that the possibility of things through concepts *a priori* obtains only insofar as they are formal and objective conditions of an *experience* in general, and that the possibility of mathematical concepts is established not merely by constructing an object corresponding to it *a priori*, but through the fact that such objects are the form of an *object of experience* (A223–4/B270–1, my emphasis). Kant is not claiming that mathematics has no objects.

that when Kant states in the Discipline of Pure Reason that algebra "wholly abstracts from the constitution of the object that is to be thought in accordance with such a concept of magnitude [i.e., *quantitas*]" (A717/ B745), he does not intend that algebra has no object at all and is only a method for solving problems. Rather, algebra has *quantitas* as its object of study, where *quantitas* is understood as the quantitative properties of *quanta* as well as the relations among those quantitative properties of *quanta*, that is, the properties of homogeneous manifolds in intuition. When algebra is employed in geometry, for example, it considers only the *quantitas* of geometrical *quanta* and the relations among the *quantitas* of those *quanta*. If this reading is correct, then there are various increasingly abstract objects that are candidates for the objects constructed in mathematics: first, a *quantum*, that is, a homogeneous manifold in intuition; second, the *quantitas* of a *quantum*, that is, the quantitative properties of a *quantum*; and third, the relations among the *quantitas* of *quanta*. In any case, however, *quanta* and their *quantitas*, understood as homogeneous manifolds in intuition and their quantitative properties, lie at the heart of Kant's account of mathematics.

There is much more that needs to be filled out and explained here; for example, the sense in which *quantitas*, as a quantitative property, can be constructed, and the role of symbols in mathematical cognition. Moreover, there is the pressing question of the relation between numerical judgments such as "5 + 7 = 12" and both *quanta* and *quantitas*. For now, I provisionally assert that the *quanta* corresponding to whole numbers are collections of homogeneous units represented in intuition, that is, discrete *quanta*, and that a whole number is the *quantitas* of such a collection, that is, its size.[39] Kant thereby assimilates arithmetic into an overarching theory of mathematics as the science of magnitudes and their measurement.

For now, it is sufficient to recognize that mathematics is about magnitudes, and that, at least in the first instance, mathematical objects are *quanta*. This places Kant's theory of magnitudes at the foundation of his theory of mathematical cognition; moreover, given the Axioms and Anticipations principles concerning magnitudes, it is at the foundation of Kant's understanding of human cognition more generally.

[39] There are passages that seem to indicate that Kant does not countenance the notion of discrete *quanta*, but this is misleading; Kant thinks of collections as discrete *quanta*. The present work provides direct and indirect support for this claim and the further point that Kant understands *quantitas* as including the "how much" of collections of discrete *quanta*, i.e., their "how many"; a more thorough and detailed argument will be given in subsequent work.

3.6 *Quanta* and *Quantitas* in the Schematism

Kant also discusses *quantitas* in the notoriously obscure Schematism of the *Critique*. A thorough treatment of the Schematism and the passage concerning mathematics will not be possible here. Nevertheless, a preliminary examination provides further indication of the relation between *quanta* and *quantitas*. According to Kant, schemata are representations that mediate concepts and intuitions. More specifically, a schema is "a rule for the determination of our intuition in accordance with a certain general concept" (A141/B180). This is the most general description of their role, but schemata play different roles for pure concepts of the understanding, on the one hand, and empirical and pure sensible concepts, on the other. Kant holds that pure concepts of the understanding have nothing in their content in common with intuition, and for that reason they cannot stand in direct relation to intuitions. The categories are pure concepts of the understanding, and schemata play the role of mediating between them and intuitions. The schemata of the categories take the form of rules for determining the time relations among representations. In Kant's account, concepts are rules of the understanding, and insofar as schemata represent rules, they share with concepts an expression of generality. As determinations of time, schemata relate to every representation that appears to us in time and hence everything that appears to us in intuition. By relating to both categories and intuitions in this way, schemata are meant to bridge the gap between them.

Like the schemata for the categories, the schemata for empirical concepts and pure concepts of sensibility are rules "for the determination of our intuition in accordance with a certain general concept" (A141/B180). There is an important difference, however. The schemata of pure concepts of the understanding can never be brought into an image (*in gar kein Bild gebracht werden kann*) (A142/B181), while a schema of an empirical or pure concept of sensibility is "a representation of a general procedure of imagination for providing a concept with its image [*Bild*]" (A140/B179–80).

Kant describes the schemata of the various categories, beginning with the categories of quantity, but rather than treat the concepts of unity, plurality, and allness separately, he introduces the concept of magnitude:

> The pure image of all magnitudes (*quantorum*) for outer sense is space; but that of all objects of the senses in general is time. The pure **schema of magnitude** however (*quantitatis*), as a concept of the understanding, is **number**, which is a representation that holds together [*zusammenbefaßt*] the successive addition of one to one (homogeneous). Number is thus nothing other than the unity of the synthesis of the manifold of a homogeneous intuition in general, through which I produce [*erzuege*] time itself in the apprehension of the intuition. (A142–3/B182)

It is clear that the concept of magnitude is in some way intended to play the role of all three quantitative categories – unity, plurality, and allness – and that the schema corresponding to these categories is number. On the other hand, the relation between *quanta* and *quantitas* is less clear. Since the concept of magnitude stands in for the three categories and Kant contrasts the image and the schema of a concept, one might think that a *quantum* and *quantitas* are the image and schema of the concept of magnitude, respectively.[40] However, both of the words "*quantum*" and "*quantitas*" appear in the genitive, so that Kant is describing the image of *quanta* in the first case and the schema of *quantitas* in the second, and hence the image and schema of different concepts. The relationship between *quanta* and *quantitas* is nevertheless close enough that the image of *quanta*, on the one hand, and the schema of *quantitas*, on the other, somehow complement each other in our cognition of magnitudes. This closeness is reinforced by Kant's habit of referring to one or both *quanta* and *quantitas* simply as "magnitude."

The passage above also reveals that the concept of *quantitas*, not the concept of a *quantum*, corresponds to the categories of quantity. Kant is giving an account of how the categories of quantity are applied by means of their schema, and his explanation describes the schema of *quantitas*. On the other hand, his reference to *quanta* is limited to its image, and he has stated that the schema of a category cannot be brought into an image. In addition, Kant follows the phrase "schema of magnitude (*quantitas*)" with "as a concept of the understanding," indicating that magnitude as *quantitas* is treated as a pure concept of the understanding, and in particular, that *quantitas* corresponds to the categories of quantity.

Since Kant holds that no image can be provided for the categories, the images of *quanta* that are space and time cannot themselves provide an image for the categories of quantity. Instead, space and time provide a homogeneous manifold in intuition, and this homogeneous manifold is determined in accordance with the categories of quantity. (See the discussion of determination in Sections 2.4 and 2.5.) It is significant in this regard that Kant does not refer to the image of a particular *quantum*; he refers to the image of all *quanta* of outer sense and all objects of inner sense. The images to which Kant refers to are not particular determinate spaces and times but space and time themselves – that is, undetermined space and time as given intuitions. Kant thereby allows that *quanta* refers to any homogeneous manifold in intuition in general, whether or not that manifold has been determined. In the Transcendental Aesthetic, moreover, he makes it clear that space and time as given intuitions,

[40] This interpretation is also suggested by the placement of "however" in the description of the schema, which separates "schema of magnitude" from *quantitas* in a way suggesting that the schema is the topic of the sentence and is equated with *quantitas*. The German is "Das reine Schema der Größe aber (*quantitatis*), als eines Begriffs des Verstandes . . ."

and hence as undetermined, are magnitudes (A25, B39). These considerations show that the concept of *quanta* applies to any homogeneous manifold in intuition in general, whether or not it has been determined. Space and time as indeterminate *quanta* serve as a "substratum" determined by *quantitas*.[41] In contrast, the schema of *quantitas* determines *quanta* and thereby generates determinate spaces and times underlying both mathematics and appearances.

Distinguishing *quantum* and *quantitas* allows a further clarification of the principle of the Axioms. The principle asserts that all appearances are, as regards their intuition, extensive magnitudes. An appearance is, with respect to this intuition, a determinate *quantum*. As noted earlier, the paragraph Kant adds to the B-edition is intended to supplement the argument that all appearances and all determinate intuitions are extensive magnitudes, and segues into that argument. Thus, when Kant says that appearances are extensive magnitudes, he means that they are extensive *quanta*. A clearer version of the A- and B-versions of the principle of the Axioms would therefore be: "All appearances, as regards their determinate intuition, are extensive *quanta*" and "All determinate intuitions are extensive *quanta*."[42] It is in virtue of their being extensive *quanta* that the categories of quantity, under the name of *quantitas*, can be employed in their cognition, and this employment in turn plays an important role in explaining the possibility of both pure and applied mathematical cognition.

3.7 Summary

Let me summarize the results. The foregoing analysis makes it apparent that the notion of magnitude plays a central role in Kant's theory of cognition more broadly and in his theory of mathematical cognition in particular. Kant holds that mathematics is about magnitudes and that the objects of mathematics are magnitudes. Moreover, Kant calls the principles of the Axioms and Anticipations mathematical principles because they explain the possibility of mathematics as well as the application of mathematics to appearances, and both principles concern magnitudes. To be a magnitude (*quantum*) is to be a homogeneous manifold in intuition, and the Axioms and Anticipations principles concern a special mathematical synthesis called composition, which is a synthesis of a manifold in intuition that is homogeneous. The homogeneity of intuition therefore plays a central role in Kant's philosophy of mathematics.

[41] See §38 of the *Prolegomena* (4:321–2). Kant also allows that we can intuit individual spaces and times as indeterminate *quanta*, a point to which I will return in the next chapter. I will discuss the sense in which space and time are a substratum for determination in Chapter 5.

[42] See Section 2.8 for including "determinate" in this more perspicuous formulation of the principles.

The close connection between *quanta* and intuition is revealed by the fact that space and time, considered in the Aesthetic in abstraction from the contributions of the understanding, count as *quanta*, while determinate spaces and times are determined by means of the employment of the concept of *quantitas* corresponding to the categories. Furthermore, the definition of *quanta* as a homogeneous manifold in intuition in general underscores the fact that what is of relevance for Kant's philosophy of mathematics is not just properties particular to space or time, but a property they have in common, that is, that each is homogeneous. These considerations indicate that intuition plays an important role in Kant's philosophy of mathematics by allowing the representation of a homogeneous manifold. We will look more closely at the role of intuition in representing magnitudes in Chapter 7.

So far, we have we have uncovered important features of Kant's understanding of magnitude, *quanta*, and *quantitas*. But Kant's larger argument in the Axioms concerns extensive magnitudes in particular, so we will now turn to Kant's distinction between extensive and intensive magnitudes, and a particularly important property of magnitudes, namely, their continuity.

4

Extensive and Intensive Magnitudes and Continuity

4.1 Extensive and Intensive Magnitudes

The Axioms and the Anticipations concern extensive and intensive magnitudes, respectively. As mentioned in the previous chapter, a circle or the space occupied by the book on my table are paradigm extensive magnitudes; the intensity of a light is a paradigm example of an intensive magnitude. The strength of forces as well as speeds are also intensive magnitudes, so that intensive magnitudes are crucial to Kant's account of the fundamental features of experience and the foundations of natural science.

It is important not to confuse Kant's distinction between extensive and intensive magnitudes with nineteenth- and twentieth-century interpretations of the distinction. For most of the latter period, neither the concepts nor the terminology were entirely fixed, but extensive magnitudes, quantities, or attributes were viewed as those that were "additive" and allowed of fundamental measurement, that is, those for which, at least paradigmatically, an ordering relation and concatenation operation can be defined that allow the generation of a standard sequence of units that measures the attribute.[1] The meaning of intensive magnitude was less stable, but by roughly the beginning of the twentieth century, it came to mean a magnitude, quantity, or attribute that was measured only on an ordinal scale.[2] The hardness of minerals on the Mohs scale is a classic example of an intensive attribute measured on an ordinal scale. Talc is assigned a hardness of 1 and diamond a hardness of 10, and other minerals assigned to 2 through 9; for example, gypsum is assigned hardness 2 and apatite is assigned hardness 5. The ordering of these minerals and any other mineral compared with them is determined by the ability of one of these minerals to scratch the other. What is important is that the scale is a mere ordering of hardness; there is no implication that diamond

[1] See Krantz et al. (1971, pp. 7, 71, and 502–3). Krantz et al. showed that those attributes that allow of fundamental measurement are not confined to those for which a standard-sequence procedure can be constructed through a concatenation operation.

[2] See, for example, Cohen and Nagel (1968, pp. 295–6) (this work originally appeared in 1934).

is five times harder than gypsum or twice as hard as apatite, for example. Intensive magnitudes in this sense are of significantly less interest than extensive magnitudes, so that the term "intensive" and the distinction between intensive and extensive attributes fell out of use.[3] According to the modern account, extensive magnitudes can be measured on an interval or ratio scale, and speed, the intensity of a light, and a force are all extensive attributes that are capable of being measured according to a ratio scale.[4]

Kant's notions of extensive and intensive magnitude were quite different; his understanding of intensive magnitudes has its roots in the medieval theory of the intension and remission of forms, a theory that treated speed in particular as an intensive magnitude.[5] Kant agrees with the modern account that magnitudes such as speed, the intensity of a light, and a force can at least in principle can be additively measured. In fact, Kant's *Metaphysical Foundations of Natural Science* concerns in no small part the way in which different magnitudes, such as the "quantity of matter," that is, mass, can be measured.[6] Thus, both extensive and intensive magnitudes in Kant's sense fall within the domain of extensive magnitudes in the modern sense.[7]

For Kant, the distinction between extensive and intensive magnitudes marks a difference in the way they are represented that has implications for whether or not they manifest their part–whole structure. Thus, he fleshes out the traditional understanding of extensive and intensive magnitudes in mereological terms. In the second paragraph of the B-edition Axioms, Kant defines an extensive magnitude as one in which "the representation of the parts makes possible the representation of the whole (and therefore necessarily precedes the latter)" (A162/B203). Kant holds that space and time are special because only determinations of space and time are extensive magnitudes; there are no other sorts of extensive magnitudes. Kant's strategy in the Axioms is to argue that all appearances contain determinate spaces or

[3] Michell (2006).

[4] See Krantz et al. (1971, chapters 1 and 3).

[5] Kant's understanding of speed as an intensive magnitude plays a particularly important role in the Phoronomy chapter of his *Metaphysical Foundations of Natural Science*. See Sutherland (2014) for more on the medieval theory of the intension and remission of forms and Kant's understanding of speed as an intensive magnitude.

[6] There is a large body of excellent work on all aspects of Kant's philosophy of natural science, far too large to survey here. For both a systematic and detailed account of Kant's understanding of the foundations of natural science, see Friedman (2013).

[7] As far as I am aware, Kant shows no interest in mere ordinal measurement or ordinal scales such as the Mohs scale. We will see in later chapters, however, that Kant is aware of the difference between an ordering that does not allow additive measurement and one that does.

times, that all determinate spaces and times are extensive magnitudes, and hence that all appearances are extensive magnitudes.

The principle of the Anticipations of Perception, on the other hand, asserts that sensations and that which is represented in the appearances corresponding to them, that is, the "real," have an intensive magnitude (A165/B207). The Anticipations of Perception concerns qualities: the Anticipations of Perception corresponds to the categories of quality; Kant refers to colors and tastes as examples of qualities of sensation (A175/B217); and he refers to the real of appearances as qualities (A174/B217, A176/B218). Kant provides an example of a real quality in the Analogies of Experience, where he describes mathematically constructing the intensity of the sensation of sunlight; in the Anticipations, he cites gravity as a reality that has intensive magnitude (A178–9/B221, A168/B210). If regarded as a cause, this reality is related to sensation as an influence (*Einfluß*) on the sense (A167/B208), but it can also be regarded as the cause of another reality in appearance, for example, a cause of alteration (A168/B210). More generally, the real is regarded as a ground of consequences; Kant describes the real as a ground, and its intensive magnitude as the amount of influence it has. In the *Metaphysik Volckmann* (1784–5), Kant again uses the intensity of a light as an example:

> the illuminating power of candlelight is intensively greater than tallow light, because with the first we can read at a distance of 2 feet, and with the latter only 1 foot, so the first is a ground of a greater consequence and the second of a smaller. (1784–5, 28:424–5)[8]

The difference in the intensity of the two kinds of candle is manifested in their effects.

Kant states that in sensation "neither the intuition of space nor of time are encountered" (B208). This is not to deny that the real in an object corresponding to sensation can have a duration, nor that the real can be represented as occupying a spatial extent. Rather, Kant is claiming that sensation itself includes no spatial or temporal intuition, and the contribution of the real to appearances is not itself spatial or temporal; instead, it is something, for example, a light or a force, that is represented as having temporal extent and in many cases spatial extent as well. Kant introduces a further example in the

[8] Note that the ordering of the intensities of the two lights is revealed through the comparison of distances at which they have the same effect, i.e., are equal in brightness. Thus, the relative brightness of two intensive magnitudes is determined with the aid of distance, an extensive magnitude. In fact, finding distances at which two intensive magnitudes have the same effect can also be used to reveal that the intensity of light varies with the square of the distance from the light source, namely, by comparison of the relative distances at which two candles are equal in brightness to one candle (of the same kind). This was a well-known example of exploiting one directly measurable magnitude to measure another indirectly.

Metaphysik Volckmann that helps to disassociate the real and its intensive magnitude from its spatial and temporal manifestation:

> If I take a kettle and a thimble full of warm water, then the first is extensively larger than the second, but if the water in the kettle is only lukewarm and the water in the thimble is boiling, then the second is intensively greater than the first. (28:425)

Thus, while the warmth of each is extended over a volume of water for a particular duration, and so appears in space and time, the warmth itself is an intensive magnitude.

Returning to the light example, Kant holds that there is a manifold contained in the light that "coalesces" to constitute the light at a given intensity. This manifold is homogeneous insofar as it belongs to one and the same quality that varies only in intensity, that is, the "how much" of that quality; hence the manifold is a magnitude.[9] Nevertheless, it is an important feature of Kant's understanding of qualities that apprehension by means of sensation "fills only an instant," and the apprehended magnitude can be apprehended only as a unity (A167–8/B209–10). As a consequence, the manifold in it can be represented only indirectly, through representing that sensation or reality as approaching to zero or as growing from zero to its given measure in a certain time. Kant uses this feature in his characterization of intensive magnitudes: "I call that magnitude which can only be apprehended as a unity, and in which the plurality [*Vielheit*] can only be represented through approximation to negation = 0, **intensive magnitude**" (A168/B210).[10]

Kant's discussion of intensive magnitudes raises questions about their relation to the *quantum*/*quantitas* distinction. As discussed in Section 3.5, these terms distinguish two senses of magnitude (*Größe*); the former connotes something relatively concrete, and the latter something more abstract, a property of the former. More specifically, Kant defines a *quantum* as a homogeneous manifold in intuition, while *quantitas* is the "how much" or amount of a *quantum*. A determinate space or time is a magnitude in the first sense, and has a magnitude in the second sense, two senses that were often conflated in Kant's time and before, and which Kant doesn't always clearly distinguish for the reader, despite drawing the distinction. How does Kant's account of intensive magnitudes relate to this distinction?

[9] See B201n for Kant's use of "synthesis of coalition [*Koalition*]" to characterize this synthesis directed to intensive magnitudes, a synthesis of the manifold he asserts is homogeneous.

[10] Note that Kant describes the manifold here as a plurality (*Vielheit*), indicating that the term applies to not just discrete manifolds but also continuous manifolds, such as the intensity of a light. This is important for understanding how Kant thinks of the category of plurality (*Vielheit*), as will be discussed in Section 4.7 below.

The principle of the Anticipations asserts that sensations and the corresponding real in appearances have an intensive magnitude, and throughout, the Anticipations focuses on the fact that a sensation or the real has an amount (no specific amount, just some amount or other). This indicates that, in contrast to the principle of the Axioms, which argues that all appearances are extensive *quanta*, the principle of the Anticipations concerns *quantitas*. This in turn suggests that sensation and the real are themselves *quanta*, which have an intensive *quantitas*, but Kant does not explicitly call them *quanta* in the Anticipations. Are they indeed *quanta*? We have just seen that sensation and the real corresponding to it contain a homogeneous manifold, one that is revealed by representing it as diminishing down to zero or increasing from zero up to its measure. On the other hand, recall that Kant defines a *quantum* as a homogeneous manifold in intuition. This might seem to exclude sensations, "in which no intuition of space or time are encountered." As we have seen, however, the real does appear in time and, at least quite often, in space, so that if we understand "in intuition" to mean not that the manifold is *of* space and time, but that it is a manifold represented *in* space and time, the real would count as a *quantum*. Moreover, even in the case of sensation, we represent it as containing a homogeneous manifold by representing it as varying in time. In fact, Kant does refer to light, warmth, and so on as *quanta*.[11] Thus, sensation and the real do in fact count as *quanta*, which have an intensive *quantitas*.

Kant's account of intensive magnitudes includes a further claim grounded in how the manifold of the *quanta* is represented:

> every sensation, thus also every reality in appearance, however small it may be, has a degree, i.e. an intensive magnitude, which can still always be diminished, and between reality and negation there is a continuous nexus [*kontinuerlicher Zusammenhang*] of possible realities, and of possible smaller perceptions. Every color, e.g. red, has a degree, which, however small it may be, is never the smallest, and it is the same with warmth, the moment of gravity, etc. (A169/B211)[12]

Kant then immediately segues into a discussion of the continuity of both intensive and extensive magnitudes, which he defines as the property "on account of which no part of them is smallest (no part is simple)." In his discussion, Kant treats the manifold contained in an intensive magnitude as parts, so that appearances in their mere perception (sensation and thus

[11] *Metaphysik Volckmann* (28:424). See also *Metaphysik Herder*, 1762–4 (28:22), *Metaphysik von Schön*, 1785–90 (28:519), *Metaphysik Dohna*, 1792–3 (28:637). See as well *Reflection* 5582 1778–9 (18:240).

[12] Kant makes virtually the same point two paragraphs earlier, indicating its importance for him (A167–8/B209–10). He uses the phrase "continuous nexus [*kontinuerlicher Zusammenhang*]" there as well.

reality), are continuous intensive magnitudes (A170/B212). Elsewhere, he also refers to the manifold contained in an intensive magnitude as parts, for example, in the *Metaphysik Volckmann* (28:425). We will return to the relation between these parts and continuity below; for now, the important point is that an extensive magnitude is distinguished from an intensive magnitude by the conditions for the representation of its parts. The former presupposes the representation of its parts in the representation of the whole, while the latter is apprehended only as a unity, and its parts can be only indirectly represented through a representation of its diminution to zero or increase from zero up to its intensity.

The Anticipations of Perception explains an important condition of the possibility of applying mathematics to phenomena that have intensity – they are homogeneous manifolds, that is, *quanta*, which have a *quantitas*. Kant does not, however, think that intensive magnitudes play a role in explaining the possibility of mathematics itself or the possibility of pure mathematics, apart from the application of mathematics to appearances. In contrast, the properties of extensive magnitudes do, which gives extensive magnitudes a priority over intensive magnitudes in explaining the possibility of mathematical cognition.

This priority of extensive over intensive magnitudes is a result of the difference in the conditions for the representation of their part–whole relations. Extensive magnitudes play a leading role in Kant's philosophy of mathematics precisely because they manifest their part–whole structure in a way that makes that part–whole structure cognitively accessible. In fact, we cannot represent an intensive magnitude as a magnitude at all, that is, represent its part–whole structure, without the aid of the extensive magnitude of time. Moreover, as Kant's candlelight and tallow light example above shows, we use extensive magnitudes to help us measure intensive magnitudes in applying mathematics to them: the intensity of the light is measured against the distance at which the two lights have equal intensity.[13] Because extensive magnitudes are primary in Kant's account of mathematical cognition, they will be the central focus of this book.[14]

4.2 The Extensive Magnitude Argument

As we saw in the previous chapter, Kant's B-edition Axioms argument that all intuitions, and hence all appearances with regard to their intuition, are

[13] See note 8 above. In fact, Kant states in one of his notes that "All intensive magnitudes must eventually be brought to extensive magnitudes" (*R5590* 1778–9, 18:242).

[14] See Daniel Warren (2001) for a penetrating account of Kant's treatment of forces, and Friedman (2013) for a detailed reconstruction of Kant's mathematical physics. For more on Kant's account of intensive magnitudes and their relation to the medieval theory of the intension and remission of forms, see Sutherland (2014).

extensive magnitudes has two stages: the first establishes that all intuitions and all appearances are magnitudes, while the second establishes that they are extensive magnitudes in particular. We will now focus on the second stage, which can be roughly reconstructed as follows:

1. An extensive magnitude is that in which the representation of the parts makes possible the representation of the whole (and therefore necessarily precedes the latter).
2. The cognition of a determinate space or time is possible only through a successive synthesis from part to part.
3. Anything cognized by means of a successive synthesis requires that the representation of the parts makes possible the representation of the whole.
4. Therefore, all determinate spaces and times are extensive magnitudes.
5. All appearances contain a determinate intuition in space and/or time.
6. Therefore, all appearances are extensive magnitudes.

Note first that the argument establishes a claim only about determinate intuitions and appearances; it cannot establish that all intuitions – in particular, space and time as given intuitions – are extensive magnitudes. Thus, a more precise statement of the A-edition principle would be: all appearances are, as regards their determinate intuition, extensive *quanta*. A corresponding more precise statement of the B-edition formulation would be: all determinate intuitions are extensive *quanta*. The structure of Kant's argument provides further support of the claim, defended in Chapter 2, that Kant's discussion of space, time, and mathematics in the Axioms contrasts with that in the Aesthetic in concerning determinate spaces and times and hence the categorial conditions of mathematical cognition.

A crucial step of this argument states that the representation of a determinate space or time is possible only through a successive synthesis. There is an obvious connection here to the claim of the Schematism that we examined in Section 3.6: the determination of intuition in accordance with the categories of quantity is guided by the schema of *quantitas*, that is, number, which is a "representation that summarizes the successive addition of one to one (homogeneous)" (A142/B182). Kant holds that the addition must be successive because the schemata mediate between the categories and intuitions through a transcendental determination of time.

We have seen that the argument inserted at the beginning of the B-edition of the Axioms concerns magnitude in general, not just extensive magnitudes, and it does not mention successiveness at all; it turns on the general conditions for determining any homogeneous manifold in intuition in general. Moreover, we have just seen that apprehension by means of sensation alone "fills only an instant," which contrasts with his claim that the representation of extensive magnitudes is possible only through a successive synthesis, a synthesis which itself determines time. Thus, the successiveness of the synthesis of extensive magnitudes is important.[15]

It is clear as well that successive synthesis in time plays an important role in Kant's philosophy of mathematics, although there have been different interpretations of its significance. Charles Parsons has argued, for example, that in Kant's account the successive construction or apprehension of marks or signs provides a model for the structure of the numbers,[16] and this aspect of Kant's understanding of number most closely corresponds to our contemporary account of ordinal numbers. Michael Friedman has argued that the successiveness of time makes it possible to represent the unlimited iterability of mathematical operations such as the construction of a line or the addition of a number;[17] Friedman also emphasizes the ordinal aspects of Kant's understanding of number.[18] We have seen, however, that successive synthesis requires that the representation of the parts makes possible and precedes the representation of a whole determinate intuition. As noted, mathematical cognition depends in particular on extensive magnitudes, so the fact that, in the case of extensive magnitudes, the representation of the parts makes possible and precedes the representation of the whole must have a special significance. I do not think that the role of this feature of the part–whole relations of extensive magnitudes in mathematical cognition has been sufficiently emphasized.[19]

Kant underscores the importance of this property in the Axioms. He asserts that successive synthesis guarantees that all appearances are represented as aggregates, that is, as collections (*Menge*) of previously given parts (B204). That is because they can be cognized only in their apprehension through successive synthesis "from part to part." I noted in the previous chapter that Kant identifies a special synthesis, the synthesis of composition, that is particular to homogeneous manifolds, and hence, magnitudes, and that Kant further distinguishes between the synthesis of composition of extensive and intensive magnitudes; the former is the synthesis of aggregation, while the latter is the synthesis of coalition (B201n). Successive synthesis entails that

[15] As Kant puts it, the successive addition of the schema of *quantitas* takes place through a generation of time itself in the apprehension of an intuition (A142–3/B182), which is an *a priori* time determination of the time series (*Zeitreihe*) (A184–5/B185).

[16] Parsons (1983b, especially pp. 139–41).

[17] Friedman (1992, chapters 1 and 2; see especially pp. 63, 87–9, 92, 116–20). Friedman further develops his view to allow for a role of intuition in allowing us to represent the translation and rotation of a point of view (Friedman (2000, pp. 191–3)).

[18] See Sutherland (2017) for an argument that Kant's conception of number includes both cardinal and ordinal elements.

[19] There are exceptions; both Kirk Dallas Wilson (1975) and Gordon Brittan (1978) have emphasized the importance of part–whole relations in Kant's account.

determinate spaces and times, and hence appearances, are represented as aggregates of previously given parts out of which they are composed. I will call this the "part–whole aggregation property."

Immediately following his assertion that appearances are intuited as aggregates, Kant adds:

> On this successive synthesis of the productive imagination, in the generation of shapes, the mathematics of extension (geometry) with its axioms is grounded, which express the conditions of sensible intuition *a priori*, under which alone the schema of a pure concept of outer appearance can arise: e.g., between two points only one straight line is possible, two straight lines do not enclose a space, etc. These are axioms that actually only concern magnitudes (*quanta*) as such. (B204)

Kant indicates in this passage that successiveness plays an important role in geometry in particular. What is crucial to geometry is not the successiveness of the synthesis itself, but that such synthesis leads to a representation of the part–whole structure of determinate spaces and times.

Despite the focus in the Axioms on determination of spaces and times and on the successiveness of the synthesis required to determine them, there is still an important question: Why is it that determinate spaces and times can be determined only through a successive synthesis? The answer appears to point to a feature of the manifold of space and time themselves; the nature of space and time imposes this constraint on how its manifold is synthesized and determined, for not all manifolds must be synthesized in this way, as we saw above in the case of intensive magnitudes.

Note that Kant's definition of extensive magnitude does not itself refer to a condition of cognition, or a condition of the determination or synthesis required for cognition, but only to representation: an extensive magnitude is one in which the *representation* of its parts makes possible the *representation* of the whole. Similarly, his definition of the concept of magnitude refers to the consciousness of the manifold homogeneous in general, "insofar as it makes the *representation* of an object first possible" (B203, my emphases). Kant's characterizations of both magnitude and extensive magnitude appear to be basic, in this sense: they concern not higher-level requirements of cognition or perception, but features of representation. On the other hand, the system of principles concerns the application of pure concepts of understanding to possible experience, and the Axioms concern the conditions for the cognition of appearances.

The focus on conditions of representation raises an issue about the scope of extensive magnitudes. Kant states in the Transcendental Aesthetic that space is represented as an infinite given magnitude (B40–1). Kant considers our representations of space itself and time itself as discussed in the Transcendental Aesthetic to be representations of a homogeneous manifold.[20] This leads naturally to a question: Are the intuitions of space and time themselves, as described in the Transcendental Aesthetic, extensive or intensive magnitudes? As we saw in Chapter 2, it cannot be that they are extensive magnitudes; Kant argues that we can represent only a single space and that the parts of space cannot "as it were precede the single all-encompassing space as its components (from which its composition would be possible), but are rather only thought **in it**" (A24–5/B39). Thus, space itself cannot be an extensive magnitude. This led to an apparent tension between the account of space and time in the Aesthetic and the Axioms of Intuition that is resolved by the observation that the Axioms of Intuition concern determinate spaces and times, while the Transcendental Aesthetic does not (see Section 2.3). Only determinate spaces and times are extensive magnitudes. One might then wonder whether the representations of space itself and time itself must be representations of intensive magnitudes. After all, they are represented as a unity and not through a successive synthesis, and thus bear some of the properties of intensive magnitudes.[21] Kant nowhere suggests that they are; moreover, the Anticipations of Perception attribute intensive magnitudes only to sensation and the real corresponding to sensation. Thus, the distinction between extensive and intensive magnitudes applies only to determinate magnitudes; the space and time of the Transcendental Aesthetic are neither extensive nor intensive magnitudes.

4.3 The Categories of Quantity and the Mereology of Magnitudes

As discussed in Chapter 2, the primary role of the Axioms in the *Critique* is to argue for and explain the applicability of the categories of quantity – unity, plurality, and allness – to objects of experience. Kant sometimes refers to these categories as the "category of magnitude." This suggests that employing the categories of unity, plurality, and allness make it possible to cognize part–whole relations of appearances and intuitions insofar as they are magnitudes: a part of a manifold, the plurality of parts, and the synthesis of those parts into a whole are cognized through the categories of quantity. That is, the employment of the categories of quantity is fundamentally mereological.

Kant does not directly discuss the part–whole relations of magnitudes, in exactly those terms, in the *Critique* or other published work. Nevertheless, Kant's lectures on metaphysics and his notes reveal that the part–whole relation corresponds to the categories of quantity outlined in the *Critique*, a correspondence found in the development of Kant's critical philosophy. As mentioned in Section 2.4, Alexander Baumgarten, a student of Wolff, wrote a

[20] See also A25, as well as the Schematism, where Kant refers to space as the pure image of all magnitudes (*quanta*) (A142/B182) and Kant's Notes on Kästner (20:419).

[21] I would like to thank Samuel Levey for raising this possibility.

metaphysics text that Kant used for many years in his lectures. Kant repeatedly worked through this Leibnizian-Wolffian metaphysics, considering and reconsidering the fundamental concepts upon which it was built; his evolving views are revealed in student lecture notes, Kant's notes in his copy of Baumgarten, and various other notes. Kant discusses various pairs and triplets of concepts and their relations, but few so frequently as one–many–one. His discussions show that he thought of these concepts as corresponding both to the concepts of unity–plurality–allness and to the concepts of part and whole. The three categories correspond to the part–whole relation by giving us the concepts of the unity of a part, the plurality of parts, and the totality of parts in a whole. In the *Metaphysik Volckmann* lecture notes, for example, Kant explicitly states that the concepts of part and whole stand under the categories of quantity (28:422, 1784–5; see also 29:803, 1782–83; compare 28:504–5, late 1780s). Thus, the three categories of quantity serve a mereological function.[22]

More specifically, these categories allow us to cognize the part–whole relations of magnitudes. As noted, Kant sometimes refers to the categories of quantity as the categories of magnitude or the category of magnitude (B115, B162, B201, B293). In the *Critique* of the Power of Judgment, he connects the cognition that something is a magnitude with the cognition of plurality and unity:

> That something is a magnitude (*quantum*) can be cognized from the thing itself, without any comparison with another; if, namely, a plurality of the homogeneous together constitute a unity. (5:248)

These passages confirm that a primary role of the categories of quantity is to allow us to cognize the part–whole relations of magnitudes, as well as confirming the central importance of the part–whole relations of extensive magnitudes in particular. They show that the categorial foundation of mathematical cognition is mereological.

We will return to the cognition of the part–whole relations of magnitudes under the categories of quantity in Chapter 7, where we will more closely consider the role of intuition. For now, we will turn our attention to a particular problem that arises for the part–whole relations of extensive magnitudes.

[22] Parsons has noted (1984, p. 112) that part–whole notions dominate in Kant's explanation of the categories of quantity, and I am indebted to his work on this point. The evidence is spread throughout Kant's notes in his copy of Baumgarten's Metaphysics, and his own notes on metaphysics (1902, vols. 17 and 18). The development of Kant's views is more obscure and complex than my summary suggests; the relations between unity–plurality–totality, one–many–one, and part–whole are also connected to Kant's views on unity, truth, and perfection, for example.

4.4 The Extensive Magnitude Regress Problem

Kant's definition of extensive magnitudes raises a deeper issue with important consequences for his entire account of both mathematical cognition and the mathematical character of the world. When Kant states that an extensive magnitude is one in which "the representation of the parts makes possible the representation of the whole (and therefore precedes the latter)," one naturally wonders which parts are presupposed and in what sense they must be represented.

These questions have bite because Kant also holds that space and time are continuous. In the Anticipations of Perception, Kant argues that all appearances whatsoever are continuous "both in their intuition, as extensive magnitudes, as well as in their mere perception (sensation and thus reality), as intensive ones" (A170/B212).[23] The first part of this argument concerning extensive magnitudes rests on a further argument that space and time are continuous:

> The property of magnitudes on account of which no part of them is the smallest (no part is simple) is called their continuity. Space and time are *quanta* continua, because no part of them can be given except as enclosed between boundaries (points and instants), thus only in such a way that this part is again a space or a time. Space therefore consists only of spaces, times of times. Points and instants are only boundaries, i.e. mere places of their limitation; but places always presuppose those intuitions that they limit or determine,[24] and from mere places, as components that could be given prior to space and time, neither space nor time can be composed. (A169–70/B211)

Kant states that space and time are continuous *quanta*, by which he means that no part is the smallest.[25] In addition, Kant follows a long tradition concerning the relation between points and extended spaces: precisely because points are unextended, no number of them composed together will result in something extended (see A439/B467). If they could compose an extended space, then space and time could have simple parts, but this is ruled out. The represented parts presupposed by the representation of an extensive

[23] Guyer and Wood translate "sowohl . . . als auch" as "either . . . or . . .." which I think is less perspicuous and misses Kant's emphasis on the continuity of *all* features of an appearance.

[24] Guyer and Wood translate this phrase as "but places always presuppose those intuitions that limit or determine them," which I think gets the limitation or determination the wrong way around. The German is grammatically ambiguous between the two readings, but the sense of the passage makes it clear that points and instants presuppose that which they limit or determine.

[25] See also (4:354) and *Reflexionen zur Mathematik, Physik und Chemie* 14:131–2, 1764–1804).

magnitude seem themselves to be extensive magnitudes, which requires a representation of their parts. Because no part is the smallest, we have an infinite regress.

As a consequence, the questions concerning extensive magnitudes – which parts are represented and in what sense they are represented – is closely related to a perennial problem in metaphysics, mathematics, and natural philosophy about the nature of continuity, problems that are at least as old as Zeno and with which Aristotle famously wrestles in his *Metaphysics*. These same problems appear in treatments of magnitude in the medieval period in the doctrine of indivisibles, and gained added urgency with the development of infinitesimal methods in mathematics and mathematical physics in the early modern period, leading Leibniz to famously refer to the seemingly intractable issues surrounding continuity as the "labyrinth of the continuum." A full treatment of Kant's understanding of continuity in metaphysical, geometrical, and infinitesimal contexts would take us too far afield;[26] I will limit myself to its relation to the nature of extensive magnitudes, and will return again to the representation of continuous magnitudes in Section 4.5 below.

There are several options available to solve the regress problem, but it is helpful first to reflect on the nature of those magnitudes Kant regards as extensive in contrast to those he regards as intensive. There is clearly some sense in which one cannot represent a line without thereby representing its parts; I cannot represent a three-inch line segment without representing its left half and its right half, for example. In contrast, my representation of the intensity of a light does not carry with it a representation of the smaller intensities which compose it, or the smaller intensities that can be represented as composing it. Because I can't represent a whole extensive magnitude without thereby representing its parts, the representation of an extensive magnitude manifests its part–whole relations. The point of Kant's drawing the distinction between extensive and intensive magnitudes is precisely to highlight this difference and this property of extensive magnitudes. The fact that the representation of extensive magnitudes necessarily manifests their part–whole structure makes that structure constitutive of our cognition of extensive magnitudes, and hence of our mathematical cognition.[27]

[26] See Sutherland (2020b) for more on Kant's treatment of infinitesimals.

[27] This raises questions about how the representations of space and time manifest their part–whole structure. According to Kant, the representation of times depends on spaces, and vice versa. In particular, we can represent time only through the drawing of a line. Kant states that in doing so we "attend solely to the action of the synthesis of the manifold" and thereby to the "succession of this determination in inner sense" (B541). It is not immediately clear to what extent Kant thinks the representation of the *part–whole structure* of time depends on the representation of the *part–whole structure* of the line. Moreover, in a lecture on metaphysics, Kant also suggests that time is essential to representing the parts of a whole space (*Metaphysik Vigilantius*, 1794–5, 29:994).

Nevertheless, the sense in which the parts of an extensive magnitude are manifest is not entirely clear. When looking at a line segment, I do not explicitly represent to myself the left and right halves as the left and right halves, unless prompted to do so – any more than I explicitly represent the thirds of the line, or the fifths of the line; or any size collection of possibly unequal parts, such as the two parts corresponding to the left 17/23rds and the right 6/23rds, or one part corresponding in length to the square root of the entire length and the other to the remainder, or perhaps some randomly determined parts. All these are parts of the line, and all are in some sense presupposed by the representation of the whole line, though they are not explicitly represented as those parts.

This fact prompts one to search for a weaker sense in which the representation of the whole presupposes the representation of the parts, one that corresponds to the sense in which an extensive manifold manifests its parts. Unfortunately, Kant's account of our cognition of extensive magnitudes involves the synthesis of composition, which appears to demand that parts be given for the act of synthesis to combine them. Moreover, Kant explicitly adds not only that the parts are presupposed by the synthesis, but that the parts precede the whole, at least suggesting a temporal dimension to the synthesis. There are many, myself included, who in at least some contexts would like to downplay the ways in which Kant's words sometimes strongly suggest temporally extended acts and temporal ordering, interpreting them as a way of expressing more abstract properties and relations among our representations. In the present context, for example, the precedence of the parts need not be interpreted temporally, but might be interpreted as simply reflecting that feature of extensive magnitudes I have been trying to articulate – a dependence of the representation of the whole extensive magnitude on representing its parts. Nevertheless, as noted in the previous section, Kant is at pains here to emphasize, and even highlight, the successiveness of the synthesis of composition of an extensive magnitude. Here, too, one could attempt to interpret this as a reference merely to an ordering of the parts synthesized, but the reading starts to feel strained. Kant immediately follows his definition of extensive magnitude with the claim that:

> I cannot represent to myself any line, no matter how small it might be, without drawing it in thought, i.e. successively generating all its parts from one point, and thereby first sketching this intuition. It is exactly the same with even the smallest time. I think therein only the successive progress from one moment to another. (A162–3/B203)

Whether or not manifesting the part–whole structure of either space or time requires both space and time; their part–whole structure is manifest in a way that is not in the case of intensive magnitudes. I discuss the part–whole relations of magnitudes further in Section 7.7 below.

In the case of the representation of time, one could chalk up the reference to moments and successive progress to the fact that what is represented is extensive magnitudes of time in particular, but Kant's claim that the representation of a line requires successive generation of parts in drawing it also sounds inescapably temporal. Moreover, Kant has already insisted on the need for drawing or describing for thinking lines and circles in §24 of the B-Deduction, and in particular the need for motion in an attached footnote (B154–5, B155n). In addition, his doctrine of the schematism explicitly argues that the schematism of the categories of quantity rests on a transcendental time determination and thereby connects the categories of quantity to the intuition of time and all our representations appearing in time.

If we take the temporal aspect of the successive synthesis seriously, however, it only underscores the regress we are trying to avoid; it seems that we can never even get the synthesis started if each part of a line must itself be represented through a synthesis of parts that are given temporally prior to the synthesis. We will return to this point, but let us set aside temporality for now, for even if we were to insist on a nontemporal reading of the successive synthesis, the threat of regress remains: the representation of the whole presupposes the representation of the parts, and these parts in turn have parts that need to be represented. The crux of a solution to the regress challenge is this: we need to find a distinction between the representation of a whole extensive magnitude and the representation of its parts that allows us to deny that the parts are themselves extensive magnitudes, and hence are subject to the same condition. Since extensive magnitudes are determinate spaces and times, the most obvious and plausible solution, and the one most strongly suggested by the texts, is that the parts are only indeterminately represented. The challenge is to explain what the indeterminate representation of parts amounts to.

To begin, recall that Kant almost exclusively employs the notion of determination for an act of the understanding by means of concepts (see Sections 2.4 and 2.5). This strongly suggests that the indeterminate representation of parts is a representation that does not directly involve concepts for the representation of those parts. Whatever views we consider, an important underlying issue will be the extent to which concepts are or are not involved in their representation. Kant's writings suggest three possible accounts of the representation of indeterminate *quanta*. I will consider and rule out two of those in the next section before turning to what I believe is Kant's actual solution to the regress problem.

4.5 Two Attempts to Solve the Extensive Magnitude Regress Problem

4.5.1 *Indeterminate* Quanta *within Boundaries*

Kant mentions the possibility of intuiting indeterminate *quanta* in the Antinomy of Pure Reason. The First Antinomy concerns whether the world has a beginning in time and is enclosed within boundaries in space. The thesis of the First Antinomy argues by *reductio* that it is. In explaining the proof of the thesis, Kant makes two claims that he endorses and are not simply part of an *ad hominem* argument. He states:

> We can think of the magnitude of a *quantum* that is not given within **certain** boundaries of every intuition* in no other way than by the synthesis of its parts, and we can think of the totality of such a *quantum* only through the completed synthesis, or through the repeated putting-together [*Hinzusetzung*] of units with each other. (A426–8/B454–6)

Here Kant makes a few comments that reinforce points we've established about his theory of magnitude. Note first that by "magnitude of a *quantum*" Kant can mean only the quantitative property of a *quantum*, and hence its *quantitas*. His linking it to totality confirms that *quantitas* corresponds to the categories of quantity and, in particular, the category of totality. Note as well that Kant affirms that a repeated putting-together in the synthesis of parts is needed in order to think the *quantitas*. Yet it is the exception Kant allows to this condition that is of particular interest; it applies only to *quanta* that are not given within "**certain** boundaries (***gewisser*** *Grenzen*) of every intuition." He appends a footnote to explain what he has in mind:

> * We can intuit an indeterminate *quantum* as a whole, if it is enclosed within boundaries, without needing to construct its totality through measurement, i.e., through the successive synthesis of its parts. For the boundaries already determine its completeness by cutting off anything further. (A426–8/B454–6n)

Here too, Kant makes comments that are helpful to understanding his views of magnitude: the thought of a *quantitas* presupposes the thought of the totality of the repeated putting-together of units, and the latter is described as constructing a totality. Kant's mention of construction connects the thought of a *quantitas* with his doctrine of construction, that is, *a priori* exhibition of an object corresponding to a concept, which is characteristic of mathematical cognition. Moreover, he describes it as constructing a totality through measurement. But what is most important for our current topic is that Kant explicitly allows that we can intuit an indeterminate *quantum* as a whole if it is enclosed within boundaries. Thus, if it were possible to represent the parts of an extensive magnitude in this way, then we could represent them as indeterminate *quanta*, and not as themselves determinate extensive magnitudes, which would stop the regress.

Kant does not here expand on how we might represent a particular space or time as an indeterminate *quantum* by representing it as a whole enclosed within boundaries, but in other contexts he does repeatedly refer to particular spaces and times as enclosed within boundaries. In the resolution to the

Antinomies, Kant refers to every space intuited within its boundaries and an external appearance (a body) enclosed with its boundaries, and in the *Prolegomena* discussion of incongruent counterparts, Kant discusses the spaces enclosed within the boundaries of a hand or an ear. Kant holds that surfaces are the boundaries of volumes, lines of surfaces, and points of lines, while moments (*Augenblicke*) are the boundaries of times (4:354; see also the *Inaugural Dissertation* (2:403), and *Metaphysik* L_2 1790–1? (28:570)). In the Discipline of Pure Reason, he refers to the concept of a triangle as the concept of a figure enclosed by three straight lines (A715/B743).[28] For Euclid and for most geometers in the tradition, a triangle is not three lines in a particular arrangement, but the area enclosed by them. This suggests that we represent the area of a triangle through enclosing its area within boundaries. Unfortunately, Kant does not say more about how geometrical figures enclosed within boundaries are represented. Hence, he does not affirm that they are represented as indeterminate *quanta*.

More importantly, Kant nowhere indicates that the representation of a determinate magnitude within boundaries solves the regress problem. Moreover, there is a good reason it would not. If the representation of an indeterminate *quanta* were to rely on the determination of boundaries – the determination of the lines enclosing a triangle, or the determination of the surface of a sphere or cube enclosing a volume, for example – then the representation of an indeterminate space would depend on the representation of a determinate space (albeit of a lower dimension). But the point of appealing to the representation of an indeterminate magnitude in the first place was to avoid a regress in the representation of determinate magnitudes.

One could perhaps argue that the representation of an indeterminate magnitude of a given dimension by means of a determinate magnitude of a lower dimension would solve the regress problem at that given dimension. The regress would then arise for the lower dimension, requiring the representation of an indeterminate magnitude by means of a magnitude at yet a lower dimension. Moreover, this regress would stop with points or instants, since these are not themselves *quanta* at all. But the solution to the regress problem has now become rather elaborate. If Kant had entertained anything like this elaborate solution, one would have expected him to say so, or at least that there were signs of his working out the solution in his own notes. There is no indication that he did.

The representation of an indeterminate magnitude by means of a determinate magnitude of a lower dimension is in fact an important feature of Kant's views to which I will return in Section 4.6.3 Nevertheless, it won't do as a

[28] See also A303/B360. Kant also refers to the impossibility of enclosing a space with two straight lines at A47/B65.

solution to the regress problem; it seems that the solution lies elsewhere. This leads us to a section of the *Critique of the Power of Judgment* that appears more promising, a section in which Kant directly addresses a regress problem involving magnitudes.

4.5.2 *The Aesthetic Estimation of Magnitude*

In the *Critique of the Power of Judgment*, Kant describes a representation of spaces other than through a synthesis of part to part; in fact, he presents such representation as a condition of all "mathematical estimation" of magnitude. Kant outlines his views in his explanation of the mathematically sublime; his broader aim in this portion of the *Critique* of the Power of Judgment is to explain aesthetic judgments of the sublime, the feeling of the sublime, and their purposiveness.[29] We will focus on Kant's account of that on which the idea of the mathematically sublime depends, that is, the estimation of the magnitude of things of nature.

Kant distinguishes between the mathematical and aesthetic estimation (*Schätzung*) of magnitude, and it is the latter that plays a key role in explaining the mathematically sublime. Kant uses the term "estimation" (*Schätzung*) of magnitude as a genus that covers mathematical as well as nonmathematical ways of representing the magnitude of something. By mathematical estimation of magnitude, Kant has in mind measurement in particular, that is, the determination of the *quantitas* of a *quantum* with respect to a unit of measure. He states that mathematical estimation is by means of numerical concepts; all logical estimation of magnitude is mathematical insofar as it depends on numerical concepts, and determinate concepts of how big something is require "numbers (or at least approximations by means of numerical series progressing to infinity), whose unit is the measure [*Maß*]" (5:251). In contrast, the aesthetic estimation of magnitude is "in mere intuition (according to the measure of the eye [*nach dem Augenmaße*])" (5:251).

What is most important for our purposes is that Kant argues for a dependence of mathematical estimation on aesthetic estimation; more specifically, the aesthetic estimation of magnitude solves a regress problem that looks like the one raised by Kant's definition of extensive magnitude that we've been considering. After stating that determinate concepts of how big something is require numbers, whose unit is the measure, Kant adds:

[29] The mathematically sublime is that which is sublime with respect to the estimation of its magnitude as absolutely great; it concerns that in comparison with which everything else is small, and it demonstrates a faculty of the mind, namely, reason, that surpasses every measure of the senses.

> But since the magnitude of the measure must still be assumed to be known [*bekannt*], then, if this is in turn to be estimated only by means of numbers whose unit would have to be another measure, and so mathematically, we can never have a first or fundamental measure [*erstes oder Grundmaß*], and hence we can never have a determinate concept of a given magnitude. (5:251)

Kant here reveals two important features of the determinate concept of a given magnitude, namely, that the determinate concept is always relative to a unit, and that the unit of measure must be "known" (*bekannt*).[30] These two features lead to a regress if the method of knowing the measure is itself mathematical, and hence relative to a further unit. I will call this the "measurement regress problem." Kant infers that our knowing the fundamental measure must be a result of a different, aesthetic estimation of magnitude:

> Thus, the estimation of the magnitude of the fundamental measure must consist simply in the fact that one can immediately grasp it in an intuition and use it by means of imagination for the presentation of numerical concepts – i.e., in the end all estimation of the magnitude of objects of nature is aesthetic (i.e., subjectively and not objectively determined). (5:251)

Kant solves the measurement regress problem he has raised by appeal to a nonmathematical means of estimating the magnitude, one that is not relative to or dependent on the representation of a further unit.

Before proceeding we should note the difference between the regress problem we raised for extensive magnitudes and the present one. The former arose because the cognition of a whole extensive *quantum* rests on a successive synthesis that presupposes a representation of its parts, which precede it. The present regress problem does not turn on synthesis, much less successive synthesis, and it concerns more than just the representation of parts; it concerns the conditions of measurement. There is, to be sure, an intimate connection between the representation of extensive *quanta* and measurement; the nature and depth of that connection will be the topic of Chapter 9. Nevertheless, measurement is a more complex cognitive achievement. Measurement is a determination of *quantitas*, the "how much" of something, not merely a determinate representation of a *quantum* as such. It requires not just determination of a unit of measure, but a means of iterating or replicating the same size unit, and counting those units to reach a number of units measured relative to the unit of measure. Thus, the regress that arises for the measurement required for the cognition of *quantitas* is at a more sophisticated level of cognition than the regress that arises for simply cognizing any

[30] A better translation for *bekannt* might be "acquainted with," to bring out the looser sense of knowing connoted by the term and to more clearly contrast it with *erkannt*.

extensive *quanta*, which concerns only its parts. The two regress problems are therefore distinct. Moreover, Kant in no way indicates that his solution to the measurement regress also addresses the extensive magnitude regress. We should therefore be wary of too hastily linking the two.

Despite these cautions, Kant's solution to the measurement regress descends to the level of intuition, and explains the aesthetic estimation of magnitude in a way that appears to provide resources that would also solve the extensive magnitude regress, distinguishing two acts. Kant states:

> To intuitively take up a *quantum* in imagination, in order to be able to use it as measure [*Maße*] or as unit for the magnitude-estimation through numbers, belongs two acts of this capacity: **apprehension** [*Auffassung* (*apprehensio*)] and **comprehension** [*Zusammenfassung* (*comprehensio aesthetica*)]. (5:251)

The former act is a taking-up of representations in intuition, while the latter is a grasping together of those representations.[31] Kant describes apprehension in temporal terms, contrasting those representations that have been taken up first with the further representations taken up, stating that at some point in the continuation of this act, the former representations partly fade before the latter have been taken up (5:252). Kant's description of apprehension in intuition bears similarity to Kant's account of apprehension in the *Critique of Pure Reason*; nevertheless, in the latter, Kant is primarily focused on determinate theoretical cognition, and in the former he does not mention a synthesis on which the apprehension is based. In the case of comprehension, Kant explicitly distinguishes *comprehensio aesthetica* from "*comprehensio logica* in a numerical concept" (5:254). The latter is the representation of a collection of units constituting a number effected through concepts, which underlies measurement. The former is a more basic holding-together of intuitive representations in one intuition.

The aesthetic estimation of magnitude provides a "standard of measure of sense [*Maßstab der Sinne*]" (5:255), one that is not relative to a further unit, and thereby solves the measurement regress problem in particular. But Kant's countenancing an aesthetic comprehension of a manifold of intuitive representations also appears to solve the extensive magnitude regress problem; if the aesthetic comprehension allows us to indeterminately represent the parts that are presupposed by and precede the representation of extensive

[31] A note on terminology: I translate *fassen* and *Fassung* as "grasping" and "grasp," since I think it aptly connotes a way of holding onto something, as with a hand. Some Kant scholars today use "grasp" in a sense that connotes an involvement of the understanding, and sometimes also employ the phrase "grasp as" interchangeably with it. I am not using "grasp" in this way.

magnitudes, then those indeterminately represented parts would not themselves be extensive magnitudes, and the extensive magnitude regress can be halted.

This solution, however, comes with substantive commitments concerning Kant's views. The commitments enter into the way in which we can represent parts both as individuals and indeterminately. We saw in Sections 2.4 and 2.5 that in Kant's view, determination of an object of cognition is an act of the understanding employing concepts. This strongly suggests that a representation of an indeterminate magnitude will be one that in some way does not involve the employment of concepts. In fact, that appears to be exactly Kant's point when he states that the estimation of the magnitude of the fundamental measure "consists simply in the fact that one can immediately grasp [*fassen*] it in an intuition" (5:251). This is the work of the faculty of imagination, which opens the door to a possible involvement of the understanding, if one understands this faculty as the productive imagination and "as an effect of the understanding on sensibility" (B152). But Kant's emphasis here is that the *quantum* is taken up into imagination "intuitively." He also refers to the aesthetic estimation of magnitude and to aesthetic comprehension, emphasizing their close connection to intuition. Moreover, aesthetic comprehension (*Zusammenfassung*) is a holding or grasping together, rather than a composition (*Zusammensetzung*), where the latter is the synthesis underlying our representation of determinate magnitudes in accordance with the categories, as discussed in Section 4.2 above. Thus, the aesthetic estimation of magnitude appears to be free of conceptual determination.[32] As a consequence, intuition can represent individual parts of an extensive magnitude, and hence objects, without the employment of concepts. In short, it suggests a nonconceptualist reading of objective representation.[33]

This is not itself an objection to this solution to the extensive magnitude regress, but it does show that Kant's theory of magnitudes and the current Kantian version of the conceptualist/nonconceptualist debate are directly connected. That is as one should expect, for if one thinks that concepts are required for the representation of objects, one would expect that even if particular sortal concepts are not required, at least the categories are. More specifically, one would expect that the categories of quantity – unity, plurality, and allness – are required, and we have seen that with regard to both

[32] Kant does not completely rule out *all* determination, however. At one point, Kant concludes that "in the end all estimation of the magnitude of objects of nature is aesthetic (i.e. subjectively and not objectively determined)" (5:251). I will return to this point below.

[33] The current framing of the conceptualist/nonconceptualist debate in Kantian scholarship concerns the objective purport of either perceptions or intuitions; I will accept this framing in what follows.

mathematical cognition and our cognition of objects, these categories play a fundamental role in the cognition of magnitudes.

The connection between Kant's theory of magnitudes and the conceptualist/nonconceptualist debate has been noted. The terms of that debate are contested and cut across different versions of conceptualism and nonconceptualism, but dividing authors in broad camps, Emily Carson and Thomas Land could be broadly classified as arguing along conceptualist lines that the categories of quantity play an ineliminable role in either cognizing or representing any magnitudes as objects, leaving no room for the representation of objects in intuition without the aid of concepts.[34] On the other side, Lucy Allais argues for a nonconceptualist interpretation of Kant according to which "intuitions do not depend on concepts to play their role of presenting us with objects" (Allais (2015, p. 149)). In her view, those objects include the presentation of distinct perceptual particulars in intuition, which she brings to bear on Kant's theory of magnitude.[35] In fact, she argues for her version of nonconceptualism from the claim that it is required to solve a combination of both the extensive magnitude and measurement regress problems (pp. 170–2). Her combining the solution to the extensive magnitude regress and measurement regress reflects the fact that she thinks the solution to the latter is also a solution to the former, which is the strategy we have been considering in this subsection.[36]

I think that Kant's theory of magnitudes has important implications for whether Kant is or is not a nonconceptualist, and I think a nonconceptual reading needs to be taken seriously. I will not attempt to settle that controversy here, however, or even the more focused question of the bearing of Kant's theory of magnitudes on it.[37] I will, however, make a few remarks about challenges facing the nonconceptualist strategy of solving the extensive magnitude regress by appeal to an aesthetic estimation of magnitude. The first concerns its place in the larger conceptualist/nonconceptualist debate. Using

[34] My understanding of these issues is indebted to discussions with Emily Carson and an unpublished paper by her entitled "Number, the Category of Quantity, and Non-Conceptual Content in Kant." I am also indebted to conversations with Thomas Land and to Land (2014), as well as his feedback on an earlier draft of this chapter. See also Land (2016). Both Carson and Land treat the issue with more care and sophistication than I can delve into here.

[35] Allais draws on a version of a direct relational theory of perception for her interpretation of Kant and describes her position in terms of the perceptual presentation of particulars, but she is careful to note that she is arguing for a nonconceptualist reading of intuition, and not of perception in Kant's sense (Allais (2015, pp. 149–50 and 150–1)). See also Allais (2016) and Allais (2017)).

[36] See Allais (2015, pp. 171n46 and 172n47), where she cites Sacha Golob (2011).

[37] Doing so would require, among other challenges, an assessment of the terms of the Kantian version of the debate. See Allais (2016) and McLear (2020) for helpful overviews.

the extensive magnitude regress to argue for a nonconceptual presentation of particulars in intuition presupposes that there is no other possible response to the extensive magnitude regress problem. I will present an alternative in the next section, and, if successful, it shows that the extensive magnitude regress does not provide an independent argument for nonconceptualism.[38]

There are also specific challenges to appealing to a nonconceptual presentation of particulars in intuition to solve the extensive magnitude regress. First, there is a threat of yet a third regress. As we have seen, the aesthetic estimation of magnitude is the result of two acts, apprehension (*Auffassung*) and aesthetic comprehension (*Zusammenfassung*), where the former is a taking-up of partial representations and the latter grasps them together.[39] But in what sense are these partial representations distinct representations that are then held together? What accounts for their particularity? If these partial representations in turn require aesthetic apprehension and comprehension, we have a regress, which I will call the "aesthetic estimation regress." On the other hand, if the partial representations do not require aesthetic apprehension and comprehension, then what is it that accounts for their being distinct representations, and whatever that might be, why wouldn't that obviate the need for the distinct acts of apprehension and aesthetic comprehension for the aesthetic estimation of magnitude? This is an issue for Kant's account of the estimation of magnitude and not just for the nonconceptualist argument, but it is all the more pressing for the latter, which purports to solve the extensive magnitude regress by appealing to it.

Because this worry is a more general puzzle about Kant's views, one that may have a solution, let us set it aside for the sake of argument. The nonconceptualist interpretation under consideration still faces a serious challenge about the nature of the merely intuitive presentation of particulars. According to Allais, we are presented with perceptual particulars in intuition without the aid of concepts, and these count as objects presented in intuition. I will refer to this as "intuitive atomism."[40] What accounts for the distinctness and unity of

[38] It should be noted that this is just one of the arguments Allais gives for her claim that intuition presents us with objects, so her position does not stand or fall with this argument. Nonconceptualists more generally appeal to a variety of considerations to support their position.

[39] Allais (2017) draws a distinction between synthesis and binding, arguing that the representation of perceptual particulars rests on the latter rather than the former. Allais does not provide a full account of binding, but it might be thought that aesthetic apprehension and comprehension could play such a role.

[40] Again, it is important to keep in mind that although Allais refers to perceptual particulars, she is arguing for a nonconceptualist reading of Kantian intuition rather than perception in a Kantian sense. I think, however, that her use of the phrase "perceptual particular" may have obscured the fact that what is at issue is not merely particulars with which we are perceptually acquainted, but particulars insofar as they are presented in

these perceived particulars? As a part of her broader interpretation of Kant's idealism, she proposes that they are grounded in perceptual features of the objects we experience. These particulars are, at least minimally, a thing which a subject singles out as a perceptual unit without the aid of concepts – a distinct bounded thing, for example, a table and an attached lamp, or just the light bulb in the lamp, or a spot of light moving on a wall, "something that the subject can pick out as a unit" (Allais (2015, p. 147n2)). Even if we grant an ability to pick out such particulars in response to features of perception, however, it will at best do as an account of empirical intuition; it cannot account for the role of intuition in mathematics. For in the case of pure mathematics, there are no perceptual features presented in intuition to "glom onto." Mathematical cognition is quite different in this respect; the manifold of intuition is uniform and homogeneous, and presents no distinguishing features. In pure mathematical cognition, concepts are required for even the representation of objects, including the representation of a part as a part. There is no alternative but to use concepts to arbitrarily determine the parts that are represented as parts; thus, appealing to perceptual particulars whose representation does not depend on concepts is not adequate in the case of mathematical cognition.

This result has direct consequences for the extensive magnitude regress. As we have seen, the Axioms of Intuition concerns pure as well as applied mathematics, and the B-edition version of the principle states that all (determinate) intuitions are extensive magnitudes, and that includes both empirical and pure determinate intuitions. Thus, any solution to the extensive magnitude regress will need to work for pure mathematical cognition as well. Hence, appealing to perceptual particulars whose representation does not involve concepts will not solve the extensive magnitude regress.[41]

Finally, there are broader considerations that militate against appealing to the aesthetic estimation of magnitudes to solve the extensive magnitude regress. The first is that, as emphasized at the outset, Kant's concern in the

intuition, and hence particulars in pure intuition as well. See Allais (2016, p. 6n10). See also note 35 above.

41 On the other hand, the fact that concepts, and in particular the categories of quantity, are required for even the *representation* of mathematical objects does not itself entail a thoroughgoing conceptualism concerning our intuition of the objects of experience – that would require further argument. This is so despite the close connection between mathematical cognition and empirical cognition I have been emphasizing throughout this book. While the categories of quantity are required for the cognition of empirical objects as extensive magnitudes, for all that has been said so far, they may not be required for the mere intuitive representation of empirical objects. I would like to thank Tyler Burge and Matt Boyle for helping me appreciate the importance of distinguishing the conditions of cognition and even intuitive representation for pure mathematical objects, on the one hand, and for the intuitive representation of empirical objects, on the other.

Critique of the Power of Judgment is with measurement in particular, that is, a determination of *quantitas*, which is at a more sophisticated level of cognition than the mere cognition of an extensive magnitude, that is, a *quantum*. Moreover, Kant's point is that – to put it in a way that he does not – the meaningfulness of a measurement ultimately depends on some direct acquaintance in intuition with the unit of measure. If I have no idea how long a Saxon league is, it will do me no good to know that a village is seventeen Saxon leagues away. Kant brings up the potential regress in the mere mathematical estimation of magnitude by means of numbers to emphasize this point; if I attempt to help you grasp the size of a Saxon league by telling you that it is 1,000 Dresden rods, you are no better off unless you are acquainted with a Dresden rod, and so on, *ad infinitum*. Kant is claiming that what is required is direct acquaintance with a unit of measure in intuition; that is why Kant refers to the fact that we need to assume that the unit is *bekannt*, that is, familiar. Kant's interest here is in the meaningfulness of measurement, *quantitas*, in relation to our faculties, and its implications for our experience of the mathematically sublime. He is not here addressing a worry about a regress in the determinate cognition of any extensive *quanta*.

Of course, Kant might nevertheless regard his account of the aesthetic estimation of magnitude as a solution to the extensive magnitude regress. But if he does, he makes no reference to it. Furthermore, this conjecture presupposes that Kant had either overlooked or saw but not taken the time to directly address the extensive magnitude regress problem in either edition of the *Critique of Pure Reason*. I think it unlikely that he would not have seen the threat of regress. Moreover, I think that although he does not explicitly address the extensive magnitude regress problem, he does in fact indicate in the Axioms of Intuition that which avoids the regress, adding more details in the Anticipations of Perception. The solution is right under our nose, where it should be, rather than in the *Critique of the Power of Judgment*.

4.6 The Solution to the Extensive Magnitude Regress Problem

To begin, I would like to reiterate an important point that emerged in the previous section. Any way of avoiding the extensive magnitude regress must work for pure mathematical cognition as well as empirical cognition. That leads to a difficulty, however. In the case of pure mathematics, we must employ concepts, and the categories of quantity in particular, to represent an object, and that includes the representation of particular parts of space. Since the representation of an object by means of concepts is a determination of that object, that would make any representation of a part of space in mathematics the representation of a determinate part. But in that case, the part represented would be an extensive magnitude after all, and off we go on the regress. We were looking for a way in which the parts of an extensive magnitude whose representation make possible the representation of the whole can be indeterminately represented in order to avoid those parts in turn counting as extensive magnitudes. That was the motivation behind both solutions considered in the previous section. Now it appears that at least in pure mathematics, we cannot represent parts of space or time indeterminately. This prompts a closer look at infinite divisibility and what it means to represent magnitudes indeterminately. Kant says more about the infinite divisibility of space and time and the representation of indeterminate *quanta* in his resolution of the Second Antinomy, to which I will turn next. Once completed, we will be in a position to appreciate what I take to be Kant's actual solution to the extensive magnitude regress.

4.6.1 The Indeterminate Representation of Quanta

Kant draws an important contrast between the First and Second Antinomy. The former concerns the totality of the composition of a world-whole, and because the world in its totality "is not given to me through any intuition, hence its magnitude is not given at all prior to the regress" (A519/B547), we cannot say that the regress is a regress to infinity (*regressus in infinitum*), but "only an **indeterminately continued regress** (*in indefinitum*)" (A518/B546). In contrast, the Second Antinomy concerns the totality of division of "a given whole in intuition." The division of the whole consists only in a progressive decomposition, so that the completed division is not contained (*enhalten*) in the whole, and we cannot say that the whole "consists [*bestehe*] of infinitely many parts." Nevertheless, because the whole is given as an intuition enclosed within boundaries, and the parts are contained (*enthalten*) within it, the parts are all given along with the whole (A523–4/B552). Because the parts are given, we can assert that the regress continues to infinity (*in infinitum*).

What is important about Kant's claims here is that there is sense in which the parts are given in the representation of the whole and yet are not given in their totality. The division consists only in the progressive decomposition, "which first makes the series actual [*wirklich*]" (A524/B552). Kant's exposition of "actual" in the Postulates of Empirical Thought requires a connection with the material conditions of experience (of sensation), while the Second Antinomy concerns the whether or not composite substances or things in the world consist of simple parts. Nevertheless, the arguments turn on the part–whole relations of intuitions. Thus, Kant's point is to distinguish between, on the one hand, actually divided parts of space and, on the other, merely potential and as-yet-undivided parts of space that are nevertheless given in the whole intuition. The former are determinate parts of space, while the latter, because they have not yet been represented through division, are indeterminate.

Kant reinforces the indeterminacy of the given parts in a passage worth quoting in full:

> But now, although this rule of progress to infinity applies without any doubt to the subdivision of an appearance as a mere filling of space, it cannot hold if we want to stretch it to cover the multiplicity of parts already detached in a certain way [*auf gewisse Weise ... schon abgesonderten Teile*] in a given whole, constituting thereby a *quantum discretum*. To assume that in every whole that is articulated (organized) [*gegliederten (organisierten)*], every part is once again articulated, and that in such a way, by dismantling the parts to infinity, one always encounters new complex parts [*Kunstteile*] – in a word, to assume that the whole is articulated to infinity – this is something that cannot be thought at all, even though the parts of matter, reached by their decomposition to infinity, could be articulated. For the infinity of the division of given appearance in space is grounded solely on the fact that through this infinity merely its divisibility, i.e. *an in itself absolutely indeterminate multiplicity of parts, is given*, but the parts themselves are given and determined only the through the subdivision – in short on the fact that the whole is not already divided up. (A526/B554; final emphasis mine)

Note first that representing a whole as having particular parts is to represent it as a *quantum discretum*. This supports the claim I made in Section 3.5 that although Kant asserts that an appearance, insofar as it is in intuition, is a continuous *quantum*, he also allows that we can represent appearances as discrete *quanta*. What is important for our present topic, however, is that Kant describes the representation of a whole into particular parts as representing the whole as *gegliedert*, that is, as articulated, structured, or jointed. Kant's main point is to argue against the assumption that these parts are themselves represented as articulated, and so on to infinity. Instead – and this is the key point – the multiplicity of parts is first given as "absolutely indeterminate."

But how is it that we can talk about the representation of a multiplicity of parts that are given absolutely indeterminately? It might seem that any representation of a multiplicity of parts presupposes a representation of the parts as parts, and hence in some sense determinately. Kant is clearly using "parts" in a way that does not carry that implication. I would like to suggest that what Kant has in mind is a representation of an indeterminate manifold akin to what would fall under a mass rather than a count noun, that is, a "stuff," and in the case of extensive magnitudes, something whose manifold is manifest and hence readily determinable into parts, but *not* in any way determinately represented as parts. When Kant refers to indeterminate parts, he means that feature of a manifold in virtue of which it counts as manifold even prior to the actual articulation of parts. Thus, when Kant refers to the representation of parts in these contexts, he is referring to the representation of parts that are as yet unarticulated and hence absolutely indeterminate.

This understanding of the representation of an indeterminate *quantum* may, I believe, be obscured by the thought that what is represented must be some particular part as a *quantum* that in some way has indeterminate properties, while what is indeterminate is the representing of the parts, so that no particular parts are represented. It is thus perhaps more perspicuous to refer not to the representation of indeterminate *quanta* but to the indeterminate representation of *quanta*.[42]

If this account is correct, it offers a solution to the extensive magnitude regress. If the parts whose representation makes possible, and therefore necessarily precedes, the determinate representation of an extensive magnitude are represented as parts only absolutely indeterminately, and hence are not yet articulated, no regress will ensue. There is, however, a serious challenge to reconciling this understanding of the indeterminate representation of *quanta* with Kant's account of synthesis. Responding to it will allow us to fill out Kant's views.

4.6.2 Continuity and Successive Synthesis

Let us return to the claim that space and time are *quanta continua*, which, together with the definition of extensive magnitudes, led to the regress. Kant defines their continuity as the fact that they have no smallest parts. As we noted in the previous section, Kant's support for this is based on the claim that the division or decomposition can continue *ad infinitum*, which Kant carefully distinguishes from the claim that the whole consists of infinitely many parts. Thus, the continuity of space and time as *quanta continua* is cashed out in terms of the possibility of articulating parts, and nothing more.

If, however, we interpret references to the representation of parts in the way suggested in the previous section – that is, in a way that does not commit to them being determinately represented as parts – it is unclear how we are to make sense of the successive synthesis of parts at the foundation of Kant's account of our cognition of extensive magnitudes. As discussed in Section 4.2, the successiveness of the synthesis plays an important role. After explicating an extensive magnitude as one in which the representation of the parts makes possible (and therefore necessarily precedes) the whole, he states:

> every appearance as intuition is an extensive magnitude, as it can only be cognized through successive synthesis (from part to part) in apprehension. All appearances are accordingly already intuited as aggregates (multitudes of previously given parts) . . . (B203–4)

[42] I would like to thank Chen Liang for helping me see this point.

How are we to understand the synthesis "from part to part" if the parts are not represented as parts; that is, if they are not articulated? Moreover, if we take the temporality of successive synthesis seriously, how can we understand that synthesis as extended over time except by representing distinct parts at different times? There seems to be a tension between the idea that indeterminately representing parts does not require the representation of particular parts, and Kant's description of the synthesis underlying extensive magnitude, which seems to require a representation of those parts.

The solution is expressed in the very same paragraph. Immediately after defining extensive magnitude, Kant states:

> I cannot represent to myself any line, no matter how small it may be, without drawing it in thought, i.e. successively generating all its parts from one point, and thereby first sketching this intuition. It is exactly the same with even the smallest time. I think therein only the successive progress from one moment to another, where through all the parts of time and their addition a determinate magnitude is finally generated. (A162–B203)

To begin, note that Kant says that a determinate magnitude of time is "finally" generated through the synthesis, reinforcing the claim that the parts of time synthesized are not themselves determinate magnitudes. Most importantly, Kant states that the synthesis in the drawing of a line begins from a point. This is highly significant: it indicates that the synthesis does not begin with some arbitrary part or other, the first of a multitude of parts to be synthesized. There is no first part; rather, the successive synthesis is itself continuous and "runs through" the continuous manifold in intuition. Moreover, Kant states that the synthesis runs through *all* the parts, that is, runs through all the indeterminate parts – all of the parts which could potentially be determinately represented. He is not envisioning the representation of each part as a particular part.

As has been well noted, Kant's invocation of motion in the generation of magnitude appears to be a reference to Newton's doctrine of fluents and fluxions.[43] In fact, Kant employs and explicitly draws attention to his use of the German phrase *fließende Größe*, which was used to translate Newton's "fluxion." Immediately following the passage cited above in which Kant claims that space and time are *quanta continua* because no part of them is the smallest, Kant states:

> Magnitudes of this sort can also be called **flowing** [***fließende***], since the synthesis (of the productive imagination) in their generation is a progress in time, the continuity of which is customarily designated by the expression "flowing" ("elapsing") ["*Fließens*" ("*Verfließens*")]. (A170/B211–12)

43 See Kitcher (1975, pp. 40ff.) and Friedman (1992, pp. 74ff.) for detailed discussions.

And in a draft of his September 25, 1790, letter to August Rehberg, Kant refers to "representing a line through fluxion, and hence as produced in time" (14:53).[44] Finally, in the collection of notes dated in 1800 and included in the *Opus Postumum*, Kant refers to the "great mathematical discovery of the fluxional calculus" (22:519).

Newton introduced the notions of fluent and fluxions to allay worries concerning the foundations of analysis, and in particular the status of infinitesimals. The latter seemed to have contradictory properties, being both extended and yet not composing a finite extension no matter how many of them were composed. The result was that sometimes they were treated as nonzero and capable of standing in ratios, while at other times they were treated as equivalent to zero. Newton attempted to avoid these difficulties by considering magnitudes as generated by a continuous motion, rather than as composed of parts. Kant appears to be appealing to Newton's theory of fluents and fluxions for the same reasons.

Michael Friedman has described in detail the significance of Kant's references to Newton's theory of fluents and fluxions for understanding his philosophy of mathematics and his foundations of natural science, and the implications of Kant's appeal to motion for his understanding of the continuity of *quanta*. He notes, for example, that the continuity underlying the theory of fluents and fluxions is stronger than the continuity presupposed by the claim that *quanta* have no smallest parts; the former entails that the generated quantities are everywhere differentiable, while the latter does not.[45] Friedman has also further argued for the significance of Kant's appeal to motion for his understanding of the continuity of *quanta* and Kant's transcendental philosophy more generally.[46] What is important for our present purposes, however, is that the continuity of fluents and fluxions based on motion in the generation of magnitudes allows Kant to accommodate the fact that continuous *quanta* have no smallest parts while accounting for the synthesis of a manifold that in some sense has parts without representing the particular parts to effect the synthesis. Moreover, the successiveness of the synthesis brings out the part–whole relation of extensive magnitudes and the sense in which a whole magnitude manifests the indeterminate parts contained in it.[47]

There are, however, reasons to be skeptical about the depth of Kant's commitment to a fluxional account of the calculus and its solution to worries about the foundations of analysis. First, Newton's theory of fluents and

[44] See Friedman (1992, p. 76n29).

[45] See Friedman (1992, pp. 78–80), where Friedman is also concerned to point out the inadequacy of the appeal to fluents and fluxions for mathematics and mathematical physics, because it does not distinguish between continuity and differentiability.

[46] See Friedman (1992), Friedman (2000), and Friedman (2020).

[47] I would like to thank Thomas Land for encouraging me to emphasize this last point.

fluxions were roughly a century old when Kant was writing; it had not been popular on the continent, where Leibniz's appeal to infinitesimals and differentials had dominated; and it was not even unequivocally supported by British natural philosophers in Kant's time. Second, Kant only once tangentially mentions fluxions and fluents in the place one would most expect to find them, namely, his *Metaphysical Foundations of Natural Science*, whose aim is precisely to account for the foundations of Newtonian mathematical physics.[48] In fact, Kant repeatedly refers in the *Metaphysical Foundations* to the ratios of ever smaller magnitudes and the ratio that they ever more closely approach. This follows another treatment of the foundations of the calculus proposed by Newton: the theory of ultimate ratios, which came to be known as the method of limits, and was eventually adopted by D'Alembert and others on the continent. Moreover, Kant sometimes even refers to differentials, which belonged to the tradition of Leibnizian analysis based on infinitesimals. All of this seems to militate against taking Kant's references to the theory of fluents and fluxions too seriously. Nevertheless, it was indeed quite important to Kant's understanding of both mathematics and the nature of experience. I cannot provide a full account here, so will merely summarize.[49] By the mid-eighteenth century, the British Newtonian and continental Leibnizian accounts of the foundations of analysis were increasingly viewed as two ways of expressing the same thing, while the theory of limits came to the fore as the best approach in mathematics and mathematical physics. Kant recognized and was influenced by these developments in the *Metaphysical Foundations*. At the same time, however, he turned to the theory of fluents and fluxions to explain the cognitive presuppositions underlying those metaphysical and mathematical foundations. In Kant's view, it is ultimately the continuous successive synthesis in the generation of the representation of a continuous *quantum* that accounts for our cognition of the continuity of those representations, that is, the fact that no part of them is the smallest.[50]

[48] Kant comes closest to mentioning them in the *Metaphysical Foundations* in the Preface, in which he refers to the "flux" (*Abflusse*) of the phenomena of inner sense according to the law of continuity (9:471). Friedman expressed to me in conversation (some years ago) the need for caution in evaluating Kant's settled views on the status of fluents and fluxions.

[49] For a more thorough treatment, see Sutherland (2020b).

[50] A note in the *Opus Postumum*, for example, strongly suggests that the "principle of fluxions" accounts for the fact that physical division can proceed to the differential magnitude (*Differentialgröße*), and hence that the principle of fluxions is more fundamental. (The note is quoted below; it is the last in the set of quotes.) See also *Reflexion* 5382 in which Kant links fluxions to differentials in his notes on the principle of continuity: "The smallest difference would be called a *differential*, but because there is no smallest, it is called 'fluxion'" (1776–8, 18:167).

Unfortunately, despite the work already mentioned as well as more recent work on Kant's *Metaphysical Foundations*, I do not think the implications of Kant's understanding of continuous motion for his broader theory of experience has been sufficiently appreciated. Recent attributions of some form of intuitive atomism to Kant, such as those based on his treatment of measurement in the *Critique of the Power of Judgment*, to solve the extensive magnitude regress are evidence of this; they overlook the role of continuous motion in the generation of our representations of extensive *quanta*. One reason may be an unfamiliarity with the meaning of "flowing magnitudes" and the importance of the theory of fluxions and fluents in Kant's thought. Another may be the fact that Kant talks about synthesis in a way that might suggest that particular parts must be represented in order for them to be synthesized, an impression I have sought to dispel in this and the previous section. Finally, one might have an aversion to taking Kant's claims concerning the necessity of successive synthesis too seriously, because it seems phenomenologically implausible that we have to draw a line in order to think it or that we "as it were" draw the shape of a house in perceiving it (in Kant's sense of perception). Despite this qualm, however, we need to take seriously what Kant says on his terms, which means recognizing the central role of continuous synthesis in his account. In the present context, that means that the solution to the extensive magnitude regress appeals to a continuous synthesis of indeterminately represented parts, and not to an intuitive atomism.[51]

In conclusion, the solution to the extensive magnitude regress rests on the continuous successive synthesis underlying the representation of extensive magnitudes. Because the manifold of intuition is synthesized continuously, the parts whose representation makes possible and precedes the representation of a whole determinate extensive magnitude are not represented determinately; that is, they are not explicitly represented as parts. Instead, they are only indeterminately represented in the way that a continuously drawn line only indeterminately represents all the parts of the line; the parts are only potentially determinately represented. Nevertheless, in representing a line, there is an important sense in which it is necessary to represent all its parts, and in drawing it, an important sense in which the representation of those parts precedes the representation of the whole. This is the sense in which an

[51] I would like to again emphasize that this conclusion does not by itself entail that we cannot be presented with perceptual particulars in the sense that Allais describes, nor does it by itself rule out a nonconceptual understanding of intuitive representation. That would require substantial further argument. It does, however, show that concepts, and in particular the categories of quantity, are required for us to have cognition of an appearance as an extensive magnitude. It also shows that a nonconceptualist reading does not find support in an argument that the extensive magnitude regress can be solved only through a form of intuitive atomism. See notes 41 and 38 above.

extensive magnitude immediately manifests its part–whole structure in intuition in a way that intensive magnitudes do not.

4.6.3 *Indeterminate Magnitudes within Boundaries Revisited*

In Section 4.5.1 I examined Kant's views on the representation of indeterminate magnitudes within boundaries. Although I argued that it is not Kant's solution to the extensive magnitude regress problem, it is clear that he makes room for it in his account of our representation of magnitudes, and it bears further comment in light of the continuous synthesis of a homogeneous manifold described above. If I draw a line from a point, I thereby generate a representation of a determinate *quanta*, namely, the line. The manifold of parts of that line are only indeterminately represented, and that indeterminate representation is contained within the boundaries of the two end points. These two end points limit and determine the line, while presupposing the representation of the indeterminate magnitude they enclose. Moreover, understood within a kinematic conception of the generation of a line, the end points are limits in a mathematical sense: they are the start and end of a series of ever-smaller parts of space.

If I now draw three straight lines in drawing a triangle, I enclose a surface within boundaries, and the parts of that surface are also only indeterminately represented. Thus, there are two instances of the indeterminate representation of parts of space in the representation of a triangle – the indeterminate parts of each line and the indeterminate surface they enclose. There is a sense in which the area of the triangle is determinately represented – the area is determinately distinguished from the surrounding area and has a determinate shape. But the manifold that composes the area is only indeterminately represented, which is what Kant has in mind in the passage from the Antinomies.[52] Finally, at the same time as the lines of the triangle determine the area they enclose, they also determinately mark off that surface from the unlimited surface of the plane outside the triangle. Although this is not a representation of the space outside the triangle as enclosed within boundaries, it does delimit that space, and there is a sense in which the manifold of space outside the triangle has been indeterminately represented by the lines of the triangle.

4.7 Continuous Synthesis and the Categories

We saw in Section 3.5 that there is an important distinction between two senses of "magnitude," *quanta* and *quantitas*, where the former is a

[52] I would like to thank Joshua Williams for prompting me to be clear on the sense in which the area is determinately and indeterminately represented.

homogeneous manifold in intuition, and the latter is the quantitative determination of the former under the categories of quantity – that is unity, plurality, and allness employed in the determination of the "how much" of something. This determination paradigmatically takes the form of a measurement relative to some unit, so that the *quantitas* is a determination of the number of units, the plurality of which taken in their totality constitute the *quantum* measured. As also noted, Kant's definition of extensive magnitude is of a determinate extensive *quantum* in particular, and the continuous successive synthesis we have been discussing generates a representation of a determinate extensive *quantum*. This suggests that we can cognize a determinate extensive *quantum* apart from determining its *quantitas*. In fact, in his notes on Baumgarten's Metaphysics, at the point Baumgarten introduces his notion of *quantitas*, Kant inserted:

> better *quanta*; that something is a *quantum*, can be absolutely cognized, how big however (*quantitas*) only relative. (17:42)

Kant seems to have in mind that we can cognize that something is a homogeneous manifold without determining how big it is relative to a unit. But cognition of a determinate *quantum* will require, in bringing it under the concept of a *quantum*, some employment of a category. If we take the role of continuous successive synthesis in the representation of extensive *quanta* seriously, the best explanation is that to cognize something as a homogeneous manifold of intuition is to bring the manifold under the category of plurality. Hence, the category of plurality is employed not only in the cognition of discrete manifolds, consisting of explicitly represented pluralities of things, but in the cognition of continuous manifolds as well.[53] The concept of plurality can thus subsume what would ordinarily fall under both a count noun and a mass noun, not just a "many" but a "much." *Quantitas*, as an answer to the question, "How big is something?" provides an answer in both cases. In the first, the answer is a determination of how many, that is, the number of discrete elements of a manifold. In the second, the answer is a determination of how much, where that is determined relative to a unit of measure, and hence how many relative to that unit.

Solidifying this interpretation of our cognition of continuous *quanta* will require more work, but we have already uncovered enough to make Kant's views intelligible. In the next chapter, we will look more closely at the role of intuition in representing magnitudes, in particular, the relation between the singularity of intuition and the representation of a continuous manifold, as well as its relation to concreteness.

[53] In Section 4.1 above, we already saw an indication that plurality (*Vielheit*) applies to a continuous manifold in the case of intensive magnitudes. See note 10.

5

Conceptual and Intuitive Representation

Singularity, Continuity, and Concreteness

5.1 Introduction

The previous chapters have revealed that Kant's theory of magnitudes is central to his theoretical philosophy more broadly, and that Kant understands a magnitude to be a homogeneous manifold in intuition. This raises questions about Kant's understanding of homogeneity as well as the role of intuition in the representation of a homogeneous manifold. We won't be in a position to properly address this issue until Chapter 7. In the meantime, however, another issue about the role of intuition emerges from our examination of extensive magnitudes. The last chapter grappled with the extensive magnitude regress, which appears to require the representation of parts of space and time as parts, which in turn motivates attributing to Kant some sort of intuitive atomism. The threat of regress is dissolved by the claim that the parts are not represented as determinate parts, but are at most indeterminately represented. I ruled out two attempts to explain what it means for particular parts of a magnitude to be represented indeterminately, and argued that the real solution rests on the indeterminate representation of parts that involves not the representation of particular parts of space and time, but a representation of an undifferentiated continuous manifold in intuition, and that the successive synthesis underlying our representation of extensive magnitudes is continuous. In the representation of a line, for example, the parts are represented indeterminately in that the drawing of a line is the result of a continuous synthesis in which none of the parts is determinately represented, yet an extent of continuous space between two points is represented. The line as a whole is a determinate space, but the parts are only indeterminately represented. Similarly, the sides of a triangle are determinate spaces, but the area they enclose is only indeterminately represented; the area is the sort of indeterminate *quantum* represented within boundaries that Kant mentions in the Antinomies and which I discussed in Section 4.5.1 above.

One might well wonder how the indeterminate representation of a continuous manifold squares with Kant's characterization of intuitions as immediate and singular, where singularity seems to consist in representing singular objects. What sense of object Kant has in mind may not be entirely clear, but at the very least, the singularity of intuition appears to consist in the representation of individuals – both individual objects of experience and individual spatial extents in geometry, that is, the representation of individual parts of space. (In what follows, I will refer to individual or singular objects in a broad sense that would include ordinary objects of experience as well as individual spatial extents, such as a length or a cube.) If we take the singularity of intuitions seriously, it seems that by its nature, intuition would have to represent individual parts of space, rather than a continuous manifold. And that appears to be inconsistent with the position for which I have been arguing. This motivates a closer look at the singularity of intuition.

This issue, however, goes much deeper than a regress problem in the representation of extensive magnitudes. Understanding Kant's views on the nature of intuitive representation is crucial to understanding his philosophy as a whole, and how we interpret the singularity of intuition will frame our understanding of both Kant's philosophy of mathematics and human cognition more generally. In the case of Kant's philosophy of mathematics, understanding the singularity of intuition as consisting in the representation of a singular object reinforces a post-Fregean focus on the status of mathematical objects, especially numbers, and the role of intuition in their representation. Asking after Kant's views concerning the status of mathematical objects is not in itself objectionable, but it can occlude Kant's theory of magnitudes and the role of intuition in representing a homogeneous manifold. In the case of Kant's theoretical philosophy more broadly, it can lead one to the view that intuitions have the capacity to represent single objects without the aid of, or logically prior to, any action of the understanding. That is, the singularity of intuition may be thought to provide another argument for a nonconceptual or non-intellectualist interpretation of Kantian intuitions, in the sense of conceptualism and nonconceptualism currently mooted among Kant scholars. That would have significant implications for understanding the role of the intellect and sensibility, and concepts and intuitions, in Kant's theory of cognition. There are therefore multiple reasons for getting clearer on the singularity of intuition.[1]

Before embarking on this investigation, I want to make three points. First, my primary aim is to argue for a view of singularity that is compatible with the representation of a continuous homogeneous manifold, and in fact accords this representation a fundamental role in Kant's account of theoretical cognition, a role that is, in a sense to be explained below, more fundamental than

[1] I am grateful to Matt Boyle, Tyler Burge, and Chen Liang for productive discussions and communications on the topics raised in this chapter, and helpful detailed comments from Matt Boyle, Tyke Nunez, Daniel Smyth, and Thomas Land, as well as the UIC graduate students attending my seminar on Kant in the fall of 2020; I hope I have done some justice to their assistance.

the representation of individual objects. The motivation for this view springs not simply from the texts that will be considered but from the broader theory of magnitudes that we have uncovered in the previous chapters. While I have not yet fully articulated Kant's theory of magnitudes, accounting for it is an adequacy condition for a successful interpretation of Kant's theoretical philosophy. Thus, we need to make sense of the role of intuition in the representation of magnitudes, and the singularity of intuition appears to raise problems. Whether or not one agrees with all the particular features of the view I argue for below, we have enough evidence of the centrality and importance of Kant's theory of magnitudes to expect some account of the singularity of intuition compatible with it.

The second point rests on the first. I will begin by outlining a quick resolution of the apparent tension between the singularity of intuition and the intuitive representation of the continuous homogeneous manifolds of space and time. As far as it goes, this first solution would meet the immediate challenge; nevertheless, I set it aside because I think there is a deeper and more interesting account of the singularity of intuition worth bringing to the fore, one that I believe more accurately represents Kant's views. What will follow is an exploration of this deeper and more interesting explication of Kant's views, an exploration that will range over the nature of conceptual and intuitive representation, take into account a distinction between intuiting and intuited, and provide an explanation of Kant's understanding of *in concreto* representation. If there are aspects of the account I give below that were deemed unsatisfactory or problematic, however, the interpretation of Kant's theory of magnitudes I have been unfolding would not fall with it. Another way of reconciling the singularity of intuition and the representation of a continuous manifold might be found.

The third point is that this chapter's investigation is far from a complete account of the role of intuition either in mathematical cognition or in theoretical cognition more broadly. First, Kant also singles out immediacy as a characteristic property of intuitions. I will not attempt to give an account of immediacy or its relation to singularity; although the immediacy of intuition is certainly important, an account of singularity is more pressing at this juncture. Even with respect to singularity, what follows is a description of a view rather than a full-blown defense, which would require a great deal more work and engagement with other interpretations than I can provide here. I argue that the singularity of intuition is best understood as a way of representing, and that understanding it this way helps us to see that the singularity of intuition is compatible with its role in the indeterminate representation of a continuous homogeneous manifold. Moreover, it is this mode of representing such a manifold that underlies our representation, cognition, and experience of individual objects – both individual spatial parts of space and objects of experience.

Section 5.2 begins with the quick solution I just mentioned to the puzzle about how to reconcile singularity and continuity, and then describes another solution that gets at something deeper in Kant's view. Arguing for the latter requires a closer look at Kant's account of the generality of concepts, which is the task of Section 5.3. It considers what the generality of concepts consists in as well as its implications for the relations between concepts and the representation of individuals; it also establishes claims about Kant's understanding of conceptual representation that will be important in later chapters. I then return to the singularity of intuition in Section 5.4, where I argue not only that singularity is consistent with intuition representing an indeterminate and continuous homogeneous manifold, but that this mode of representing makes the representation of individual objects possible, although in the case of mathematical objects at least, fulfilling that possibility also requires the employment of concepts. Section 5.5 then discusses the implications of this new interpretation of the singularity of intuition for three aspects of Kant's philosophy: his philosophy of mathematics, the current Kantian version of the conceptualist/nonconceptualist debate, and the extensive magnitude regress.

The singularity of intuition is also directly connected to what Kant calls "representation *in concreto*," which contrasts with representation *in abstracto* by means of concepts. Representation *in concreto* plays a central role not just in mathematical cognition but in the representation of any individuals. Section 5.6 clarifies the different ways in which Kant uses "concrete" and "abstract" and how they relate to the singularity of representing *in concreto*. Doing so will also allow us to compare Kant's understanding of the concrete/abstract distinction with our modern way of raising questions about the ontology of abstract and concrete objects. This section clarifies the sense in which our cognition of mathematical objects is concrete and singular, despite the fact that our concepts of them are in an important sense less determinate and more general than objects of experience. Finally, I first addressed the sense in which *quanta* and *quantitas* were concrete or abstract in Section 3.5 above. Section 5.7 employs the analysis of abstractness and concreteness given here to further clarify the sense in which *quanta* and *quantitas* are concrete or abstract.

5.2 A Quick Solution and an Alternative

Before beginning, I would like to clarify some terminology. I refer in this and in the previous chapter to the representation of particular parts of space. This is our contemporary way of using the term "particular." Kant uses "particular" in a different sense, and uses the term "singular" (*einzelne*) only for what we usually mean by "particular." I will sometimes use "particular" in our sense in what follows, as well as "singular" or "individual."

There appears to be a fairly straightforward solution to the apparent problem reconciling the singularity of intuition and the role of intuition in representing a continuous homogeneous manifold. The singularity of intuition essentially consists in the representation of individual objects, including the representation of particular parts of space. But this is compatible with the role of intuition in the representation of a continuous manifold, because the latter is characterized in terms of the former. More specifically, what it is for spatial intuition to represent a continuous manifold in space is for the nature of spatial intuition to be such as to place no constraints on which individual parts of space can be represented, and furthermore, for it to be always possible to represent further parts of any part of space. The point is that the representation of a continuous magnitude is wholly characterized by the possibilities governing the representation of singular parts of space. This allows us to maintain that it is the nature of intuition to represent singular objects – in this case parts of space – while appealing only to the possibility of the representation of these parts. There is therefore no need to carve out a way for intuition to represent a continuous manifold independently of the possible representation of parts.

This explanation of the representation of continuity also explains the representation of an indeterminate manifold. Because it is the nature of intuition to represent singular objects, intuition by its nature represents parts of space. As a consequence, only the representation of particular parts constituting a manifold, that is, a manifold in which we represent at least some parts, is a representation of a manifold at all. To represent an indeterminate manifold is to represent something that could be represented as having particular parts. As in the case of continuity, indeterminacy is characterized in terms of the possible representation of parts, so that there is no need to propose that intuition somehow represents an indeterminate manifold singularly in a way that avoids the representation of singular parts of space.

While I think this solution solves the puzzle with which we started, I believe it leaves something out. I think it is part of Kant's view that we can both represent and have cognition of an indeterminate homogeneous manifold in intuition in a manner that does not reduce to a possible representation of parts; that is, intuition makes it possible to *directly* represent a genuine indeterminate manifold without either representing a manifold of parts, or reducing what it is to represent an indeterminate manifold to the possibility of presenting parts. I believe that in Kant's view, intuition makes it possible to represent and have cognition of an indeterminate extent without parts, an extended undifferentiated "stuff," if you will, and that this is fundamental to his understanding of intuitive representation. This is exemplified in our representation of a line or the area enclosed by the sides of a triangle; I represent the indeterminate and undifferentiated manifold of the line or area. This is not to say that we can cognize an indeterminate manifold as an indeterminate manifold through intuition alone, without the aid of concepts. Cognition of it as an indeterminate manifold requires the category of plurality, that is, the concept of a manifold, where that encompasses both discrete and continuous manifolds (see Section 4.7). The point is rather that in having a cognition of an indeterminate magnitude, we have a cognition of an indeterminate magnitude as indeterminate, that is, a cognition of an extent without any need for either the representation of its parts or a reduction of its indeterminacy to the mere possible representation of parts. Moreover, as I will argue below, this representation of an indeterminate manifold made possible by intuition is consistent with the singularity of intuition.

Nevertheless, to say that intuition allows us to represent an *indeterminate* manifold in this way is not to say that it allows us to represent a *continuous* manifold as continuous. I draw a sharp distinction between the intuitive representation of an indeterminate manifold and what is required to represent a continuous manifold as continuous. In representing an indeterminate extent of space, we are representing something that is continuous, but not as continuous; that is, we do not directly intuit nor cognize that it is continuous. For magnitudes to be continuous, according to Kant, is for no part of them to be the smallest. To come to know this requires concept-employing reflection on the conditions for the representation of parts of space, that is, that they can be represented only between boundaries, and hence as again extended parts of space; this reflection on the conditions for the representation of parts grounds our knowledge that space is continuous. This much is in agreement with the first solution proposed; continuity is a property that cannot be *directly* represented, and can be represented only by appeal to the possible representation of parts. My proposal disagrees with the first solution in claiming that it is the nature of intuition to make it possible to directly represent an indeterminate spatial manifold and to cognize an indeterminate spatial manifold as such – a manifold that we can come to understand is continuous.

The representation of an indeterminate manifold is like the representation of a "stuff," that is, a genuine representation of a "muchness" in contrast to a many-ness and that does not include a representation of its parts.[2] It is the representation of something that would be referred to by means of a mass noun rather than a count noun. I am assigning a primacy here to the direct representation of an indeterminate manifold over the representation of particular parts of it. And to miss this feature of Kant's views not only obscures the role of intuition in the representation of magnitudes and hence his philosophy of mathematics; it can also mislead one into thinking that it

[2] Just to be clear, by "muchness" I do not mean a particular amount, a "*how* much," but rather muchness in contrast to a many-ness or a distinguished, articulated, discrete plurality. Thanks to Thomas Land for prompting me to make this clear. "Stuff" and "muchness" are intended to suggest to the reader the indefiniteness of an undistinguished plurality.

follows from the nature of intuition alone that it can represent only singular objects, including spatial parts.

The interpretation I have outlined requires an account of that in which the singularity of representing an indeterminate and continuous homogeneous manifold consists, and moreover explains the relation between this sort of singularity and the singularity of individual objects of cognition. While I do not believe it is strictly necessary to do so to support this interpretation, I think it is helpful to first back up and consider the nature of conceptual and intuitive representation and distinguish between representing and what is represented. For if we regard intuitive representation as fundamentally a way of representing, I think it is easier to see how what might seem like two sorts of singularity are at root one and that same. I begin with the nature of conceptual representing, because Kant's views are clearer in that case, before returning to intuitive representing in Section 5.4

5.3 The Generality of Conceptual Representation

5.3.1 Concepts as a Mode of Representing Generality

To begin, recall the famous *Stufenleiter* passage in Kant's discussion of the concepts of reason in the *Critique*:

> The genus is **representation** in general (*representatio*). Under it stands the representation with consciousness (*perceptio*). A **perception** [*Perception*] that refers to the subject as a modification of its state is a **sensation** (*sensatio*); an objective perception is a **cognition** (*cognitio*). The latter is either an **intuition** or a **concept** (*intuitus vel conceptus*). The former relates immediately to the object [*Gegenstand*] and is singular; the latter mediately, by means of a mark, which can be common to several things. (A320/B376–7)[3]

The passage places Kant's contrast between intuitions and concepts in the context of cognition, and following his classification, characterizes cognition as an objective representation with consciousness. Kant goes on to say that an intuition relates immediately to the object (*Gegenstand*), and concepts relate to things (*Dingen*), so that by objective representation, he means a representation insofar as it relates to some intentional content as its object of representation, a representation of an object with consciousness. In the *Jäsche Logik*, moreover, Kant glosses cognitions in the following way: "All cognitions, that is, all representations related with consciousness to an object, are either intuitions

[3] I depart from the Guyer and Wood translation in Kant (1998) rendering *mittelbar* in adverbial form. This is closer to the German, and reinforces my emphasis below on distinguishing conceptual and intuitive representation by the way they represent. (I would like to thank Daniel Smyth for pointing this out.)

or concepts" (9:91). In the context of discussing cognitions, Kant emphasizes the relation of an intuition to an object.

More specifically, Kant states in the *Stufenleiter* passage that concepts represent mediately, by means of a mark that can be common to more than one thing (*was mehreren Dingen gemein sein kann*), while intuitions represent immediately and are singular. It is the nature of concepts to represent by means of a common mark; as a consequence, they can represent more than one thing in virtue of those things possessing the mark.[4] Kant continues the *Jäsche Logik* passage just cited by contrasting intuitions and concepts as singular (*repraesentatio singularis*) and as universal (*allgemein*) (*repraesentatio per notas communes*) or reflected (*repraesentatio discursiva*). The universality of a concept consists in it representing by means of a common mark; it is the way it represents that makes it a universal representation, and hence a concept.

This way of putting the point might still sound as if there is a kind of thing, a common mark, which can be common to more than one thing, and that a concept employs a specific common mark to refer to things belonging to a specific kind. But this doesn't quite get to the heart of it. In Kant's view, the generality of a concept consists in the act of considering a mark of a thing as something that could be common to several things. The *Stufenleiter* passage might be interpreted as suggesting that marks are inherently general, but as Houston Smit has pointed out, they are not: Kant refers to intuitive marks as well, which are singular.[5] A concept differs from an intuition in that it considers a mark as a representation that can be common, and taking this representation as a ground of cognition of a thing.

To fully appreciate this point, we need to look more closely at Kant's account of fundamental acts of the understanding. Considering a mark as a representation that can be common to more than one thing is a logical act of the understanding that Kant calls reflection. That is the significance of Kant's claim in the *Jäsche Logik*, quoted above, that a concept is a "universal representation (*repreasentatio per notas communes*) or reflected representation (*repreasentatio discursiva*)"; it is through the logical act of reflection that a mark is considered as a common mark.[6] Sections 4–6 of the *Jäsche Logik* elaborate on the nature of concepts and on the conditions for constituting or generating a concept. Kant states that the latter requires three logical acts of the understanding (*logische Verstandes-Actus*): comparison, reflection, and abstraction. The last is only a negative condition of generating a concept, for it leaves something out; only the first two acts are positive conditions of

[4] This is not to say that every concept can only represent more than one thing. I return to this issue below.

[5] See Smit (2000).

[6] See Longuenesse (1998) for a thorough and helpful account of reflection and the origin of concepts. The account I give here is focused and simplified.

generating a concept.[7] Comparison allows us to distinguish marks from one another in one consciousness, but it is reflection that constitutes the generality of a concept. This is not quite as clear as it might be if one reads only Kant's description of the logical act of reflection: "Reflection [*Reflection, d.i. Überlegung*], as to how various representations can be conceived in one consciousness" (9:94). Kant does not employ the word "common" (*gemein*) or "universal" (*allgemein*) here, and it might sound as if reflection is the mere holding-together of representations in one consciousness. Nevertheless, the broader context of §§4–6 of the *Jäsche Logik*, as well as the *Logik Dohna-Wundlacken* (from the early 1790s), make it clear that reflection constitutes the generality of a concept. Let's begin with the *Logik Dohna-Wundlacken*. Without labeling them, Kant distinguishes three acts that do not correspond to the acts adumbrated in the *Jäsche Logik*. Kant states that the making of a concept occurs:

> (1) through the fact that something is considered as [*betrachtet als*] a partial representation, which can be common to several, e.g., the red color. (2) When I consider the partial representation as a *nota*, as a ground of cognition of a thing, e.g., I cognize blood, a rose, etc., through red. The 3rd action is abstraction, to consider this partial representation as ground of cognition, insofar as I ignore all other partial representations. A concept is thus a partial representation, insofar as I abstract from all others. (24:753)

In both the *Logik Dohna-Wundlacken* and the *Jäsche Logik*, abstraction is the third act, and in both, Kant emphasizes that abstraction is only a negative condition, a leaving-out of consideration. But in the *Logik Dohna-Wundlacken*, there is no act corresponding to comparison. Instead, the first act corresponds to reflection, while the second regards a partial representation as the ground of cognition of a thing. What is important for our present purposes is that the act of reflection is clearly the act that constitutes the generality of a concept, which is the result of considering a partial representation as capable of being common to several.

Let's return to the *Jäsche Logik*. Kant's account of the form and matter of a concept also emphasizes the generality of concepts and its dependence on the act of reflection: "The matter of concepts is the object [*Gegenstand*], their form universality [*Allgemeinheit*]" (9:91). Kant states that, in logic, the generation of concepts whose conditions we've been considering concerns their logical origin, which is their origin with respect to their mere form (9:94), that is, their universality.

In this context, Kant appears to also use the term "reflection" in a broader sense. He states that the logical origin of a concept consists in reflection and that

[7] We will return to Kant's views on abstraction and the abstract in Section 5.6 below.

logic considers "only the difference in reflection in concepts" (9:592) – right before differentiating the three logical acts of combination, reflection, and abstraction. This suggests a broader use of "reflection" that encompasses comparison, reflection (in the narrow sense), and abstraction. This only further underscores the fundamental role of reflection in generating the logical form of a concept, which is its generality.

The upshot of this closer examination of the generation of concepts with respect to their form is this: the generality of a concept is constituted by an act of the understanding whereby a mark of a thing is considered as something that could be common to several things. Most importantly, to describe a concept as general is to say something about the way it represents, that is, how the understanding considers a mark. It is thus a part of the nature of concepts to be a general mode of representing; that is, a concept "represents generally."

5.3.2 *Generality and the Relations among Concepts*

The generality of conceptual representations is also manifested in the relations among concepts, the study of which belonged to general logic. Kant's understanding of the relations among concepts is familiar, but is worth reviewing because it bears on the role of intuition in representing individuals and hence the singularity of intuition, and will be useful for later reference.

One concept can be related to one another by containment or inclusion: one concept contains another if it includes all the marks of another and more. For example, the concept <horse> contains, say, the concepts <animal> and <ungulate>. A concept also has an extension or sphere, which comprises concepts that fall under it; for example, the concept <horse> falls under the concept <ungulate>. Kant sometimes expresses this by saying that the sphere of the concept <horse> is part of the sphere of the concept <ungulate>. It is important to note that Kant's notion of extension differs from our modern notion; for Kant, the extension of a concept comprises concepts that fall under the concept, while our modern notion of extension refers to objects falling under a concept. In Kant's view, an object can also be said to fall under a concept, but when it does so, it is in virtue of the object possessing the marks contained in the concept – those marks that constitute the concept under which the object is cognized. "Trigger" falls under the concept <horse> because the marks that constitute the grounds of cognition of Trigger are present in Trigger.

The intension and extension of a concept are reciprocally related: a concept is in the extension of another concept just in case the former contains the latter; the concept <horse> is subordinate to (falls under) the concept <ungulate> if and only if it the concept <horse> contains the concept

<ungulate>. The relations of inclusion and subordination can be used to order concepts into a genus–species hierarchy.[8]

It is an important feature of Kant's view that the extension of a concept is not limited to concepts of actual objects; it comprises all further possible determinations of a concept by additional marks. This leads to a genus–species hierarchy of possible concepts completely contained in one another *ad infinitum* with no *infimae species*, that is, no lowest species. This is a feature of the relations among concepts that rests on their nature as general representations: because they refer generally, that is, by means of a mark that can be common to more than one thing, any concept makes it in principle possible to find a further mark to further differentiate among the possible things to which it refers.

Kant holds that the fundamental generality of concepts is consistent with their employment in our cognition of individuals. Although it is the nature of concepts to refer generally, and there are an infinite number of possible concepts that fall under a concept, it is possible to employ the concept to refer to just one individual; Kant explicitly allowed for the employment of concepts in singular and particular as well as universal judgments. He nevertheless denies that this employment makes the concepts themselves singular, particular, or universal. In two notes at the beginning of the *Jäsche Logik*, Kant states:

> Note 1. A concept is opposed to intuition, for it is a universal [*allgemeine*] representation, or a representation of what is common [*gemein*] to several objects, hence a representation insofar as it can be contained in various ones [*verschiedenen*].
>
> 2. It is a mere tautology to speak of universal or common concepts – a mistake that is grounded in an incorrect division of concepts into universal, particular, and singular. Concepts themselves cannot be so divided, but only their use. (9:91)[9]

Note that Kant describes a concept's being universal (*allgemein*) in terms of its representing what is common (*gemein*), drawing on the similarity of word form. The employment of a concept can also be universal (*allgemein*), but it need not be, and whether a concept is used in a universal, particular, or a singular judgment, it is nevertheless essentially general. That is why it is a mere

[8] The containment and falling under relations obtain only when one concept is completely contained in another, so that the ordering relations are only partial. Whether there is a canonical hierarchy in which all concepts find their proper place is a further issue; we are here interested in the possible relations among concepts. For an in-depth exploration of the role of the hierarchy of concepts in Kant's view, see Lanier Anderson (2015)

[9] For a helpful discussion of the singular use of concepts, see Thompson (1972).

tautology to speak of universal (*allgemein*) or common (*gemeinsammen*) concepts; it is by its nature a representation that refers generally.[10]

Kant's view that the hierarchy of concepts includes no *infimae species* entails that, in virtue of their general mode of representation, no finite number of determinations added to a concept would yield the representation of an individual or a single object, for the marks are general and hence can in principle be common to several things. Neither concepts nor combinations of concepts represent singular individuals, without the aid of intuition. Kant reinforces his claim that there are no *infimae species* of concepts in his treatment of the principle of thoroughgoing determination:

> The proposition **Everything existing is thoroughly determined** signifies not only that of every **given** pair of opposed predicates, but also of every pair of **possible** predicates, one of them must always apply to it.... What it means is that in order to cognize a thing completely one has to cognize everything possible and determine the thing through it, whether affirmatively or negatively. Thoroughgoing determination is consequently a concept that we can never exhibit *in concreto* in its totality, and thus it is grounded on an idea which has its seat solely in reason ... (A573/B601)

Kant thus rejects our representation of individuals by means of a mere combination of general representations.[11] In sharp contrast to the inherent generality of conceptual representation, intuitions are singular, and this gives intuitions a role in the cognition of individual objects. I will return to this role for intuition in Section 5.6.

Kant's view is that thoroughgoing determination of an individual through concepts alone, and hence complete cognition of an individual through concepts alone, is not possible, while intuition makes it possible to cognize individuals. The enduring debates in the history of Western philosophy, beginning with Plato and Aristotle, about the status of universals and their relation to individuals includes various attempts to account for the singularity of the latter. Kant's appeal to intuition supplies a principle of singularity and

[10] Kant does not further develop an account of how concepts are employed in singular and particular judgments, nor whether it is possible to define a concept in such a way that only one object could fall under it – for example and most importantly, the concept of God (as pointed out, for example, by Parsons (1983b, p. 113)). It does seem that such concepts are incompatible with Kant's doctrine that there are no *infimae species*. I will not try to resolve this issue here. For a different view of the generality of concepts, see Daniel Smyth (2014).

[11] Kant's claim assumes that a concept cannot contain an infinite number of determinations; see B40. I will not explore Kant's justification for this claim. See Friedman (1992, pp. 68f.) for an argument that it is based on the limitation of monadic concepts, and Daniel Smyth (2014) for an argument that it is based on the limitations of discursive representation.

even of individuation.[12] This fact about intuition, however, does not by itself imply that intuitions on their own and without the aid of concepts allow us to cognize or even represent singulars or individuate them without the employment of concepts at all. I will return to this role for intuition in Section 5.5 below, after presenting an account of the singularity of intuition.

5.4 The Singularity of Intuition Explained and Defended

Let us return to the issue raised in Section 5.1. Kant contrasts intuitions with concepts as singular and immediate representations, and the singularity of intuitions appears to consist in representing singulars or individuals, such as tables or parts of space. And this seems to be contrary to my claim that intuition can directly represent an indeterminate homogeneous manifold, that is, directly represent a "stuff" or a "muchness" that is a genuine and direct representation of a manifold without a representation of its parts, and does not reduce the representation of its manifoldness to the mere possibility of representing its parts. As I understand Kant, however, there is a common notion of singularity that encompasses both the direct representation of an indeterminate homogeneous manifold and the representation of singular objects. To flesh out this view, I take a cue from Kant's understanding of the generality of conceptual intuition as a mode of representing generally and regard intuition as a mode of representing singularly. The mode of representing intuitively grounds the singularity of what is represented through intuition. On the other hand, our knowledge of this singular mode of representing is acquired by reflecting on what is represented.

5.4.1 *Intuition as a Mode of Representing Singularly*

To begin, let me emphasize again that in the context of the theoretical cognition of objects, a central role for intuition is in making it possible to represent singular objects. Moreover, in the context of such cognition, we individuate intuitions by the individual objects they allow us to represent. Kant freely uses the phrase "the intuition" or "an intuition" to refer to an intuition whose employment allows us to represent a single object: an intuition of a house, for example, or the intuition of a triangle in geometry each of which is an intuition employed in the representation of a single object. Moreover, the *Critique of Pure Reason* focuses on the conditions that make the cognition of objects possible. The focus on the representation of objects is

[12] Thus, in this respect, intuition plays the role of a principle of individuation akin to the role of matter in Aristotle or haecceity in Scotus. I will not attempt to draw out the parallels or important differences between Kant and Aristotle (or the rest of the tradition) on this point here.

found even in the Transcendental Aesthetic, where Kant broaches the topic of intuition through its role in representing objects, and sensibility is characterized as the capacity to acquire representations through the way in which we are affected by objects (A19/B33). Furthermore, Kant holds more generally that our knowledge of the nature of our cognition rests on philosophical reflection on the nature of our cognition of objects.

Despite the central role of intuition in the cognition and representation of objects, however, intuition plays a more fundamental role in representing the indeterminate manifold of space and time. The singularity of intuitions consists in representing singularly, and to represent singularly is for intuition to only ever represent the manifold of space or time singularly, even when representing an indeterminate extent of space or time. This requires an account of what it means for intuition to represent an indeterminate manifold singularly.

The singularity of spatial intuition can be understood through analogy with the properties of maps, paintings, pictures, or other images that represent an individual spatially extended thing.[13] Each of these kinds of spatial representation can represent a continuous spatial extent, and each part of a map or image represents only a single, individual part of the spatial extent represented, and no other; each part of the map represents a distinct part of space. In that sense, each part of a map or image represents singularly. Moreover, one can regard the map or painting as continuous, and whatever arbitrary part of it one might chose to isolate, and no matter how small the extended part of the map or painting, that part corresponds to a single and unique part of what is represented, distinct from other parts. It is in this sense that the map or painting represents singularly. This singular mode of representation is antithetical to Kant's understanding of conceptual representation, which, as we saw in the previous section, by its nature considers a mark as something that can be common to more than one thing, and hence represents generally.

In the case of a map or picture, the mode of representing singularly depends on there being a one-to-one correspondence between the possible parts of the spatially extended two-dimensional surface of the map or picture and the spatial extent it depicts.[14] In contrast, the spatial intuition of a house is not itself spatial. Intuition is a kind of mental representation, which is itself something nonspatial, by means of which we represent spatial extents in a singular way. We call spatial intuition "spatial" in virtue of what it represents,

[13] I will focus on the paradigm case of spatial intuition in what follows. The same points will apply for temporal intuition, although there are complications.

[14] At least in paradigm cases. I am not claiming that it is essential to maps, etc. to represent singularly, but rather using the analogy to exemplify the compatibility of representing singularly and representing a continuous extent.

not because it is itself spatial.[15] That means we cannot explain its singularity as a one-to-one correspondence between the spatial parts of an intuitive representation and what is represented. We can understand this mode of representing only by analogy with a map, and, as I will argue below, by reflection on the nature of what is intuitively represented. It follows from the nature of intuitive representing that it represents any possibly distinct part of space uniquely. (And a parallel point holds for time.) As will be explained in Chapter 7, it is an important feature of Kant's view that the manifold represented is homogeneous; that is, any possible part of space is qualitatively indistinguishable from any other; the numerical distinctness of the parts of space is purely intuitive. For now, the essential point is that it follows from the nature of the intuitive mode of representing that it represents any possibly distinct part of space uniquely.

Before expanding on this further, let us first return to the context of our cognition of objects. My cognition of a house is of a single object in virtue of the individual spatial and temporal extent it occupies. But my cognition of the singularity of this object as a house includes several factors: intuition singularly representing continuous spatial and temporal extent, and the unity of the house determined by the concept of the house, which includes the concept of the determinate spatial and temporal extent it occupies. Depending on one's understanding of intuition, one might also hold that intuition presents us with a singular something, an object, which we come to cognize as the house. Or one might hold that the presentation in intuition of any kind of object already presupposes, at the very least, the categories, and perhaps also sortal concepts.[16] Regardless of whether or not one thinks that intuition can present objects without the aid of either sortal or categorial concepts, my claim is that the singularity of whatever is represented is made possible by the underlying capacity of intuition to singularly represent an indeterminate manifold in space. And once something, such as a house or a triangle, has been cognized as an individual, we can then individuate the intuition contributing to it; we can then speak of the intuition of the house or of the triangle, or more generally an intuition of an object of experience, such as a house, or a determinate space, such as a triangle.[17]

[15] I am indebted to Chen Liang for discussions on these topics in this and the following section.

[16] Thus, I am at this point maintaining neutrality in the Kantian conceptualist/nonconceptualist debate regarding the objective purport of intuition. I will return to this debate below.

[17] Another way of putting this point is that the sorts of things that count nouns refer to depend on an underlying representation of a kind of thing that would be referred by means of a mass noun.

Unfortunately, Kant doesn't clearly state that, in the first instance, the singularity of intuition consists in the way it represents. Moreover, there are texts in which Kant seems to use term "intuition" for something that is itself singular. In the *Stufenleiter* passage, for example, Kant states that a concept *relates* to objects by means of a mark that can be common to many things, and he states that intuition *relates* immediately to its object, but adds that the intuition *is* singular. Kant thus attributes singularity to the intuition, not merely to its mode of representing, or the way it represents. In light of what I've been arguing, one might argue that this could just be a slip, or that Kant could be speaking loosely. But it is also possible that Kant uses "intuition" in two senses of representation, which might well be in play in the *Stufenleiter* passage. In the following section, I will attempt to clarify these two possible senses and bring out their implications for understanding the singularity of intuition.

5.4.2 *Intuition as a Mode of Representing and as Represented*

Kant seems to use "intuition" not only for that by means of which we represent a spatial or temporal extent, but also for what is thereby represented.[18] More specifically, Kant sometimes seems to use intuition to refer to a represented object insofar as it is represented in intuition. For example, the principle of the Axioms of Intuition in the B-edition states that all (determinate) intuitions are extensive magnitudes, while the A-edition states that all appearances are extensive magnitudes. Since being an extensive magnitude is a property of a represented object, not a mental act or a representing, it appears Kant is using intuition to refer to something represented in intuition.[19]

In the case of spatial intuition, this second sense of intuition as intuited refers to a spatial object, or more precisely, to a spatial object insofar as it is represented by means of intuition, and more generally to a part of space represented by means of intuition, which I will refer to as a "spatial extent."[20]

[18] See Sellars (1968, pp. 7ff.), who draws a distinction between intuiting and intuited. Allison also draws a distinction between intuition as intuiting and intuited, where the former is the act of representing a particular, and the latter is the object represented. In addition, he distinguishes both from intuition in the sense of a particular kind of representation. See Allison (2004, p. 82).

[19] One could perhaps resist this claim, and assert that by intuition Kant always means an act of intuiting. In that case, one can set aside what I say below, for the important point is that intuition in its primary meaning is to be understood as a mode of representing, and singularity as representing singularly. In what follows, however, I will allow that Kant also uses intuition in the sense of intuited.

[20] I'd like to thank Chen Liang for helpful discussions on this point. I am here setting aside the uses of "intuition" to refer to space as a whole; I restrict what follows to

If we use "intuition" in this second way, then it is quite appropriate to say that the intuition is spatial. One might even be tempted to say that the intuition of the window is to the left of the intuition of the door: the window insofar as it is represented in spatial intuition is to the left of the door insofar as it is represented in spatial intuition. This dual use of the term "intuition" makes it quite easy to slide between properties of intuition as means of representing and properties of intuition as object represented. This ambiguity is perhaps in play in the *Stufenleiter* passage, when Kant says that intuition *relates* to its object immediately, but also says that intuition *is* singular.

Despite the potential ambiguity of "intuition" and the confusions it can cause,[21] as well as Kant's apparent sliding between them, I believe Kant himself was not confused on the most important point: he did not commit the fallacy of confounding the properties of what is represented and properties of that by means of which it is represented.[22]

My focus in the present chapter is on intuition in the first sense, which I take to be the more fundamental sense of intuition in Kant. It is more fundamental in that it hews more closely to Kant's description of intuition as a form; it is also more fundamental in that it grounds the properties that objects have insofar as they are represented in intuition, that is, intuitions in the second sense. Despite the fact that intuition in this first sense is more fundamental, we come to know the properties of human intuition through reflection on the cognitions they make possible, and in particular by reflection on our cognition of objects, such as appearances, insofar as they are represented by means of intuition.

I would like to emphasize a point already mentioned above. Just as we can come to know about properties of intuition as a mode of representing through

representations of spatial objects and of parts of space, because it is the role of intuition in representing their singularity that is my focus.

21 There are further potential grounds of confusion. As mentioned in the previous note, I am here setting aside the intuition of space as a whole. I am also setting aside the difference among various descriptions of intuition such as pure intuition, form of intuition, and formal intuition.

22 Thus, as a point of reference, I am arguing that Kant avoids an intuitive version of what, much later, Ryle (1949) identifies as a confusion concerning sensations and U. T. Place (1956) called the "phenomenological fallacy." According to Place, it is a common fallacy to attribute to sensations the properties they allow us to represent. A green sensation is called "green" on account of what it represents, not because it is itself green; to attribute green to the sensation itself is to be fallaciously mislead by the way we individuate sensations and phenomena such as after-images. In a critique of Kant's notion of space as the form of outer sense, Sellars employs a similar distinction between representings of outer sense and the represented (see Sellars (1968, p. 8)). One can nevertheless make some legitimate inferences about the properties of the intuitive mode of representing from the properties of that which is represented, as I argue below.

reflection on the things cognized by means of intuition, such as appearances, we can also individuate the former by appeal to the latter. I can talk of the intuition of a window, for example, and mean by that the window insofar as it is represented in spatial intuition. But I can also individuate the intuitive representing of the window, picking out an individual act or means of representing, or component of an act or means of representing, corresponding to the represented window. The dependence of both on our knowledge and our individuation of intuitive means of representing on what is thereby represented further obscures the distinction between the two possible senses of intuition.

I am claiming that in Kant's view, we learn about our intuitive mode of representing by reflection on the objects we thereby represent. What can we say more specifically about how we gain knowledge of our intuitive modes of representing and what we thereby learn? Focusing again on spatial intuition, recognizing that the representation of the spatial properties of objects is made possible by the spatial intuitive mode of representing requires distinguishing those spatial properties from other properties whose representation is made possible by pure concepts of the understanding, and concepts more generally. There is no reason to assume that this is an easy task. Kant makes a related, more general point in the B-Introduction, where he anticipates the philosophical work to come in identifying the *a priori* elements of our cognition:

> it could well be that even our experiential cognition is a composite of that which we receive through impressions and that which our own cognitive faculty (merely prompted by sensible impressions) provides out of itself, which addition we cannot distinguish from that fundamental material until long practice has made us attentive to it and skilled in separating it out [*zur Absonderung ... geschickt gemacht*]. (B1)

Similarly, there is no reason to think it is easy to distinguish the properties that belong to the pure forms of intuition from those that belong to the pure concepts of the understanding. Doing so requires philosophical reflection on the nature of our faculties themselves, reflections such as those with which Kant begins the Transcendental Aesthetic, as well as further work separating out various elements of our cognition. Thus, at the close of the introduction to the Transcendental Aesthetic Kant describes the method of the Aesthetic, which reveals that there are two forms of sensibility, space and time; the method is to isolate sensibility by separating off everything that the understanding thinks and detaching everything that belongs to sensation (A22). Kant gives a more specific characterization of this method in the case of spatial intuition immediately prior to this passage, where he says that we separate from the representation of a body that which the understanding thinks about it, as well as that which belongs to sensation, and find that what is left is extension and shape (*Ausdehnung und Gestalt*) (A20–1/B35).[23] These are the properties of objects of cognition the possibility of whose representation is attributed to a specific mode of intuitive representation we call spatial. Similar reflections presumably lead to attributing the possibility of temporal properties of objects of cognition to a distinct mode of intuitive representation we call temporal. Having found that there is a mode of intuiting that allows us to represent objects as spatial, and more specifically, to represent the extension and shape of objects, further reflection on the representation of the extension and shape of everyday objects informs us further about what this mode of intuiting makes possible. Moreover, because geometry concerns the extension and shape of objects represented in intuition, reflection on the representation of geometrical figures will also inform us about the nature of the spatial mode of intuiting.

In addition, the identification of a spatial mode of intuiting quite naturally leads to reflection on this mode of intuiting more generally and prompts a closer examination of our concepts of space as a whole and our concept of time as a whole, that is, their "exposition." The exposition of the concepts of space and time is a distinct representation of that which belongs to the concepts, which proceeds by reflection on the properties of that which is represented by means of them as well as on the nature of the mode of conceptual representing. It constitutes the core of the Transcendental Aesthetic.[24]

What can such philosophical investigations teach us about the spatial and temporal modes of intuiting? First, to reiterate the point made above, I have argued for the view that Kant does not commit the fallacy of simply attributing properties of the intuited to the mode of intuiting through some confusion; he does not attribute spatial and temporal properties to the modes of intuiting themselves. What properties can we safely attribute to them? Kant clearly thinks they have the property of allowing us to represent the spatial and temporal properties of objects, respectively. But what more can we say about the spatial and temporal modes of intuiting themselves? The Transcendental Aesthetic argues that our original representations of space and time are *a priori* intuitions, and that their being so is a condition for the possibility of

[23] One might raise concerns about the method of abstraction Kant seems to endorse in these passages. While I think Kant's employment of abstraction as a component of philosophical reflection is defensible, I will not argue for this claim here. My primary point is to sketch how Kant thinks we can attribute properties to the mode of intuitive representing in general as well as to particular intuitive representings of, e.g., a house.

[24] The relationship between our representation of everyday spatial objects and spatial extents, on the one hand, and our representation of space as a whole, on the other, is crucial to understanding the Transcendental Aesthetic and Kant's account of intuition more generally. The distinctions I am employing here play an important role in understanding that relationship, but I cannot attempt a full treatment here.

geometry and the synthetic *a priori* cognition found in the general theory of motion, respectively.

These properties concern the spatial and temporal modes of representing in general. However, I claimed that in Kant's view, we can also individuate intuitions in the first sense, that is, as modes of representing, by the spatial objects or extents they, in combination with concepts, allow us to cognize. As I noted above, I can speak of the intuiting by means of which I represent the window of a house and the intuiting by means of which I represent the door, for example. This is a perfectly legitimate way of individuating intuitions as a means of representing. The individuation of these individual means of representing is derivative, because it relies on the cognition of objects represented, which in turn depends on both spatial and temporal intuition and on pure concepts of the understanding – and in this case on the empirical concepts of window and door as well.

Is there a derivative sense in which we can attribute other properties to the individuated spatial and temporal modes of representation themselves? I will again focus on spatial intuition. We might well say that the intuition of the window is to the left of the intuition of door, if by that we mean intuition in the second sense, that is, these appearances insofar as they are represented through spatial intuition. It would not be legitimate, however, to take this to mean that intuitions in the first sense – as means of representing individuated by the objects represented – stand in these spatial relations, for acts of intuiting are not spatial at all. Similarly, I can well say that the intuition of the window is a part of the intuition of the house, if I mean intuition in the second sense. But if I mean intuition in the first sense as act of intuiting, and I mean a spatial part, that would be illegitimate, for again, they are not even spatial.

Nevertheless, there does seem to be a difference between relations such as left-of, which are inherently spatial, and part–whole relations, which might be regarded as having not merely a spatial meaning, but a more abstract meaning. Kant holds that the intuition of time allows for part–whole relations, as does any intensive magnitude, such as the intensity of a light or a force, indicating that he had a more abstract understanding of the part–whole relation than merely spatial, even when we are considering only the part–whole relations of magnitudes. Furthermore, as noted in Section 5.2.2, there are part–whole relations among the intensions of concepts and among the extensions of concepts, which indicates an even broader application or understanding of the part–whole relation.[25] If we understand the part–whole relation in a more abstract way, there is room to say that the intuiting of the window is a part of

[25] This does not imply that it is not possible to use spatial or temporal intuition to represent more abstract part–whole relations. Kant appeals to spatial intuition when he uses Euler circles to illustrate the relations among concepts, for example. In fact, it may be that in some cases we *must* use spatial intuition or temporal intuition to represent part–whole

the intuiting of the house, where that means that the intuitive representing of the whole house includes the intuitive representing of the window, and more. As with the individuation of intuitions as means of representing, here too the attribution of part–whole relations is derivative – it is based on the spatial part–whole relations of objects or spatial extents insofar as they are represented in space. Thus, there is a correlation between the abstract part–whole relations attributed to intuitions as means of representing and the spatial part–whole relations among the spatial objects or spatial extents they represent. Furthermore, because the mode of representing is more fundamental and grounds the properties represented, Kant may hold that the more abstract part–whole relational structure of the intuitive mode of representing is what makes it possible to represent and grounds the spatial part–whole relations of objects. That is, properties of what is intuited may be the *ratio cogniscendi* of the properties of intuiting, while the latter are the *ratio essendi* of the properties of what is intuited. And while I have focused on spatial intuitions, the same points hold for temporal intuitions.[26]

This exploration of Kant's understanding of the relation between the means of representing and what is represented has implications for understanding the Transcendental Aesthetic, Kant's claims about the nature and limits of conceptual and intuitive representing, and even more particularly, the part–whole relations among intuitive representings and among conceptual representings. But the main point for our present discussion is its implication for our understanding of the singularity of intuition, to which we now return.

5.4.3 *Representing Singularity and the Represented Singular*

I am arguing for the view that the singularity of intuition fundamentally consists in the mode of representing singularly. This mode of representing accounts for the map-like representation of an indeterminate spatial manifold described in Section 5.4.1. It is also what makes it possible to represent singular objects, and in the context of our theoretical cognition of objects of experience, the latter is perhaps the most important role for intuition.

The representation of an indeterminate spatial manifold and representing a singular object of cognition are intertwined. On the one hand, the fact that spatial intuition represents a determinate spatial manifold singularly explains

relations. As we saw in Chapter 4, Kant holds that we cannot even represent an intensive magnitude as a magnitude without the aid of the intuition of time.

[26] I note again that the representation of temporal extents is complicated by the dependence of their representation on the representation of spatial extents. In fact, there is a mutual dependence of our representations of spatial and temporal extents, for Kant holds that I cannot even think a line without drawing it in thought. I set these further complications aside here.

the possibility of representing singular spatial objects of cognition such as a house. That is, a representation of a house is a spatial singular in virtue of its occupying, and in fact in part being, a singular spatial extent, that is, an extensive magnitude, and the singularity of that extensive magnitude presupposes and is grounded in the singular mode of spatial intuiting. On the other hand, in my initial appeal to maps and pictures to explain the map-like representation of spatial extent, I stated that each part of a map or picture represents exactly one part of what is depicted, and that there is a one-to-one correspondence between the parts of the map or picture and what is thereby represented. Thus, while I can represent an indeterminate manifold as such, my understanding of the singularity of what is represented appeals to the representation of arbitrary parts. In short, even in the case of the representation of an indeterminate continuous manifold of space, we learn what it is to represent singularly by reflection on the singularity of the parts of space represented. It is nevertheless the case that our singular representation of the parts of space is made possible by the singular mode of intuiting an indeterminate manifold.

There are two related questions one might reasonably have about the account I have been giving. First, what is gained by shifting attention to the nature of intuition as a mode of representing singularly if what we can know of it is derived from philosophical reflection on what is singularly represented? Second, what do representing an indeterminate manifold singularly and representing an individual object singularly have in common?

My answer to both of these questions is this: by focusing on intuition as a mode of representing singularly, we come closer to Kant's understanding of intuition as a form of sensibility, a feature of our cognitive capacities. But more importantly for present purposes, we move away from the mistaken belief that the singularity of intuition consists solely in the representation of individual objects or individuals or perceptual particulars, which makes room for understanding how it is that intuition can represent an indeterminate manifold in a way that can be described as singular. The latter is not easy to characterize. I have resorted to an analogy with map-like representation, where there is a one-to-one correspondence between arbitrary parts of a map, no matter how small, and the spaces they represent. It is only an analogy, since intuition as a mode of representing is not itself spatial, and we can attribute only parts of an act or mode of representing standing in a one-to-one relation to potentially determinate parts of an indeterminate manifold of space in a nonspatial sense of parts and can individuate those parts only derivatively. Nevertheless, once we see that intuition can represent an indeterminate manifold in a manner consistent with the singularity of intuition, we can appreciate what the intuitive representation of an indeterminate manifold and the representation of an individual object have in common. As a singular mode of representing, spatial intuition represents an indeterminate field of singularity, as it were, in which individual spatial objects can be represented. It is in this sense that representing an indeterminate manifold singularly makes the representation of singular objects possible.

According to this interpretation, the representation of the indeterminate manifold space in a singular way can be thought of as representing a kind of substratum for conceptual determination, a substratum of singularity. There are two ways, however, in which one must be careful in articulating the sense in which it can be regarded as a substratum. First, Kant most frequently employs the notion of a substratum in his First Analogy account of the principle of the persistence of substance. Kant describes that which persists as the substratum of the empirical representation of time itself, and as a condition of all time determination (A182–6/B226–9). Kant also appeals to the notion of a transcendental substratum as a ground of the thoroughgoing determination in our reason, the idea of an all of reality of all possible predicates of things (A575/B603). But Kant also describes intuition as a substratum for conceptual determination. The *Prolegomena* provides a revealing example. In a discussion of the role of the understanding in prescribing *a priori* laws to nature, Kant states:

> Space is something so uniform and so indeterminate with respect to all specific properties that certainly no one will look for a stock of natural laws within it. By contrast, that which determines space into the figure of a circle, a cone, or a sphere is the understanding, insofar as it contains the basis for the unity of the construction of these figures. *The mere universal form of intuition called space is therefore certainly the substratum of all intuition that can be determined in particular objects*, and admittedly, the condition for the possibility and variety of those intuitions lies in this space; but the unity of the objects is determined solely through the understanding . . . (4:21-2; my emphasis).

Kant begins this quoted passage by drawing attention to the uniformity and indeterminateness of space. These features of space are reflected in Kant's exposition of the concept of magnitude as a homogeneous manifold, and the need for intuition to represent such a manifold. We will look more carefully at homogeneity in Chapter 7; for now, my suggestion is that in light of his emphasis on uniformity and indeterminateness, what intuition offers is an indeterminate substratum of singularity that accounts for the uniqueness of every represented region of space. Note that Kant applies the term "substratum" to "the universal form of intuition called space"; this draws attention to intuition as form, which on my interpretation draws attention to intuition as a mode of representing. At the same time, he describes space as the substratum "of all intuition that can be determined in particular objects," emphasizing that its role as substratum is to make the representation of singular objects possible.

There is also a second way in which one must be careful in articulating the sense in which intuition is a substratum of conceptual determination. I have emphasized throughout that the singularity of intuition is to be understood as a mode of representing singularly, while talk of a substratum might sound as if we represent an indeterminate manifold as an object prior to conceptually determining it, as if it involved the representation of a kind of blank slate. While I do believe that in Kant's view, we indeterminately represent the interior of a triangle, for example, as an extent, I do not think Kant is committed to the view that we represent a substratum as an object prior to or independent of any determination whatsoever. Most importantly, Kant's references to space as a substratum do not undermine the claim that the singularity of objects represented in space rests on the mode of representing singularly. As noted, when Kant refers to the "universal form of space as the substratum of all intuition that can be determined in particular objects," he is connecting the notion of substratum to the way in which the form of intuition represents, namely, singularly.

Let me recapitulate. I have been arguing for the view that just as it is in the nature of concepts to represent generally, it is in the nature of intuitions to represent singularly, and that properly understood, this is compatible with intuition representing an indeterminate and continuous homogeneous manifold and explains the sense in which this manifold is itself singular. Moreover, the fact that the mode of intuitive representing represents singularly makes it possible to represent singular objects.

Nevertheless, the fact that intuition makes it possible to represent singular objects does not mean that the singularity of intuition *consists in* the representation of singular objects, despite what the *Stufenleiter* passage has often been taken to imply. I noted above that some may have been misled by Kant's explication of intuition in the *Stufenleiter* passage, which is an explication in the context of cognition, understood as the representation of an object with consciousness. Kant's remarks in that context, and his focus throughout the *Critique* on the conditions of the cognition of objects of experience, might lead one to believe that the singularity of intuition consists in its representing a single object. This misses the mark, however. The interpretation of singularity for which I have been arguing has implications for several aspects of our understanding of Kant's philosophy worth emphasizing.

5.5 Three Implications of This Interpretation of Singularity

5.5.1 Singularity and Mathematical Cognition

Understanding the singularity of intuition as consisting in the representation of a singular object can mislead accounts of Kant's philosophy of mathematics. In his classic and influential article "Kant's Philosophy of Arithmetic," Charles

Parsons cites several passages in which Kant refers to the singularity and immediacy of intuition and then states:

> What is meant by calling an intuition a singular representation seems quite clear. It can have only one individual object. The objects to which a concept "relates" are evidently those which fall under it, and these can be any which have the property the concept represents, so that a concept will only in exceptional cases have a single object. Thus far, the distinction corresponds to that between singular and general terms.[27]

According to Parsons, the singularity of an intuition is tantamount to representing one individual object. Now, Parson's primary focus is on the immediacy of intuition, arguing that it plays a substantive phenomenological role in Kant's account. In contrast, Hintikka argues that singularity is the primary and philosophically most interesting property of intuition in Kant's theory of mathematical cognition, downplaying any phenomenological role for immediacy. Hintikka nevertheless also understands the singularity of intuition as consisting in the representation of an individual object:

> we may say that Kant's notion is not very far from what we would call a singular term. An intuition is for Kant a 'representation' – we would perhaps rather say a symbol – which refers to an individual object or which is used as if it would refer to one.[28]

My goal in this chapter is not to adjudicate between Parsons and Hintikka on the role of immediacy, nor to give a full account of singularity and immediacy, but to make room for the idea that intuition singularly represents a homogeneous manifold, and to point out that a mistaken understanding of the singularity of intuition focuses attention on the status of mathematical objects and can thereby obscure Kant's theory of magnitudes. Obscuring that theory distorts our understanding of both Kant's philosophy of mathematics and his account of the world of experience.

5.5.2 *Singularity and the Current Kantian Nonconceptualism Debate*

A second way in which misconstruing the singularity of intuition can mislead has implications for understanding Kant's theoretical philosophy as a whole. Taking the singularity of intuition to consist in the representation of individual objects can lead one to think that intuitions have the capacity to represent single objects without the aid of, or logically prior to, any contribution of the understanding. This is a line of thought present in at least some of the current

[27] Parsons (1983, p. 112).
[28] Hintikka (1969, p. 43).

nonconceptualist readings of Kant.[29] It can lead to the further thought that the contribution of intuition to cognition is to provide objects for our concepts, and in particular, objects that fall under the categories.

There are various philosophical reasons one might find this view appealing and there certainly are texts which seem to support it.[30] There is neither space nor time to adequately address this issue here, but if what I have said is correct, it is wrong to think that the nature of the singularity of intuition consists in the representation of singular objects. It would therefore be wrong to infer from this characterization of the nature of intuition that intuition allows us to represent singular objects without any aid from concepts.

There are several important points to note about this conclusion, however. First, undermining this *prima facie* argument for a nonconceptualist reading of Kantian intuition does not rule out nonconceptualism; there may be other good arguments for the position. Moreover, for all that has been said so far, it might be that the intuitive mode of singularly representing a homogeneous manifold makes it possible for intuition to represent perceptual particulars without any involvement of the intellect or employment of concepts.

One might think that the position I've outlined in this and previous chapters already forecloses the possibility that intuition can represent particulars without concepts. In fact, I long thought so.[31] For I have argued that the intuitive representing of an indeterminate extended manifold of space or time is what makes the representation of singular individuals in intuition, including parts of space, possible, and I have also argued that the synthesis of composition of a homogeneous manifold in intuition described in the Axioms of

[29] This view can be found in the work of Lucy Allais, to cite one interesting and sophisticated example we already discussed in the previous chapter. See Allais (2015), where singularity is cashed out in terms of presenting particulars in some minimal sense (p. 147), that there is a particular thing which intuition represents (p. 154), and furthermore that "if intuitions did not involve the presence to consciousness of the objects they represent, they would not be singular and immediate" (p. 154). A similar view can be found, for example, in McLear (2015), who refers to an episodic consciousness of sensory qualities of an object (e.g., color, shape, and location) that on their own are not representations *as* objects, but are individual representational states that are representations *of* distinct, individual features of objects.

[30] As noted in the previous chapter, there is an extensive literature on whether Kant is conceptualist or nonconceptualist, complicated by a variety of interpretations of what conceptualism and nonconceptualism consists in, which I will not enter into. However, one article that bears directly on the singularity of intuition and the representation of objects is Land (2013).

[31] Others have also seen an argument for a form of conceptualism that appeals to the fundamental nature of the synthesis of composition at the heart of the Axioms of Intuition. For example, Emily Carson's unpublished manuscript, "Number, the Category of Quantity, and Non-Conceptual Content in Kant" (unpublished), and Thomas Land's work, in particular, Land (2014) and Land (2016).

Intuition is quite fundamental. It does not presuppose the presentation of individuals or perceptual particulars in intuition; on the contrary, it seems to be what makes the representations of individuals in intuition first possible. Since that synthesis is governed by the categories of quantity, we seem to have a quite strong argument for the necessity of the categories for the representation of any individuals in intuition.

This line of thought gains apparent additional support from further claims for which I have argued. The Axioms of Intuition concerns not just applied mathematics but also pure mathematics, and the synthesis required for the representation of determinate spaces and times in mathematics is the very same synthesis required for the apprehension and perception of appearances. Since mathematical cognition rests on the construction of concepts, and in particular construction of the concept of magnitude under the guise of the categories of quantity, these categories are a condition of any apprehension and perception of appearances.[32]

I think these considerations certainly suggest a line of argument for Kantian conceptualism, one that I do not think nonconceptualists have adequately acknowledged or addressed. Nevertheless, these considerations are not on their own sufficient to establish a version of Kantian conceptualism.[33] More argument would be needed to counter a three-part nonconceptualist response. First, the nonconceptualist can concede that in the case of mathematical cognition, the categories are required not just for cognition, but even for the representation of the determinate spaces and times underlying mathematical cognition. For in the case of pure mathematical cognition, we are not given an object; we instead generate an object through an elective (*willkürlich*) construction of concepts, and that presupposes an activity of the understanding grounded in the categories of quantity. That does not, however, entail that we cannot be given an object – a perceptual particular, for example – in intuition. Second, the nonconceptualist can agree that the apprehension and perception of appearances rests on the same synthesis that underlies the representation of

[32] A prior version of myself had believed that these considerations gave strong support to the view that employment of the categories of quantity was required in order for intuitions to gain any objective purport. At the same time, I retained the belief that my interpretation of Kant's theory of magnitudes and the role of intuition in it, and in particular the nature of intuitive singularity and the role of intuition in the representation of a homogeneous manifold (to be discussed in Chapter 7), provide support for a nonconceptualist reading of intuitive content in Kant's philosophy, albeit in a different sense of nonconceptual content than currently employed in the Kantian conceptualist/nonconceptualist debate. Unfortunately, I cannot pursue this line of thought further here.

[33] I would like to thank Tyler Burge and Matt Boyle for helping me see that Kant's theory of magnitudes, its role in mathematical cognition, and the interpretation of the Axioms of Intuition I have argued for do not, on their own, establish a form of Kantian conceptualism.

determinate spaces and times in mathematics, but insist that in the case of the apprehension and perception of an appearance, this is a synthesis carried out in a way responsive to perceptual particulars given in intuition. Finally, to reiterate a point alluded to above, the nonconceptualist could abandon the claim that it is the nature of intuition to represent singular objects (in some minimal sense of object), while maintaining that intuitions can represent objects without the employment of concepts or the activity of the understanding. For she can agree that what is fundamental to the nature of intuitive representing is the possibility of singularly intuiting an indeterminate homogeneous manifold, and agree that this makes it possible for intuition to present us with perceptual particulars, while at the same time insisting that the latter is independent of concepts or the intellect.

In conclusion, the interpretation of Kant's theory of magnitudes, the Axioms of Intuition, and the singularity of intuition for which I have argued do not on their own provide an argument for conceptualism as that is currently understood in the Kantian conceptualist/nonconceptualist debate. At the very least, more argument would be needed. In any case, I believe that any resolution of that debate will need to more closely consider the implications of Kant's theory of magnitudes and the nature of mathematical cognition.

5.5.3 *Singularity and the Extensive Magnitude Regress*

The final implication of the account of the singularity of intuition I have given brings us back to the issue that prompted this exploration of the singularity of intuition: how to understand the part–whole relations of magnitudes, and in particular, extensive magnitudes. Kant's definition of extensive magnitude appears to generate a regress that can be stopped only by the representation of particular parts, and some have taken this as committing Kant to some kind of intuitive or perceptual atomism. If one is inclined to think that the singularity of intuitions consists in the representation of individuals, then it will seem natural to appeal to intuitive particulars to stop the regress: the parts of extensive magnitudes are represented as particular parts in intuition. We have seen, however, that the actual solution to the extensive magnitude regress is an appeal to a continuous successive synthesis of a manifold of undifferentiated, and hence at most indeterminately represented, parts (see Section 4.5 above). The account of singularly intuiting a homogeneous manifold in intuition explains how this can be reconciled with the role of intuition in making it possible for us to represent singular objects.

Up to this point, I have attempted to sketch out a new way of understanding the nature of the singularity of intuition, one that both includes the role of intuition in representing an indeterminate homogeneous manifold as such and accounts for the singularity of individual objects represented. Even as an outline of an interpretation, however, it is incomplete. It has not explained the homogeneity of the manifold represented in intuition, which I will address in Chapter 7. It also has not explained the relationship between singular representation and representation *in concreto*, which plays a prominent role in Kant's account of our cognition of individuals. We are now in a position to address this latter topic.

5.6 Abstract and Concrete Representations and Objects

Kant uses the concrete/abstract distinction in various ways, which can obscure his views. Moreover, we think about the distinction in a different way today. Most present-day discussions of the distinction concern the ontological status of concrete and abstract objects; this is the case in metaphysics generally, but also in the philosophy of mathematics and set theory. In the early modern period, however, the concrete/abstract distinction was applied to ideas or concepts, and was applied only secondarily, when at all, to objects. Kant follows suit in applying the distinction to concepts. But he also invokes the notion of concreteness in describing a way of representing that depends on intuition, in contrast to the way concepts represent, that is, representing *in concreto* in intuition, which is tantamount to representing singularly. Sorting out Kant's views helps further explain the unique contribution of the singularity of intuition, allows us to relate Kant's views to our contemporary understanding of the abstract and concrete, and clarifies the sense in which relatively abstract concepts of magnitudes represent something concrete.

5.6.1 Kant on the Concrete and the Abstract

We saw in Section 5.3.1 that Kant appeals to abstraction in his account of concept formation. In the early modern period more generally, the term "abstract" was applied to ideas or concepts, and Kant, like most in the early modern period, primarily employed the distinction in the context of representation. Nevertheless, he rejected the phrase "abstract concept" for several reasons. In his view, all concepts depend on abstraction, that is, the leaving-out of some marks in forming a concept. Hence, the notion of an abstract concept, which suggests that there might be other concepts that do not depend on abstraction, is misleading. Moreover, concepts are not generated through abstraction alone; a concept is generated through comparison and reflection, which are two positive conditions of concept formation, in addition to abstraction, which is a final, merely negative condition of concept formation. As Kant puts it, we should say not that we abstract a representation, but that we abstract from representations (24:754).

Kant nevertheless uses the concrete/abstract distinction to distinguish two ways of *regarding* a concept as well as to distinguish two *uses* of a concept.

To regard a concept either abstractly or concretely is not to use it as a ground of cognition of represented objects but to consider its relationship to other concepts. To regard a concept abstractly is to consider it in relation to higher concepts; for example, to consider the concept <horse> in relation to the concept <ungulate>. To regard a concept concretely is to consider it in relation to lower concepts; for example, to regard the concept <horse> in relation to the concept <Clydesdale>.

On the other hand, to *use* a concept abstractly is to use it as a ground of cognition in thinking about all that falls under the concept; for example, to think about all horses in the judgment that all horses are ungulates. To use a concept concretely is to use it as a ground of cognition to think of an individual or some individuals that fall under the concept, for example in the judgments that this horse is a mustang or that some horses are palominos.[34]

Despite Kant's objection to the phrase "abstract concept," Kant does sometimes tolerate and even employ the terms "abstract" and "concrete" to describe concepts themselves, that is, with respect to their content relative to other concepts. This use of the phrases "abstract concept" and "concrete concept" is closely related to the two ways of regarding a concept. In this sense, concepts are relatively abstract or concrete, depending on whether they have fewer or more marks or determinations included in them; a more abstract concept is a broader and higher concept, while a more concrete concept is a narrower and lower concept. For example, the concept <ungulate> is more abstract than the concept <horse> falling under it, and the concept <palomino> is more concrete than the concept <horse>.[35] But he still resists this way of expressing the idea. He makes the following claim in the *Jäsche Logic* about repeated abstraction:

> my concept has in this way become still more abstract. For the more differences among things that are left out of the concept, or more determinations from which we abstract in that concept, the more abstract the concept is. Abstract concepts, therefore, should really be called abstracting concepts (*conceptus abstrahentes*), i.e., ones in which several abstractions occur. (9:95)

[34] In the case of the particular judgment, Kant has in mind that in forming the judgment "some horses," we think that at least a part of the sphere of the extension of objects falling under the concept, and hence at least a subset of the objects falling under the concept, have a property. A particular judgment does not entail that there are some objects falling under the concept that lack that property (see 4:301n).

[35] A concept being more abstract, broader, and higher than another concept, and similarly a concept being more concrete, narrower, and lower than another concept, is a relation that is defined only when all the marks of one concept are contained in the other. There is no sense to be given to whether the concept of Clydesdale is more or less abstract than, say, the concept of Western Tanager. This fact limits an ordering of concepts with respect to their relative size to a partial ordering, a point to which I will return in Section 9.7.

Kant describes a concept with fewer marks as having fewer determinations than a concept with more marks, and it is in that sense less determinate, and conversely, a concept whose content contains more marks as having more determinations, and in that sense more determinate. Although Kant sometimes directly modifies "concept" with "abstract" and "concrete," he generally avoids it. I will follow suit in avoiding applying the terms "abstract" and "concrete" to concepts with respect to their content, unless I make it clear that I am doing so. I will instead refer to less determinate or more determinate concepts or to more general and more specific concepts when describing a concept with respect to the number of marks contained in its content relative to other concepts.

Most of Kant's comments concerning the concrete/abstract distinction concern the ways of regarding concepts or the use of concepts, but Kant also employs the distinction to contrast concepts and intuitions. Using the concrete/abstract distinction to distinguish the roles of concepts and intuitions emerged in the development of Kant's critical philosophy beginning in his *Prize Essay* of 1763–4, *Inquiry Concerning the Distinctness of the Principles of Natural Theology and Morality*. Kant there systematically investigates the differences in the certainty and distinctness of philosophical and mathematical cognition, and one of the contrasts is between their use of signs. Philosophy examines the universal by means of signs *in abstracto*, while mathematics examines the universal by means of signs *in concreto*. As examples of the latter, he refers to signs employed in general arithmetic (i.e., algebra) and number arithmetic, as well as geometrical constructions, where the signs are, for example, a drawn circle (2:278–9). Kant has not yet distinguished between the faculty of sensibility and the faculty of understanding, or the distinctive nature of intuitive and conceptual representations belonging to each, but Kant states that signs are sensible means of cognition (*sinnliche Erkenntissmittel*), and he also states that indemonstrable propositions, such as the proposition that space has three dimensions, "can be explained if they are examined *in concreto* so that they come to be cognized intuitively" (2:281). By 1770, when his *Inaugural Dissertation* appears, Kant distinguishes the sensitive faculty and the faculty of the understanding (2:393ff.), contrasts intuitions and concepts, and describes intuitive representation as representation *in concreto*. Kant states that the sensitive faculty allows us to represent at least certain concepts "to oneself *in concreto* by a distinct intuition," while some "abstract ideas" (*ideas abstractas*) "cannot be followed up *in concreto* and converted to intuitions" (2:389). The connection between *in concreto* representation and intuition is further developed in the *Critique of Pure Reason*, especially with respect to mathematical cognition, which is described as rational cognition through the construction of concepts. To construct a concept is to exhibit *a priori* the intuition corresponding to the concept, and Kant underscores that this intuition, as an intuition, is an "***individual** object*" (**einzelnes** *Objekt*;

Kant's emphasis) (B741). Thus, in mathematical cognition, representing *in concreto* by means of intuition is what allows us to represent an individual object. As I interpret this passage, the intuition exhibited corresponding to the object is an intuited individual, an individual object, and in light of what we have seen above, the representation of this intuited individual depends on an *in concreto* representing, that is, an intuiting that represents singularly.

The same point holds for empirical intuitions. They are not pure, but empirical intuiting represents *in concreto* and hence represents individual objects. This is especially clear in Kant's discussion of synthetic judgments, which go beyond what is contained in a concept to either pure or empirical intuition:

> I can go from the concept to the pure or empirical intuition corresponding to it in order to assess it *in concreto* and cognize *a priori* or *a posteriori* what pertains to its object. The former is rational and mathematical cognition through the construction of the concept, the latter merely empirical (mechanical) cognition . . . (A721/B749)

Thus, both empirical and pure intuitions represent *in concreto*, and in doing so, make the representation of an individual object possible. (In this passage, Kant refers to "assessing" a concept; nevertheless, in both cases, representing *in concreto* is a manner of representing made possible by intuition.)

Drawing on the results of Section 5.4 above, it is the fact that intuitions represent singularly that accounts for and explains *in concreto* representing. Furthermore, the use of concepts *in concreto* in a cognition ultimately depends on intuition; even the meaning and sense of concepts requires that they have a use *in concreto*, by which Kant means that they "apply to some intuition or other, by means of which an object is given for them" (4:282).[36] In summary, it is the nature of intuitions to represent singularly, that is, *in concreto*, and in the context of cognition, that means that intuitions allow us to represent individual objects.

5.6.2 *From Representing* in concreto *to Concrete Objects of Representation*

We have seen that Kant employs the concrete/abstract distinction in various ways, but all apply the distinction to concepts and intuitions rather than to the object represented. But to use a concept *in concreto* as the grounds of cognition of an object, and hence to employ intuition *in concreto* in its representation, has implications for the object represented, namely, that the object is an

[36] See also §6 of the *Prolegomena*, where Kant takes up the question of how pure mathematics is possible. In §7 he claims that it must be grounded in pure intuition, "in which it can present, or as one calls it, *construct* all its concepts *in concreto* yet *a priori*" (4:281).

individual, singular object. We can thus, in a borrowed sense, apply the term "concrete" to the object of intuitive *in concreto* representing. To be the object of an *in concreto* representing is to be something concrete in the sense of being singular. Note that there is an important asymmetry here with concepts. As noted, it is the nature of concepts to represent by means of marks that can be shared by one thing; if they are used *in concreto*, they are used to cognize a singular object. Intuitive representing, however, is always an *in concreto* representing.

Extending the notion of concreteness from the mode of intuitive representing to the object represented allows us to compare and contrast Kant's understanding of the concrete/abstract distinction to current treatments of it. Our contemporary account of the distinction between abstract and concrete objects is unsettled and there are interesting difficult cases, but the articulation of the distinction usually focuses on two features of objects: spatiotemporal particularity and causal interaction.[37] Paradigm examples of concrete objects are spatiotemporal particulars that causally interact with other spatiotemporal particulars. Particular physical objects such as my desk are paradigm examples of concreta. Paradigm examples of abstracta are not undisputed, but sets, numbers, propositions, types (such as the letters of the alphabet), concepts, and universals have all been candidate abstract objects, that is, objects without spatial and temporal location that are not causally efficacious.[38] There are interesting in-between cases concerning objects that lack one or more of these properties without lacking all of them. An absence of spatiality and causal efficacy is often regarded as sufficient for abstractness, for example. But there are also concerns about how to articulate what the casual efficacy condition requires – whether abstract objects can be effects of concreta, even if not causes of concreta or changes in properties of them, for example. And there are also concerns about how to spell out what it means not to be spatial. Is it sufficient for it not to occupy a determinate volume or particular location? What do we mean by a determinate volume or a particular location? Most accounts would appeal to the way in which paradigm examples of concreta, such as my desk, have both a determinate volume and a particular location: a whole concrete object is not instantiated in – that is, it does not as a whole occupy – more than one location.[39] And in that sense, it is singular.

In Kant's account, the objects of intuitive *in concreto* representation share something in common with our contemporary views of concreta: they are

[37] In bringing Kant's views in relation to contemporary accounts, I follow custom by shifting back to using the term "particular" to mean a singular or an individual.

[38] For a helpful overview, see Rosen (2017).

[39] This does not rule out that *parts* of a concrete object might be distributed over distinct and separate spatial locations.

singular. Moreover, because the form of our intuitive representing is spatial and temporal, they are spatial and temporal, or at least temporal, singulars.

What is important to note, however, is that in Kant's account, this is all that it would mean to be concrete. In particular, standing in causal relations is not required. Thus, a triangle represented in pure intuition in the course of a Euclidean proof is concrete in the sense that it is the result of an *in concreto* representing in intuition, despite its lack of causal interaction with the physical world. Furthermore, the objects of empirical intuition are also concrete, but only in virtue of being singular objects or individuals. While standing in causal relations with other such objects is a crucial additional condition of being an object of experience, those causal relations do not contribute to making them concrete, nor do they make them more concrete than geometrical objects. They are already fully concrete in being objects represented singularly in intuition. Or to put the point more carefully, the concreteness of objects does not come in degrees in Kant's view. Something is either singular and concrete, or it is not, and both objects of experience and geometrical objects are fully concrete.

There is a further use of the concrete/abstract distinction in current metaphysical debates that bears mentioning. In some accounts, a property is a more abstract entity than the thing or substance in which it inheres; moreover, it is sometimes held that a relational property is more abstract than a monadic property. This way of understanding the distinction allows for something to be both abstract and singular, if a property is identified with its instantiation; that is, it allows for what are called "abstract particulars." For example, the whiteness of this piece of paper to my left and the whiteness of that other piece of paper to my right could be regarded as distinct abstract particulars. Note that in the Kantian sense of concreteness extended to the objects represented, concreteness simply consists in singularity. Thus, in Kantian terms, an "abstract particular" or an "abstract singular" would be an oxymoron. This is not to say that he rules out tropes, only that he would not consider them to be abstract in his sense of abstract. If Kant were to countenance the individuating of properties by their particular instantiations, they would count as concrete, not as abstract.

5.6.3 *Singularity, Representing* in concreto, *and Mathematical Objects*

I have argued for a view according to which the singularity of intuition is to be understood as a mode of representing singularly, a mode that makes it possible both to represent an indeterminate homogeneous manifold and to represent individual objects in intuition, a view that moreover grounds the representation of the latter in the former. Now that we have shifted our attention to the representation of individual objects in intuition and clarified the sense in which they are concrete, we should briefly address a further feature of Kant's view. There is an apparent difference between the singularity of mathematical objects and of objects of experience, for the former are less determinate and are in an important sense general in a way that objects of experience are not. I say "briefly," because this raises issues that we will not be able to fully address here – in particular, the relationship between the necessary role of individual intuitions in justifying general mathematical claims. It also draws us into questions about the individuation of both mathematical objects and objects of experience. A thorough treatment of these issues would deflect us too far from the primary goal of this book, and my present aim is only to further elucidate the sense in which both mathematical objects and objects of experience are singular.

To start, it is helpful to first revisit the discussion of the conceptual representation of individuals in Section 5.3 above. One might have thought that to be an individual object requires that it be fully determinate, in the sense that an individual object must be determinate with respect to every possible predicate or property expressed by a concept, so that, with respect to every (nonvague) property P, the individual either is P or is not P. Kant does in fact hold this principle,[40] but only as articulating an idea of a thoroughly determined thing, an idea for which we can give no corresponding object. As we saw in Sections 5.3.2 and 5.3.3, it is the nature of concepts to represent generally, and the possible determination of concepts by further possible concepts never ends, a claim expressed in Kant's doctrine that there are no *infimae species*. Kant's discussion of the principle of thoroughgoing determination makes it clear that the notion of a thing determined by every pair of possible predicates is a mere idea of reason, and cannot be exhibited *in concreto*:

> Thoroughgoing determination is consequently a concept that we can never exhibit *in concreto* in its totality, and thus it is grounded on an idea which has its seat solely in reason . . . (A573/B601)

Thus, the representation of an individual by means of a thoroughgoing conceptual determination is not possible for us. In the *Jäsche Logik*, however, Kant distinguishes a determination through concepts, which in this context he

[40] As Kant formulates it: "Every **thing**, . . . as to its possibility . . . stands under the principle of thoroughgoing determination, according to which, among **all possible** predicates of **things**, insofar as they are compared with their opposites, one must apply to it" (A571–2/B599–600). See Section 5.3.2 above for a closely related passage. Recall that Kant uses "predicate" to refer both to that which expresses or refers to a property and to the property itself. Vagueness was not a concern for Kant and his contemporaries; setting it aside assumes that each genuine property and corresponding predicate is not vague.

calls "logical determination," from the representation of a thoroughly determinate cognition as an intuition:

> The highest, completed determination would yield a thoroughly determinate concept (*conceptus omnimode determinatus*), i.e., one to which no further determination might be added in thought.
>
> Note. Since only individual things or individuals [*einzelne Dinge oder Individuen*] are thoroughly determinate, there can be thoroughly determinate cognitions only as intuitions, but not as concepts; in regard to the latter, logical determination can never be regarded as completed (§11, Note). (9:99)

Intuition is what allows us to have thoroughly determinate cognitions, and hence cognition of individuals. Kant ends his note with a reference back to his note to §11 of the *Jäsche Logik*, which concerns the doctrine of no *infimae species*. After articulating that doctrine, he adds:

> For even if we have a concept that we apply immediately to individuals, there can still be specific differences in regard to it, which either we do not note, or which we disregard. Only comparatively for us are there lowest concepts, which have attained this significance, as it were, through convention, insofar as one has agreed not to go deeper here. (9:97)

This is further support of the point that concepts on their own do not allow us to represent individuals. What is of further significance here is its relation to determination through concepts. Although we take the individual to be governed by the principle that one of every opposed pair of possible predicates must apply to it, our representation of an individual is in an important sense always incomplete: our representation of them does not include a specification of every property an object has. There are two important points to note about this. First, the incompleteness of the logical determination of a concept is hardly surprising given the finitude of human cognition, but it underscores the fact that in Kant's account of cognition, intuition plays the role of allowing us to represent individuals *in concreto* despite an incompleteness in our concepts of them. Neither mathematical objects nor objects of experience are represented in a fully determinate way through concepts. Relative to the conceptual determination of its properties, my cognition of the Norway maple in front of my window is also general; in this respect, the difference between my cognition of it and my cognition of a triangle is only the many further, yet finite number of conceptual determinations I employ in my cognition of the maple. Intuition is not only necessary to represent an individual; it is also sufficient in the case of both objects of experience and mathematical objects, and in both cases, the objects are singular in Kant's sense of singularity.

The second noteworthy point is that there is a connection between the notion of concreteness applied to concepts and the *in concreto* representation of individuals made possible by intuition. The fact that there are no *infimae species* means that the logical determination of concepts can lead to more and more relatively concrete concepts, but on its own this will never lead to the representation of a concrete individual. In other words, there is no absolutely concrete conceptual representation; that would be tantamount to representing *in concreto*, which only intuition can do.

In Section 5.4.3, I discussed Kant's claim that intuition plays the role of substratum for conceptual determination, and suggested that what Kant has in mind is that the form of intuition represents singularly, which makes it possible to represent a uniform indeterminate manifold singularly as well as singular objects in that manifold. In light of its role of in the representation of individuals we've been discussing here, intuition can be described as a "principle of individualization"; that is, in the case of spatial intuition it is what makes possible the representation of unique parts of space and individual spatial objects.[41]

The above analysis of representing *in concreto* by means of intuition in the context of incomplete conceptual determination makes room for the thought that the representation of a particular *quantum*, a particular homogeneous manifold in intuition, is as singular and concrete an individual as any object of experience; the representation of a triangle, for example, is an individual represented *in concreto*, despite the fact that there are many more properties with respect to which it is indeterminate than a Norwegian maple.

On the other hand, this does not mean we can employ intuition to represent an individual *in concreto* corresponding to any arbitrarily chosen concept while omitting any further conceptual determinations we wish; the representation of an individual object possessing some particular properties often commits one to representing further properties. I can think of color in general, but I cannot represent an individual object as colored without representing it as having some color or other. Significantly, I also cannot represent a triangle *in concreto* without thereby representing it as either equilateral, isosceles, or scalene, and also representing it as either obtuse or acute, as well as in a particular orientation. As is well known, Berkeley appealed to such facts to argue against Locke's abstract general idea of a triangle, arguing that all ideas are particular, and Kant formulates a response that attempts to account for

[41] To identify intuition as such a principle, however, is *not* to say that the representation is possible without the employment of concepts, and in particular the categories of quantity, or with some involvement of the intellect. That would be a further claim that would need to be defended. Similarly, the negation of this further claim would need to be defended, that is, the claim that intuition cannot represent unique parts of space and individual spatial objects without the employment of concepts or involvement of the intellect. Thus, the claim I am making cuts across the current Kantian version of the conceptualist/ nonconceptualist debate, although this role of intuition as a principle of individualization would need to be taken into account in resolving this debate.

both the singularity of the intuitive representation and the generality of the conclusions that can be drawn from it (A713–14/B741–2).[42] I will postpone a thorough discussion of this issue for a more detailed consideration of Kant's philosophy of mathematics. For now, the point is simply that the representation of an individual comes with constraints on what conceptual determinations must be employed. That should not, however, obscure the fact that it is intuition that makes it possible to represent an individual at all. Most importantly, my representation of a triangle in intuition is every bit as much an individual object represented *in concreto* and every bit as singular as any object of experience, even if my cognition of the latter involves many more conceptual determinations.

5.7 The Concreteness and Abstractness of *Quanta* and *Quantitas*

We are now in a position to make some clarifications about Kant's notions of a magnitude, and more specifically the notions of *quantum* and *quantitas*. As we saw in Section 3.4, Kant explicates the concept of magnitude (*quantum*) as "the consciousness of a homogeneous manifold in intuition, insofar as it makes the representation of an object first possible" (B202). Note first that this is a concept of a manifold in intuition. That which the concept represents is thus the concept of a concretum. As I was at pains to emphasize in Chapter 3, all that Kant's explication of the concept of *quanta* requires is that a *quantum* is a homogeneous manifold in intuition; any homogeneous manifold in intuition counts as a *quantum*. Thus, the concept itself is relatively indeterminate; that is, it is general and broad, because it includes relatively few determinations or marks. I noted earlier that the term "abstract" is sometimes applied to concepts to describe a concept that is general in the sense of including relatively few marks. If we used the term in this way, then the concept of a *quantum* is a relatively abstract concept, that is, includes relatively few marks, while its object is concrete, that is, singular.

What about *quantitas*? As I described in Section 3.5, *quantitas* is, at least in the first instance, the quantitative determination of a *quantum*, that is, a determination of the size of a *quantum* relative to a unit through a representation of a plurality of these units as a totality. It is a more involved and hence more determinate concept than that of a *quantum*. If so, it would count as a less abstract concept than the concept of a *quantum*. What can we say about that which the concept represents? Is it concrete or abstract? That depends on whether the quantitative determination of a specific *quantum* is individuated as such. If the eleven-inch length of this piece of paper is regarded as distinct

[42] Sensitivity to the relation between what are today called "determinables" and "determinates" is arguably already found in Aristotle. See Wilson (2017) for an overview of the topic.

from (although equal to) the eleven-inch length of that other piece of paper, then each of them is singular and hence concrete. Recall my point above in Section 5.6.2, that in Kant's terminology, an abstract particular or an abstract singular would be an oxymoron, which is not to say that he rejects tropes, only that if he did countenance tropes, they would count as concrete in his sense. If, however, the *quantitas* is the property of being eleven inches long, a property shared by both pieces of paper (and quite a few others), then what *quantitas* represents will not be concrete in the Kantian sense of concreteness, when that notion is extended to represented objects. I will have to postpone fully addressing this issue, which will require a more in-depth exploration of *quantitas* than I can give here. For now, it is sufficient to note that the concept of a *quantum* is less determinate and in that sense more abstract than the concept of a *quantitas*. At the same time, that which is represented by means of the concept of a *quantum* is something fully concrete, and whether that which is represented by means of the concept *quantitas* is concrete or not is left for further investigation.[43]

I have focused on the singularity of intuition in order to explain what singularity consists in, which I have described as "representing singularly," an understanding of singularity that is compatible with the direct representation of indeterminate manifolds. The singularity of intuition plays the role of allowing us to represent and cognize individual objects, but the nature of the singularity of intuition does not consist in playing this role. Moreover, I have argued that the role of intuition in representing indeterminate manifolds as manifolds underlies its role in representing individual objects, both geometrical objects and objects of experience. I would like to emphasize again at this point that Kant's theory of magnitudes and the role of intuition in the representation of magnitudes must be accounted for in any adequate interpretation of Kant's theoretical philosophy. Thus, whether or not one agrees with all the claims of this chapter, some account that accommodates and incorporates Kant's theory of magnitudes will be required.

I have also explained Kant's employment of the concrete/abstract distinction, how we can extend Kant's concrete/abstract distinction to objects represented, and the role of the singularity of intuition in representing objects *in concreto*. Extending the notion of concreteness to the objects represented allows us to see that in Kant's view, mathematical objects are no less concrete

[43] I have been explicating Kant's understanding of the concrete/abstract distinction and relating it to contemporary accounts. We are, of course, entitled to our own understanding of the concrete/abstract distinction in formulating what we take to be important philosophical questions, for example, the sense in which Kant's understanding of number is concrete or abstract. A closer investigation of Kant's understanding of number in future work will take into account the understanding of concrete and abstract I've described here and help us relate Kant's views to contemporary questions.

than objects of experience. The analysis of Kant's understanding of abstract and concrete also allows us to explain the sense in which *quanta* are concrete, and what it would mean for *quantitas* to be concrete or abstract.

This exploration of the role of intuition in the representation of magnitudes is not exhaustive. In fact, we have not yet addressed one of the most important features of intuitive representation – its ability to represent a manifold that is homogeneous. To understand Kant's views of homogeneity and to deepen our understanding of his theory of magnitudes, however, it is necessary to first back up and consider why Kant might think that mathematical cognition rests on the cognition of magnitudes in the first place. For there was a very long history of thinking of mathematics as concerning magnitudes that shaped Kant's views. The theory of magnitudes found in Euclid and the Euclidean tradition will be the topic of the next chapter. With that background in place, we will then turn to Kant's adoption, and adaptation, of that theory.

INTERLUDE

The Greek Mathematical Tradition as Background to Kant

6

Euclid, the Euclidean Mathematical Tradition, and the Theory of Magnitudes

6.1 Introduction

What follows is a highly selective summary of a few features of Euclid's *Elements* and the history of Euclidean geometry relevant to understanding Kant's views.[1] Some of those features are about the *Elements* as a whole, but the most important concern the Eudoxian theory of proportions it contains and the understanding of magnitude and number on which that theory rests. The encapsulation given here is intended for those not already familiar with Euclidean geometry and the theory of proportions, or those who would appreciate a refresher; it will also serve as a reference for later chapters. Developments in mathematics from the early renaissance through the early modern period challenged this Greek mathematical tradition, and the Greek theory of magnitudes underwent changes. Despite these changes and further transformations and rapid developments in the field of mathematics in the seventeenth and eighteenth centuries, however, its influence persisted. It set both the background and a basic framework for thinking about the nature of mathematics, particularly with respect to standards of rigor and certainty, and more broadly with regard to what mathematics is about and the nature of mathematical knowledge. It also deeply influenced Kant's attempt to explain the foundations of mathematical cognition and its relation to our cognition of experience.

As I touched on in Chapter 1, the Euclidean tradition that lay in the background of eighteenth-century theorizing about mathematics differs deeply from our modern understanding after the "arithmetization" of mathematics in the nineteenth century. Broadly speaking, we now think of the foundation of all mathematics as beginning with the natural numbers (and ultimately set theory and logic). Fractions and real numbers are in turn founded on the naturals. Magnitudes are generally thought of as those continuous properties of things to which real numbers are subsequently applied, or as the real

[1] I am indebted throughout this chapter especially to Heath's 1956 translation and Mueller (1981); the latter has influenced my understanding of Euclid more than any other. I am also grateful for conversations about Euclid's theory of proportions with Mueller, and for frequent discussions and correspondence with Vincenzo De Risi over the years.

numbers that result from their application in the measure of those properties. In contrast, the Euclidean tradition passed down to the eighteenth century viewed mathematics as a science of magnitudes and their measure. From our modern point of view, this description of mathematics might sound as if it would be more accurately understood to be a matter of applied mathematics that follows pure mathematics, and perhaps an anticipatory description of the ultimate aim of mathematics. In fact, however, pure mathematics was itself viewed as a science of magnitudes and their mathematical relations. The brief overview of Euclid and the Euclidean tradition offered in this chapter aims to draw out this very different way of thinking about mathematics.

The *Elements* comprises thirteen books, beginning in Book I with some definitions and foundational principles, and culminating in Book XIII with the construction of the Platonic solids, that is, the five regular convex polyhedra: the tetrahedron, cube, octahedron, dodecahedron, and icosahedron. In each case, Euclid both describes how to construct the solid and determines the ratio of the length of an edge to the diameter of the circumscribed sphere – an impressive achievement. The *Elements*, however, does not seem to be solely directed at achieving these results, establishing other results along the way and surveying fundamental geometry at the time it was completed, c. 300 BCE, some twenty years after Aristotle's death. It is a compilation and synthesis of the work of various Greek geometers, and was probably the fourth such work written by Greek geometers since the second half of the fifth century BCE. The extent to which Euclid modified his sources is unknown, but an analysis of the *Elements* provides plenty of evidence that Euclid did not iron out all the differences between them and only imperfectly unified them into a whole. This feature of his work raises various puzzles about how different parts of the *Elements* fit together. Those puzzles are directly bound up with an important theme in the Greek mathematical tradition right through to eighteenth-century mathematics: whether there is a unified account of mathematics that governs all the parts of Euclid's *Elements*, and whether some parts are properly considered as prior to other parts. A primary aim of this chapter and of this book as a whole is to draw out this theme, for it plays an important role in the eighteenth-century understanding of the foundations of mathematics and is key to interpreting Kant's views. It also lays the groundwork for a reappraisal of Euclidean geometry and of Kant's philosophy of mathematics in Chapter 9, which argues that a unifying foundation rests on a general theory of measurement. The next two sections will describe the parts of the *Elements* and their relations in order to flesh out this theme.

6.2 Organization of the *Elements*

Books I and II cover rectilineal geometry in a plane, starting with the construction and properties of triangles, and proceeds to constructions involving parallel lines and parallelograms, and squares in particular. There is an emphasis throughout the *Elements* on constructing figures of equal area to another, for example, a parallelogram equal in area to a given triangle. Book I famously culminates with the Pythagorean theorem, but as for the *Elements* as a whole, Book I sometimes establishes propositions that are not needed for the proof of this theorem. Book II further extends rectilineal plane geometry to constructing figures meeting various conditions, and establishes equalities covering the addition and subtraction of squares and rectangles. The fact that these equalities can be expressed in algebraic form led some later historians of geometry in the nineteenth century to describe this work as "geometrical algebra," in explicit contrast to algebraic geometry, since Book II articulates these equalities entirely in geometrical terms, instead of using algebraic notation to express geometrical relations. Books III and IV expand plane geometry to circles and regular polygons, respectively. Up to this point, Euclid develops plane geometry as far as he reasonably can using the relations of equality and congruence, without appeal to the relation of similarity between rectilinear figures or to proportionalities. At this point the *Elements* takes a turn.

Book V presents the theory of ratios and proportions among magnitudes, which allows him, in Book VI, to define similarity in terms of proportion and to further develop plane geometry using the theory of proportions. As Ian Mueller has emphasized, Euclid appears to avoid the theory of proportions as long as he can and sometimes even seems reluctant to employ it after its introduction.[2] The theory of proportions presented in Book V almost certainly derives from a work by Eudoxus, and the seemingly conflicted manner in which Euclid uses the theory may in part result from how he incorporated it into the *Elements*. The theory of proportions among magnitudes played a central role in the subsequent history of mathematics.

Books VII–IX take another turn, this time to the theory of number. Book VII defines number as a multitude composed of units and covers their elementary properties (such as being even and odd, prime and composite), and the relations among them, including propositions governing their ratios and proportions. Together, these so-called arithmetical books make it possible, in Book X, for Euclid to employ numbers in describing geometrical ratios and to distinguish between two sorts of incommensurable ratios. This is in turn required for determining the ratio between the edge of an icosahedron and the diameter of a sphere circumscribing it in Book XIII, where the Platonic solids are constructed. The arithmetical books most likely derive from a work by Theatetus, and are developed independently of what precedes them. This is true in particular of the laws of proportions governing numbers, even when those laws mirror those for proportions among magnitudes found in Book

[2] See Mueller (1981, pp. ix–x).

V. This raises a question about how to understand the relation between magnitudes and numbers in the Euclidean sense, with reverberations that persisted into the eighteenth century.

With the arithmetical books and their employment in geometry in place, Book XI begins solid geometry, while Book XII introduces the method of exhaustion to establish proportionalities among geometrical volumes, areas, and lengths. Book XIII, as noted, constructs the Platonic solids, and determines the ratios between a side of these solids and the diameter of a circumscribing sphere.

What is most important for our present purposes is that the *Elements* contains distinct parts treating not just plane and solid geometry, but also the Eudoxian theory of proportions in Book V and the Theatetus' elementary theory of number in Book VII, and the relations among these parts were a source of debate throughout the long tradition of Euclidean geometry that followed.

6.3 The Deductive Structure of the *Elements*

The oldest copies of the *Elements* date from centuries after it first appeared, and since that time, there have been debates about which and how many definitions and fundamental principles the original contained. It is thought that Book I began with about twenty-three definitions. Most of the other books begin with further definitions. I list a selection of the definitions of Book I to give a sense of those first definitions and allow for later reference:[3]

I.D1: A point is that which has no part.

I.D2: A line is a breadthless length.

I.D3: The extremities of a line are points.

I.D4: A straight line is one which lies evenly with the points on itself.

...

I.D10: When a straight line set up on a straight line makes the adjacent angles equal to one another, each of the equal angles is right, and the straight line standing on the other is called a perpendicular to that on which it stands.

...

I.D14: A figure is that which is contained by some boundary or some boundaries.

[3] Here and throughout, I follow Ian Mueller's translation of Euclid's definitions, postulates, common notions, and propositions; see Mueller (1981, p. 317ff.). I will indicate definitions and propositions with a Roman numeral indicating the Book followed by the number of the definition or proposition as listed in Mueller. I will not include the Book number for postulates and common notions, which were found only in Book I. Note that these numbers sometimes differ from Heath, a reflection of disagreements about which definitions, postulates, common notions and propositions properly belong to the *Elements*.

I.D15: A circle is a plane figure contained by one line such that all the straight lines falling upon it from one point of those lying inside the figure are equal to one another.

. . .

I.D20: Of trilateral figures, an equilateral triangle is that which has its three sides equal, an isosceles triangle that which has only two sides equal, a scalene triangle that which has its three sides unequal.

The definitions of Book I were followed by a list of fundamental principles. There has also been considerable debate about which and how many of these were contained in the original *Elements*, but a rough consensus of about ten emerged, with some arguing that there were as few as eight. The principles were divided into two sets: postulates and common notions, the latter often called axioms (a division found already in Aristotle). The postulates and common notions were:

P1: Let it be postulated to draw a straight line from any point to any point,
P2: and to produce a limited straight line in a straight line,
P3: and to describe a circle with any center and distance,
P4: and that all right angles are equal to one another,
P5: and that, if one straight line falling on two straight lines makes the interior angles in the same direction less than two right angles, the two straight lines, if produced *ad infinitum*, meet one another in that direction in which the angles less than two right angles are.
CN1: Things equal to the same thing are also equal to one another.
CN2: And if equals are added to equals the wholes are equal.
CN3: And if equals are subtracted from equals the remainders are equal.
CN4: And things which coincide with each other are equal to each other.
CN5: And the whole is greater than the part.[4]

Unlike definitions, which appear at the beginning of almost all the books of the *Elements*, these are the only explicitly stated principles for the whole of the *Elements*. Commentators debated the basis for the division between the two sets of principles. One interpretation dominated, but a minority interpretation that did not prevail is worth discussing first, since it emphasizes a feature of the *Elements* that had a lasting influence on geometry.

Some argued that the distinction between postulates and common notions corresponded to a distinction between those principles asserting that something could be done or found, that is, constructed, and those asserting some

[4] Note that the Euclidean sense of "part" is of a proper part. In all that follows, I will take "part" to mean "proper part." Books V and VII employ a quite different notion of part, as we shall see in Section 6.4, and as I will discuss in detail in Chapter 9.

truth. On this understanding, the last two postulates, P4 and P5, were classified (or reclassified) as common notions. Even for those who rejected this way of distinguishing postulates and common notions, the first three postulates were referred to as "construction postulates."

The distinction between, on the one hand, an assertion of something that can be done or found, and on the other, some truth that holds, runs through the *Elements*; the propositions proved from the fundamental principles were divided into "problems" that asserted the possibility of a construction, for example, to construct an equilateral triangle on a given straight line; and "theorems," for example, that two sides of any triangle are greater than the remaining side. But even the proofs of theorems depended on constructions, so that the *Elements* is at its very core "constructionist," that is, constrained by what could be constructed in accordance with the construction postulates. When considered from the perspective of the tools of a geometer, it is apparent that the construction postulates correspond to the use of a straight-edge without a ruler, and a collapsible compass, that is, a compass that maintained its shape during construction, but collapsed when lifted from the plane. That so much can be established from this modest constructionist starting point is astounding.

Commentators analyzed the structure of Euclidean proofs and distinguished those parts of a proof that involved constructions. Construction was at the heart of the understanding of geometry through the eighteenth century and was replaced by a "static" understanding of geometry only at the end of the nineteenth century. The role of construction in Euclidean geometry is worth emphasizing, because it clearly influenced Kant's view that mathematical cognition rests on the construction of concepts. It also influenced "constructivist" theories of mathematics in the nineteenth and early twentieth centuries and their references to Kant's philosophy of mathematics; it was also central to a debate about the role of intuition in the construction of mathematical concepts in the resurgence of interest in Kant's philosophy of mathematics beginning in the late 1960s.

Although the importance of construction in the *Elements* inspired an interpretation of the distinction between postulates and common notions corresponding to it, the more widely accepted understanding viewed the distinction as corresponding to those principles particular to a special science (namely, geometry) and those principles that concern any mathematical science whatsoever, that is, those that are "common to all investigation which is concerned with quantity and magnitude."[5] This distinction is also

[5] The quote is Proclus' way of putting the point. See Heath's commentary in Euclid (1956, vol. 1, pl. 123). Note that CN4 is an exception, since coincidence is a relation that specifically applies to continuous spatial magnitudes. Nevertheless, the notion of *equality* employed in CN4 could still be regarded as a quite general relation that would apply to any

fundamental to the *Elements*, and suggested the possibility of a science of mathematics more general than geometry, which reinforced the idea that there is a universal mathematics common to continuous spatial magnitudes and numbers.

The *Elements* only imperfectly satisfies the mathematical, scientific, and philosophical ideals it inspired. Euclid's conception of foundations and standards of rigor appears to have been significantly looser than those of later geometers. To begin with, his definitions often play little or no role in the proofs; I.D2, the definition of a line as a breadthless length, is an example. The definitions often seem to be ostensive or clarifications of what he assumes we already sufficiently understand. Nor are his definitions complete or systematic; he provides a quite clear definition of similarity in terms of proportions. On the other hand, the relation of equality is assumed in the earliest definitions and is nowhere defined, as is evident in the sample of definitions given above, although he does provide common notions governing equality that contribute to implicitly defining it. In contrast, congruence is neither defined nor are there common notions governing it, despite its fundamental importance to his proofs.

Euclid also assumes without definition some notions that he seems to think are familiar enough to employ either without or prior to explanation. In the early books, for example, Euclid helps himself to the notion of number as a collection of units and the properties of such collections, but only articulates these notions once a fuller theory of their properties is needed in Book VII. This further complicates the relation between the different parts of the *Elements*, an issue to which we will return in Chapter 9. A similar point applies to his principles; Euclid did not seem to share the idea that rigor required that every proof indicate all the principles on which it depends. He sometimes implicitly relies on a common notion without citing it, despite having articulated it. More importantly, he also does not articulate all the principles on which he implicitly relies. For example, the first three postulates, the construction postulates, are sufficient for plane geometry, but Euclid's solid geometry assumes the possibility of further, more complex constructions; it assumes, for example, that one can construct a sphere by rotating a semicircle around its diameter and that one can construct a greatest circle by passing a plane through the center of a sphere.[6] Even in plane geometry, Euclid makes a crucial assumption in the proof of I.4; he helps himself to the notion of applying one figure to another, that is, rigidly moving one figure onto another so that their parts coincide. Appeal to such

continuous magnitudes as well as collections. Some thought that CN4 is a later interpolation and not one of Euclid's common notions, perhaps partly because it isn't fully general.

[6] Mueller (1981, p. 29).

motion was viewed by later geometers as problematic in its own right; it also seems to undercut the need for the clever but involved construction for replicating a line segment of a given length at a different location, which he established in I.2. More generally, Euclid appears to adopt an inconsistent standard of rigor in his proofs, while increasing standards of rigor in the Euclidean tradition that followed him required that more and more assumptions be articulated.

Despite its defects, however, Euclid's *Elements* is a remarkable intellectual achievement that employed a modest foundation to ground an impressive body of knowledge, and inspired ideals of clarity, rigor, and certainty that reverberated beyond mathematics into our understanding of human knowledge in general. Moreover, the combination of its achievements and its faults stimulated a rich research program in mathematics that lasted nearly two millennia. The next two sections will describe Euclid's theory of proportions for continuous spatial magnitudes and his account of number and their proportions in more detail.

6.4 Magnitude and the Euclidean Theory of Proportions

As noted in Section 6.2, Euclid establishes as much of plane geometry as he can in Books I–IV by appeal to the relations of equality and congruence; only then does he insert the Euclidean theory of proportions of Book V, which allows him to define similarity in terms of proportions and apply the theory of proportions in Book IV. The Euclidean theory of proportions proves propositions governing the ratios and proportions among magnitudes. Book V begins with eighteen definitions, but Euclid does not define "magnitude," assuming that this notion is understood and employing it to define other terms. It is clear that in Euclid's view, magnitudes include lines, areas, and volumes, as well as angles; the notion of magnitude is intended as a general term covering different kinds of continuous spatial magnitudes and Book V is a general treatment of them.

As also noted in Section 6.2, the theory of proportions found in Book V derives from Eudoxus, and is commonly referred to as the Eudoxian theory of proportions. More carefully, however, it should be called the Euclidean-Eudoxian theory, because we do not have Eudoxus' original version and Euclid altered it when he incorporated it into the *Elements*. Eudoxus preceded Aristotle and was a teacher at Plato's academy when Aristotle was there. Aristotle states in the Posterior Analytics that at one point an important law of proportions had been proved separately for numbers, lines, solids, and times, but that they were no longer treated separately and that the proposition was subsequently proved universally. Aristotle appears to be referring to the innovative work of Eudoxus, with whose work he would have been acquainted. Aristotle adds that numbers, lines, solids, and times were considered "qua having a particular character which they were assumed to have universally."[7] Aristotle's examples of kinds of things treated in a theory of proportion do not explicitly include areas or angles, though they are most likely assumed. Note that his examples also include times, which are not spatial. Finally, they also include numbers. As I will discuss in the next section, the Greek conception of number is of collections of discrete units corresponding to the whole numbers; numbers are not by their nature spatial; nor are they continuous, but discrete. Thus, Aristotle seems to have had a broad conception of the scope of the theory of proportions.

Throughout the Greek mathematical tradition, there is a question of what sorts of things are treated as magnitudes, and what the term "magnitude" was explicitly used to cover. Despite Aristotle's reference to different kinds of things subject to the universal propositions concerning proportions, he does not use "magnitude" (*megethos*) or any other term to describe the genus to which his examples belong. It is possible that no term had yet been coined, or that he preferred to reserve the term for spatial magnitudes in particular.[8] Regardless, Euclid appears to use the term "magnitude" for continuous spatial magnitudes alone.

Let us look more closely at the definitions that open Book V. Definition 5, which defines sameness of ratios among magnitudes, is the most important; I will consider it separately in the next section. I list the first six definitions, omitting Definition 5 until next section.

V.D1: A magnitude is part of a magnitude, the less of the greater, when it measures the greater.

V.D2: The greater is a multiple of the less when it is measured by the less.

V.D3: A ratio is a kind of relation with respect to size between two homogeneous magnitudes.

V.D4: Magnitudes which, when multiplied, can exceed one another are said to have a ratio to one another.

V.D5: [Definition of sameness of ratio]

V.D6: Let magnitudes having the same ratio be called proportional.

The notion of part invoked in V.D1 is not that of the generic part–whole relation corresponding to CN5; it is a special part–whole relation commonly called "aliquot part," for which the parts are even multiples of the whole. We will consider it more closely in Chapter 9, along with the notion of multiple and the notion of measure he assumes in defining them. For now, our focus is on magnitude. What is important to note is that Euclid does not define

[7] *Posterior Analytics* A.5.74a17–25. Quoted in Mueller (1970b), on which this and the following paragraph rely.

[8] Mueller (1970b, p. 1).

"magnitude," and that magnitudes can stand in some sort of part–whole relation and in the relations of greater-than and less-than.

The characterization of ratio in V.D3 assumes not just the notion of magnitude but of homogeneous magnitude. Euclid does not define homogeneity either, but it is clear he has in mind a division of spatial continuous magnitudes into their kinds – lines, areas, volumes, and angles. Magnitudes of the same kind are homogeneous, reflecting the fact that only magnitudes of the same kind can stand in ratios with each other; that is, a line can stand in a relation with respect to size to another line, an area to an area, and a volume to a volume, and an angle to angle; in contrast, a line and an area cannot stand in a relation with respect to size. Greater-than and less-than are also relations with respect to size, and only magnitudes of the same kind stand in these relations; one area can be greater in size than another area, for example, but no sense can be attached to an area being either greater or less in size than a volume.

Since Euclid does not explicitly define magnitude, it might be tempting to take V.D1 through V.D3 to implicitly define magnitude. In an important sense they do, for we can surmise what Euclid had in mind only by looking at how he employs the notion in the definitions and subsequent propositions. It is nevertheless important to note that Euclid assumes that the notions of magnitude and homogeneity are sufficiently well understood to use them to characterize other notions, such as ratio in V.D3.

While V.D3 characterizes a ratio as a kind of relation with respect to size, V.D4 states a condition on two magnitudes having a ratio; two lines are homogeneous, for example, but for them to have a ratio, each must be able to be multiplied to exceed the other. This property identified in V.D4 is commonly referred to as the Archimedean property, and it is usually interpreted as a condition on being a nonzero element of an ordered field; that is, it is used to rule out elements that are infinitely small or infinitely large in relation to each other. This condition is usually interpreted as an additional condition on being a magnitude; that is, Euclid's V.D4 is often interpreted as part of an implicit definition of either homogeneous magnitudes or magnitude itself.[9] Note, however, that V.D4 does not state that magnitudes that do not have this property are not magnitudes, or that they are not homogeneous; it states only that two magnitudes that have this property relative to each other have a ratio. Following immediately on V.D3, it is implicitly a claim about homogeneous magnitudes. While being homogeneous is a necessary condition of two magnitudes having a ratio, meeting the Archimedean condition is a

[9] I am indebted to Marco Panza for discussion on this point; he endorses the view that V.D4 makes the Archimedean property a condition on being a homogeneous magnitude. See also Mueller (1970b, p. 2); Mueller incorporates the Archimedean property into the meaning of homogeneity.

sufficient condition of two magnitudes having a ratio. Hence, those magnitudes that meet the Archimedean condition are subject to the propositions proved in Book V. For all the Archimedean definition requires, however, an infinitely small line could count as a magnitude homogeneous with a finite line, even if they would not stand in a ratio to each other.

The fact that the Archimedean property, as Euclid states it, is not a condition on homogeneity or on being a magnitude is an important point, because it concerns what is and is not essential to being a magnitude – in short, how much is packed into the concept of magnitude. That is important for understanding the relation between the Euclidean notion of homogeneity and Kant's own account of it. This is not to deny, however, that the definitions together do contribute to describing the sort of magnitudes in which Euclid is interested: magnitudes are the sorts of things that are capable of standing in ratios if they are homogeneous and satisfy the Archimedean property.

On the other hand, V.D4 seems to assume magnitudes are capable of being multiplied. That is, they are capable of being composed with other homogeneous magnitudes to constitute larger magnitudes homogeneous with them. More particularly, they are capable of being composed with other homogeneous magnitudes that are equal in size to it. Thus, in addition to standing in relations of greater-than and less-than, they also can stand in relations of equality. Moreover, the sort of composition at issue is, like the possibility of standing in a ratio, restricted to homogeneous magnitudes: lines can be composed to make larger lines, areas with areas, volumes with volumes, and angles with angles.[10] In summary, what is special about homogeneous magnitudes is that at least some of them can stand in ratios, that they can be composed to make more of the same kind of magnitude, and more particularly that they can be multiplied to make more of the same kind of magnitude. Moreover, whether two homogeneous magnitudes stand in a ratio is characterized by appeal to relative size relations of greater-than, less-than, and equality and by appeal to composition. These are fundamental features of homogeneous magnitudes.

[10] Angles are a special case that raise difficulties. (I would like to thank Stephen Menn for first calling my attention to the unusual status of angles.) For the other spatial magnitudes, more composition always leads to more of the same kind of magnitude. An angle, however, can increase until it returns to zero and increases again. I do not know whether this issue was addressed by ancient Greek geometers, but one might try to preserve the idea that composed magnitudes always increase in size by considering, for example, an angle of 375° to be strictly speaking distinct from and 360° larger than an angle of 15°. That entails that they are not equal, however, so one would have to postulate another way of capturing the sense in which they are equivalent.

6.5 The Definitions of Sameness of Ratio and Similarity

The Eudoxian definition of sameness of ratio is the heart of the theory of proportions. It states:

> **V.D5** Magnitudes are said to be in the same ratio, first to second and third to fourth, when equal multiples of the first and third at the same time exceed or at the same time are equal to or at the same time fall short of equal multiplies of the second and the fourth when compared to one another, each to each, whatever multiples are taken.

This is a mouthful. An anachronistic use of modern symbolic algebraic notation helps to articulate the condition expressed. This anachronism can lead to misunderstandings, an issue we will address in Chapter 9, but for now, it will serve us well. For any four magnitudes *a*, *b*, *c*, and *d* and any two positive integers *m* and *n*,

$$a : b = c : d \text{ iff for all } m, n,$$
$$ma > nb \ mc > nd,$$
$$ma = nb \rightarrow mc = nd,$$
$$ma < nb \rightarrow mc < nd.$$

In other words, one pair of magnitudes stands in the same ratio as another if the comparative size of the first pair is the same as the comparative size of the second pair under all equimultiple transformations.

This appears a convoluted way of defining sameness of ratio, but there is a good reason for it. As is well known, the Pythagoreans had believed that all relations of size between two magnitudes of the same kind could be expressed by a ratio of whole numbers; those whole numbers might themselves be large, 6785:7843 for example, but in principle, a unit could always be found, in this case 1/7843rd of the second magnitude, that would allow one to "measure" and express the ratio in whole numbers. It was subsequently discovered and proved that this was not so for all continuous spatial magnitudes, for example, for the ratio between the diagonal of a square and its side. Such ratios are incommensurable in the sense that no common unit of measure could be found that would measure both magnitudes, no matter how small that common unit. Since they cannot be expressed as a ratio of two whole numbers, they came to be called "irrational." This fact meant that ratios among continuous magnitudes included more ratios than those between numbers. It also required an account of when two ratios were the same that did not appeal to numbers to express the ratios.[11] V.D5 is a work-around to this problem.

[11] Recall from above that the Greek conception of number was of a collection of units. The multiplicities mentioned in the definition, *ma*, *nb*, *mc*, and *nd*, would count as numbers in the Greek sense. Numbers are therefore employed in this definition, but they are not employed to *directly* express the ratios *a*:*b* and *c*:*d*. We will further examine the significance of their indirect employment in Chapter 9.

It is easy to see that if two ratios are the same, the trio of conditionals that are included in the condition should hold for any m and n; after all, if two ratios are truly the same, they should behave in the same way under all transformations. It may be less obvious why it will always be the case that two ratios that differ will, for some m and n, violate at least one of the conditionals, so that, for example, the first term will be greater than the second after the two are multiplied m and n times, while the third will be less than the fourth after the same multiplication. It is helpful to consider a case to see how this works. The point of the definition is to hold for all ratios among continuous magnitudes that are irrational, but we will use ratios between whole numbers to make the example clearer. Consider the following case:

Let $a = 7$, $b = 10$, $c = 8$, $d = 11$. Clearly $7{:}10 \neq 8{:}11$, so that for some m and n, the condition should be violated; that is, at least one of the conditionals should fail to hold.

Note first that if $m, n = 1$, then $7 < 10$, and $8 < 11$, so that the condition is not violated.

Suppose we try $m = 3$ and $n = 2$. In that case, we'll have $ma = 3 \cdot 7 = 21$ and $nb = 2 \cdot 10 = 20$, while $mc = 3 \cdot 8 = 24$, and $nd = 2 \cdot 11 = 22$. Since $21 > 20$ and $24 > 22$, once again the condition will not be violated. In fact, for most arbitrarily chosen values of m and n, the condition won't be violated.

However, if $m = 55$ and $n = 39$, we have $55 \cdot 7 = 385$, $39 \cdot 10 = 390$, $55 \cdot 8 = 440$, and $39 \cdot 11 = 429$. Since $385 < 390$, while $440 > 429$, the third conditional of the condition is violated. Hence, $a{:}b \neq c{:}d$.

One way to regard how this works is that although in many cases most arbitrarily chosen values of m and n will not differentiate $a{:}b$ and $c{:}d$, the right values of m and n tease out the difference between the ratios, no matter how small the difference might be, and amplify that difference in ratios to generate different relative size relations between the members of each pair. While this is speculation, one can well imagine some such line of thought led Eudoxus to his discovery.

What's important, and in fact quite remarkable, is that for Eudoxus' approach to generalize and work for all continuous magnitudes, it only needs to be the case that each of the magnitudes that constitute the members of the ratios can be multiplied indefinitely many times and that homogeneous magnitudes stand in greater-than, equality, and less-than relations. We now have a general definition of sameness of ratio that can be employed in a theory of proportions covering all continuous spatial magnitudes including those that stand in irrational ratios. This is an ingenious way of defining sameness of ratio without appealing to numbers to directly express the ratios involved; it is one of the highest achievements of ancient Greek thought for which Eudoxus rightly earned a place among the pantheon of Greek mathematicians.

There is an additional feature of the definition that should be emphasized. We saw that Euclid defines ratios as a kind of relation with respect to size between two homogeneous magnitudes. Thus, *a* and *b* must be homogeneous and *c* and *d* must be homogeneous. There is no requirement, however, that *a* and *b* be homogeneous with *c* and *d*. Thus, the area of two figures, for example, can stand in the same ratio as two lines or two volumes, or two angles. This greatly expands the strength of the theory of proportions.

Book V establishes twenty-five propositions concerning ratios and proportions among magnitudes. V.16, the law of alternation, mentioned by Aristotle in his discussion of the theory of proportions, is an example:

V.16 If four magnitudes are proportional, then they will also be alternately proportional.

In other words, for any four homogeneous magnitudes *a*, *b*, *c*, and *d*, if $a{:}b = c{:}d$, then $a{:}c = b{:}d$. Such propositions provided powerful tools for expressing the mathematical relations among spatial magnitudes in Euclidean geometry.

There is a point to make here about terminology. We saw in the previous section that after defining sameness of ratio, V.D6 defines "proportional" to mean magnitudes having the same ratio. Since a ratio is a relation between two *magnitudes*, and having the same ratio is a relation between two *ratios*, Euclid's intent is that "proportional" means that two *magnitude pairs* have the same ratio. This is clear enough as it stands. However, it should be noted that at some point later in the Euclidean tradition, "proportion" was sometimes used as a synonym for ratio as well as for sameness of ratio. In that case, one must determine from the context which sense of "proportion" is intended.

A final note before turning to Book VII and Euclid's understanding of number. As was stated in Section 6.2, Euclid uses the theory of proportion presented in Book V to define similarity in Book VI. That definition is as follows:

VI.D1 Similar rectilinear figures are such as have their angles equal one by one and the sides around the equal angles proportional.

While equality of the angles is sufficient to guarantee the similarity of triangles, proportionality of the sides is also required for rectilinear figures with more than three sides, as reflection on the dissimilarity of a square and a rectangle makes apparent; they have equal angles but the proportions of the sides are not equal.[12]

[12] In a Euclidean space, equality of angles for triangles as well as other rectilinear figures, even with the added condition of proportionality among sides, will not guarantee that two similar rectilinear figures can be superimposed unless one adopts a notion of superposition that allows the translation and rotation of the figures in three-space. Similarity of

6.6 Euclid's Definition of Number and the Euclidean Theory of Proportions

Euclid's definition of number influenced the understanding of number until the late nineteenth century, and also influenced attempts to find a universal mathematics that covered continuous magnitudes as well as number. As noted above, Books VII–IX are referred to as the "arithmetical" books, and they are clearly meant to be foundational. What is striking in comparison to modern foundations of arithmetic is not simply the relative lack in sophistication, which is after all to be expected, but how the topic is approached, that is, what Euclid explains and how he explains it. Modern foundations do not merely characterize what natural numbers in general are, but recursively define each specific number, beginning with 0, and define the arithmetical operations on them. Euclid defines number in general, but not each specific number. He instead divides numbers into different kinds, such as odd and even and prime and composite, and treats their properties insofar as they belong to these general kinds. And rather than defining the arithmetical operations, he proves propositions concerning the relations among number, and more specifically propositions governing their ratios and proportions.

What is perhaps most striking is how differently Euclid and other Greek mathematicians conceived of number itself from the way we do today. Book VII begins by defining number (*arithmos*) as "a multitude composed of units," which reflects the ancient Greek understanding of number as a collection of individuals regarded as units. The first five definitions of Book VII are:

VII.D1 A unit is that with respect to which each existing thing is called one.

VII.D2 A number is a multitude composed of units.

VII.D3 A number is part of another number, the less of the greater, when it measures the greater.

VII.D4 But parts when it does not measure it.

VII.D5 The greater (number) is a multiple of the less when it is measured by the less.

The Greek notion of *arithmos* had different possible senses, and the way in which numbers were understood by Greek mathematicians and philosophers (in particular, by Plato and Aristotle) is quite complex and a matter of considerable debate. The most basic sense of *arithmos* seems to have been nothing more than a particular collection of things, such as the sheep on a particular hill or the shoes in front of my door. Each member of the collection is regarded as a unit that together with other units composes the *arithmos*. Euclid's definition of number implies that distinct collections are distinct numbers, even when they are of the same size. Thus, the number of sheep

solids does not guarantee the possibility of superposition. I consider this issue more carefully in Sutherland (2005a).

on this hill is a different number from the sheep on that hill, even if the size of the collections is the same.

Note that according to this definition of number, one is not a number, because it is not a collection. This was a common view, also expressed by Aristotle, who held that one is not itself a number, but was instead the principle of number. This leads Euclid to sometimes provide a separate proof for the case of collections of units and for the unit. But in other places, he treats the unit as if it were a number.[13] Later mathematicians in the Greek mathematical tradition followed suit, and eventually regarded one as a number. In effect, they treated it as a degenerate collection.

The most basic notion of number is simply a multitude composed of units – something closer to what we think of as a set than to what we think of as number. But Euclid did not have our set–membership relation; he instead describes the relation between the units and the multitude they compose in terms of part–whole relations. Euclid's most basic notion of number is mereological.

There is a closely related second point. Greek mathematicians and philosophers did not carefully distinguish, as we do, between the members of a set and the set itself, much less regard the latter as an abstract object, as many would today. The Greek notion of number is of a collection of individuals regarded as units, with an emphasis on the units that together make up the collection. And with regard to treating the number one as a degenerate kind of collection, they did not distinguish, as we would today, between a thing and its unit set.

In practice, the notion of *arithmos* relies on a concept corresponding to a count noun, such as sheep or shoe, under which each of the distinct units falls, though in the most general sense, the individuals need only fall under the concept of a unit.

Another closely related sense of *arithmos* was the number that corresponds to the results of enumerating the members of a collection and is sometimes called the "counting number" of the counted collection. In practice, it presupposes the choice of a unit, which is often called a "counting-concept," such as sheep or shoe, or the concept of a unit consisting of other units, such as pairs of shoes. Interpretation is difficult, but *arithmos* in this sense might conceivably belong only to the particular collection counted, so that the six of these sheep would be distinct from the six of those sheep.[14] However, *arithmos* as counting number might be thought of as a species under which collections of a

[13] See Mueller (1981, p. 58).

[14] Rogers Albritton first suggested to me that considering the Greek notion of *arithmos* in this way (and more specifically as "abstract particulars" to use our modern parlance) is at the very least an interpretative possibility that requires serious consideration. Klein seems to at least make room for this reading; see Klein (1992, pp. 46–7). See also Stein (1990, pp. 163–4) .

particular size fall, so that the six of these sheep is the same six as the six of those sheep and the six of the shoes in front of my door. On this view, particular number words such as "six" can be viewed as referring to something common to different collections or as a name common to collections of a particular size.[15]

There is much more to be said concerning the Greek notion of *arithmos* that we will have to pass over, but I would like to draw attention to three points that had a persisting influence on the understanding of number and eventually impacted the way Kant thought about number. The first is implicit in what we've already seen, but is worth making explicit. The notion of *arithmos* corresponds to whole numbers (allowing that one is a limiting case of number), and more particularly corresponds to what we think of as the cardinals, that is, numbers thought of as in some way representing collections and the sizes of collections.

Second, even where there was a notion of number as abstracted from and common to different collections, numbers were in the first instance thought of as multitudes composed of concrete individuals, regarded as units. In this sense, Euclid has a relatively concrete conception of number.

Finally, it should be noted that Euclid's theory of number as a multitude of units can be interpreted to be very general, since to be a unit is simply "that with respect to which each existing thing is called one." Nevertheless, the point of Euclid's introducing the theory of number into the *Elements* is to apply it in geometry, and the units he has at least primarily in mind are spatial.

Let us now turn to the complex relations among Euclid's theory of number and his geometry and theory of proportions. As we have noted, geometry appears to be a more general theory, insofar as it concerns the relations among continuous magnitudes, which include ratios that cannot be expressed by the ratio between two whole numbers. On the other hand, geometry presupposes number in the sense of collections of units. At the most superficial level, the definition of a triangle, for example, appeals to three sides. At a deeper level, the measurement of spatial magnitudes presupposes conceiving of them as composed of collections of units, and, as emerged in the previous section, the definition of sameness of ratios presupposes collections of units. This is an important point to which we will return in Chapter 9.

It was, however, the relation of both geometry and number to the theory of proportions that held particular historical significance. We saw above that Aristotle regarded numbers as falling within the purview of the theory of proportions. This is understandable, given that numbers can stand in ratios and proportions. In contrast, Euclid includes a separate treatment of the ratios and proportions of number in the arithmetical Books VII–IX. These books

[15] See again Stein (1990, pp. 163–4), and Tait (2005, pp. 238–9).

contain propositions that concern only numbers, for example, propositions concerning primes. Nevertheless, Book VII of the *Elements* proves many of the same propositions for numbers that are proved for continuous spatial magnitudes in Book V. Indeed, propositions VII.12–VII.14 are identical to propositions in Book V with "number" simply substituted for "magnitude." VII.13, for example, proves the law of alternation for numbers, and corresponds to V.16.[16] These parallels strongly suggest that there is a general theory common to continuous spatial magnitudes and number; as noted in Section 6.4 above, this is reflected in Aristotle's reference to their "universal character." Moreover, Euclid's definition of part and multiple for numbers, VII.D3 and VII.D4, which appeal to greater than, less than, and measure, mirrors the definition of part and multiple for magnitudes. (I will also examine these notions more closely in Chapter 9.) Furthermore, Euclid's common notions concerning equality, CN1–CN3 listed in Section 6.3 above, appear to apply to both continuous spatial magnitudes and numbers; in the latter case, equality consists in two collections being "equimultiple."[17] Finally, the very fact that numbers stand in ratios and proportions at all suggests that numbers are a kind of magnitude. In fact, recall Euclid's definition of ratio, V.D3, listed in Section 6.4 above: a ratio is a kind of relation with respect to size between two homogeneous magnitudes. Since numbers stand in ratios, this suggests that numbers themselves are some kind of magnitude.[18] The fact that Euclid does not define ratios for numbers in Book VII only reinforces the impression that numbers ought to be assimilated to magnitudes. So why doesn't Euclid provide a theory of proportions that covers both continuous spatial magnitudes and numbers?

We do not know what Euclid's actual motivations were, but canvassing some of the possible reasons helps reveal the relation between the ancient

[16] Cited on p. 176 above.

[17] The Greek manner of expressing "equimultiple" used a "ὅσα . . . τοσαῦτα" construction to express equality. The equality is asserted of two things with respect to multitude (πλῆθος). Thus, "equimultiple" roughly corresponds to "equality of multitude." This reinforces the impression that there is a common notion of equality governing both continuous magnitudes and multitudes. I thank Marco Panza for discussion, and in particular Ken Saito for very helpful correspondence in April 2017 on this issue.

[18] This is, in fact, how I interpreted Euclid at the time of writing Sutherland (2006). It seemed to me that Euclid was compelled by his own definition of ratio to regard numbers as magnitudes. I no longer think Euclid felt this compulsion, in part because of Euclid's looser attitude toward foundations described above, and in part for further reasons I will mention in what follows. Euclid's definitions are often characterizations of notions he seems to think are already sufficiently familiar to fix reference, and are not intended to be exhaustive. Having characterized ratios in a manner sufficient for the theory of proportions in Book V, he thinks the notion is sufficiently clear to appeal to it without further definition in Book VII. (This is not to deny that the different sources of Book V and Book VII also likely influenced his exposition.)

Greek understanding of magnitudes and numbers and the relation of both to the theory of proportions; it also helps explain a fundamental tension in the *Elements* that proved consequential in the history of the Euclidean tradition. The first reason concerns Euclid's sources. As noted above, Euclid most likely compiled the *Elements* from other works, and there are differences between chapters, anomalies in the way the *Elements* is structured, the way chapters fit together, and repetitions that are best explained as the influence of Euclid's compilation from different sources – in particular, the fact that Book V derived from a work of Eudoxus and Book VII is most likely a descendant of a book by Theatetus. And despite overlaps, there are also differences in the content and aims of each book. Rather than completely rework both, Euclid allows each to stand on its own, despite the strong parallels.

The deductive structure of the *Elements* might also have recommended the separate treatment of geometry and arithmetic. Book V introduces the theory of proportions required for rounding out plane geometry in Book VI, and it is only well after this point that the theory of proportions for numbers is needed. An analysis of the deductive structure of the *Elements* appears to show that Euclid wanted to develop geometry as far as possible without appeal to the theory of proportions.[19] Euclid's apparent minimizing and postponement of the theory of proportions will later become a point of dispute for those who thought geometry would benefit from a thorough reform. But if this were Euclid's aim, then including the separate accounts allows him to postpone what he needs from the theory of proportions until he needs it.

These are merely accidental historical considerations or structural considerations that need not prevent a unified theory of magnitudes covering both continuous magnitudes and number. There are, however, general mathematical reasons for presenting them separately that may have motivated Euclid. First, the spatial magnitudes covered by Book V are continuous, whereas numbers as the ancient Greeks conceived of them are collections of units that are discrete, and this appears to be a fundamental difference. Second, homogeneity is a requirement of the combination and comparison of continuous spatial magnitudes, while no such distinction among collections of units, regarded merely as units, seems to be required. Or rather, homogeneity is satisfied for any units; insofar as each existing thing is considered with respect to being one, it is homogeneous with every other individual thing. Third, the ancient Greek emphasis on construction in geometry may have been missing in arithmetic; at the very least, the arithmetical books derived from Theatetus do not describe the "construction" of collections in the way that the construction of geometrical objects are described. I noted in Section 6.3 that only Book I contained principles in addition to definitions; although collections of units

[19] See Mueller (1981, pp. ix–x).

were thought to fall under the common notions governing equality, there were no special principles for collections of units, and in particular no construction postulates. Such collections are treated as if they were found rather than constructed.[20] These differences between continuous spatial magnitudes and number might have led Euclid to present the two separate treatments of proportions inherited from Eudoxus and Theatetus.

Finally, there were additional, more purely mathematical reasons for dividing the theory of proportions. Since there are no incommensurable ratios among numbers conceived as collections of units, it is possible to give a much more straightforward definition of sameness of ratios for numbers; the innovative and complex definition of same ratio given for continuous magnitudes in V.D5 is not required. This would not in principle rule out simply assimilating the theory of proportions for numbers into a more general and more complex account; as the example I gave in the previous section demonstrates, V.D5 works for numbers as well as for continuous magnitudes. Nevertheless, the fact that a much simpler account of same ratio for numbers is available may have spurred Euclid to present the theory of proportions among numbers separately.

A further mathematical reason is a direct consequence of the greater generality of continuous magnitudes. Some of the Eudoxian theory of proportions for continuous magnitudes assumes the existence of a fourth proportional; that is, for any homogeneous magnitude pair x and y and any other magnitude w, there is a magnitude z homogeneous with w such that $x{:}y = w{:}z$. Suppose, for example, that line b is half the length of some line a and that an area g is given. Then there is an area h that is half as large as g. One might well assume the existence of a fourth proportional for continuous magnitudes, but it is ruled out for numbers conceived as collections of units, since that conception limits numbers to whole numbers. For example, there is no whole number that could serve as the fourth proportional in the proportional $3{:}1 = 1{:}z$. The assumption of the existence of the fourth proportional for continuous magnitudes means that some of the results of Book V do not apply to numbers.

Despite these possible grounds for separating the theory of proportions for magnitudes and for numbers, the fact that the two theories have so much in common strongly suggests some general, overarching theory that would encompass both magnitudes and number. The tension between Euclid's separate treatments of continuous spatial magnitude and number and the promise of a common theory played an important role in the history of the Greek mathematical tradition.

[20] See Mueller (1981, p. 60).

Before moving on to that later tradition, it bears mentioning that there was another influential Greek approach to arithmetic distinct from the Euclidean. It was called "logistic" (*logistike*), and its contributors included Archimedes (ca. 287–ca. 212 BCE) and Nichomachus (ca. 60–ca. 120 CE), and its most important proponent was Diophantus (ca. 200–284 CE). *Logistike* focused on techniques of calculation. *Logistike* played a significant role in the complex evolution of views of number as well as algebra.[21] We can set aside that history in this work, however, and focus on the relation of number to magnitudes and the theory of proportions in Euclid.

6.7 The Euclidean Geometrical Tradition

It is impossible to overstate the influence of Euclid's *Elements* on the history of Western thought concerning mathematics, science, and philosophy. It supported the idea of the pursuit of "pure" sciences detached from immediate practical applications, exemplified a large body of knowledge flowing from a much smaller set of fundamental principles or "elements," and provided a model and ideal of rigor, clarity, and certainty in the pursuit of knowledge. At the same time, the shortcomings in the deductive structure of the *Elements* outlined in Section 6.3 also inspired those who followed Euclid, for it spurred Greek, Arabic, Persian, and subsequently European mathematicians to improve it. These two features of the *Elements* set the stage for a research program that lasted two millennia. It is, of course, impossible to do justice to two millennia of mathematics in a few short pages. What follows will necessarily be brief and oversimplified, but will help highlight a few themes that will be important in the following chapters.

By and large, most commentators that followed Euclid stuck fairly closely to the overall structure of the *Elements*, although ever greater departures with more radical reforms began to appear in the late Middle Ages and Renaissance. Some of those reforms sometimes included reordering the major parts of the *Elements*; some, for example, began with arithmetic and then the theory of proportions before turning to geometry, reflecting a belief in their priority over geometry.

Long before these reforms, however, increasing standards of rigor prompted the articulation and addition of ever more axioms. To cite just a few examples, one of the earliest additions by an unidentified commentator is the principle that two straight lines do not enclose a space; the tenth-century Persian mathematician and astronomer An-Nayrizi added a principle stating that a surface cuts a surface in a line; and in 1645 the French geometer Claude

[21] See Klein (1992, pp. 26–36). I discuss this tradition and its relation to Kant in more detail in Sutherland (2006), although my views have evolved since writing that article.

Richard added line–circle and circle–circle continuity principles that assert the existence of their points of intersection.[22]

While these examples concern geometrical figures in particular, mathematicians also added principles governing magnitudes, principles governing numbers, and principles concerning ratios and proportions to support the Eudoxian theory of proportions in Books V and the arithmetical Books VII–IX. In a recent survey he says is not exhaustive, Vincenzo De Risi identifies 318 additional axioms articulated by various geometers from the early twelfth century and into the eighteenth.[23] Many of these are variations of similar axioms, so there is considerable overlap, and no one geometer listed more than about fifty principles; nevertheless, they are a testament to the amount of work and careful thought that contributed to the Euclidean geometrical tradition.

Two schools of thought that emerged late in the tradition and concern fundamental mathematical principles are particularly worth noting, as they mark out two end points of a spectrum of views on the number of such principles. Beginning in the seventeenth century and carrying into the eighteenth century, some French mathematicians adopted the view that any clear and self-evident mathematical principle counts as an axiom; while they would list a number of them, they held that there were in fact an infinite number of axioms. A second school of thought held that any purported axiom can and ought to be derived from definitions, a view that spread in eighteenth-century Germany. There were important German mathematicians who disagreed, including Abraham Kästner, who wrote perhaps the most influential mathematical texts of the second half of the eighteenth century. Nevertheless, the view that axioms should be eliminated by appeal to definitions appears in the geometry texts of Christian Wolff, Andreas Segner, and Wenceslaus Karsten.[24] Kant was familiar with all of these texts.

Let us return to the beginning of the Euclidean tradition and the theme of a universal mathematics. There were quite early attempts to expand and unify the theory of magnitude and the theory of number found in Books V and VII, despite the reasons Euclid might have had for treating them separately. There were two parts to this. First, there is a generalization of the theory of continuous spatial magnitudes to include other sorts of magnitude, in particular, time and motion. Second, there were also attempts to broaden the theory of magnitudes to include number. Proclus, a fifth-century Neo-Platonist, wrote an extremely influential commentary on Book I of the *Elements*; he was a strong proponent of a unified theory of proportion that governs continuous spatial magnitudes as well as motion, time, and number (Proclus (1970, p. 6)).

[22] See De Risi (2016, pp. 6, 9, 15, 39, and 41). I'm indebted here and in all of what follows to this article, as well as to multiple conversations with De Risi.

[23] See De Risi (2016).

[24] De Risi (2016, pp. 9–11).

Extending the notion of magnitude to include both nonspatial continuous magnitudes such as time and motion as well as number were in service of a universal mathematics based on a theory of proportions. Proclus attributes to Eratosthenes (c. 276–c. 194 BCE) the view that the unifying bond of all mathematical sciences is proportion (Proculus (1970, p. 36)).[25] Thus, if Proclus is to be trusted, the idea that the theory of proportions provided a general theory of mathematics was promulgated only very shortly after Euclid finished the *Elements*.

Broadening the notion of magnitude to include time and motion would prove to be particularly important in the medieval theory of the intension and remission of forms; the influence of that theory on Kant, already noted in Chapter 4, makes it worth a brief description. This theory was first advanced in the early fourteenth century at Merton College, Oxford, and then in Paris; at the latter location, it was developed especially by Nicole Oresme. It revised the prior understanding of motion and other qualities in Aristotelian metaphysics, allowing a mathematical treatment of motion by treating it as a quality or form of a body that could vary in intensity and hence as a quality having an intensive magnitude. This made it possible to analyze different kinds of motion, culminating in the mean speed rule governing a uniformly accelerating movement starting from rest, an important milestone in understanding the mathematical character of the world. Kant's way of drawing the distinction between extensive and intensive magnitudes and his view that speed is an intensive magnitude trace back to this medieval tradition.[26]

Let us return once again to the fifth century CE and the relation of the theory of proportion to numbers. Proclus was a strong proponent of a unified theory of proportion that governs not just continuous spatial magnitudes, such as motion and time, but also number. Any such move would encourage treating collections of numbers as themselves magnitudes. They nevertheless differ from continuous magnitudes in being collections of discrete units, which suggests conceiving of such collections as discrete magnitudes. There was, in fact, precedent for thinking of number as a kind of magnitude or quantity. As early as chapter 6 of the Categories, Aristotle divides quantity into the continuous and the discrete; according to his classification, number and language (i.e., spoken syllables of language) are considered discrete quantities.[27]

Proclus holds that there are theorems common to arithmetic and geometry that are derived from "general mathematics." As examples, he refers to the

[25] This is the same Eratosthenes who was a librarian in Alexandria and estimated the diameter of the earth.

[26] See Section 4.1 above; I discuss the intension and remission of forms and its relation to Kant in Sutherland (2014).

[27] See Aristotle (1990, p. 12).

principles governing alternation, conversion, composition, and division of ratios, that is, the theory of ratios and proportions. At the same time, his manner of speaking reflects a continuing tension with regarding numbers as a species of magnitude, for he often uses the term "magnitude" to mean continuous magnitudes in contrast to number.

It is worth noting that while Proclus supported the idea of a common theory of magnitudes covering both continuous magnitudes and numbers conceived as collections, he also indicates that if we differentiate axioms from postulates as those that state the way things are and those which state what can be constructed, then the general principles of general mathematics will be so divided, and that there will also be such a division for specifically geometrical principles and for specifically arithmetical principles. For the latter, Proclus mentions the axiom that "Every number is measured by the one," and the postulate of construction, "A number can be divided into least parts." This is an indication that at least by the time of Proclus, and in contrast to Euclid's Books VII–IX, geometers already in the early Greek mathematical tradition were willing to consider axioms and postulates of arithmetic, and to consider construction postulates for arithmetic.[28]

As the Euclidean tradition progressed, so did the analysis of the foundations of the theory of proportions. Various mathematicians added principles concerning magnitudes that articulated their mereological properties and the operations of mereological addition and subtraction, and these principles were appealed to in the theory of proportions. Some mathematicians also introduced similar principles for number. Many, however, regarded the principles of magnitudes as governing both continuous and discrete magnitudes.[29] I cannot provide a full survey of these views, but the acceptance of both continuous and discrete magnitudes is illustrated by the famous Renaissance mathematician Luca Pacioli. He published his *Summa de Arithmetica, Geometria, Proportioni, et Proportionalita* in 1494 (and again in 1523). On the very first page, Pacioli begins by stating that there are two kinds of quantity (i.e., magnitude): the continuous and the discrete.[30]

The point of proposing principles for a common theory for continuous and discrete magnitudes was to support a universal mathematics based on the theory of proportions; the importance of proportions is reflected in the title of Pacioli's work. Pacioli went on to publish *Divina Proportione*, a highly influential Renaissance work on mathematical proportions in art and architecture. Unfortunately, the contents of Book V had been poorly transmitted through

[28] See De Risi (2016, pp. 21–22) for a discussion of this point.

[29] De Risi (2016, p. 15).

[30] As reported in Abraham Kästner's 1796 *Geschichte der Mathematik*, p. 66. This shows that Pacioli's views on discrete quantity were known in Kant's time and almost certainly by Kant himself.

the medieval period, and it wasn't until the Renaissance that its contents were regained; even then, it took some time for mathematicians to fully appreciate the subtlety of Book V, in particular V.Def.5.[31] Despite these challenges, the theory of proportions was regarded as a universal mathematics, one that would make it possible to represent the mathematical character of natural phenomena. Galileo was the foremost proponent of this view; while his reworking of the theory of proportions was ultimately not included in his 1638 *Discourse and Mathematical Demonstrations Relating to Two New Sciences*, it was known and inspired further work in seventeenth-century Italy, above all the work of Toricelli and Borelli. The theory of proportions was up until this time absolutely central to the understanding of the nature of mathematics.

Algebra emerged against this background. It grew out of the Greek logistical arithmetic tradition mentioned at the close of Section 6.6 and assumed an increasingly important role in mathematics during the Renaissance. Pacioli's *Summa*, for example, incorporates algebraic techniques of problem-solving. The history of the interaction of algebra and the theory of proportions is complicated, and I can only drastically simplify it here. In the seventeenth century, as algebra was successfully applied to solving geometrical problems, above all under the influence of Descartes' 1637 *Geometry*, attention began to shift from the theory of proportions to algebra as a universal approach to mathematics. Nevertheless, the connection between them was close, as algebraic symbols were interpreted as standing for magnitudes and the relations among them.

This extremely condensed summary of the Euclidean tradition and its role in the history of mathematics is of course quite incomplete; in particular, it does not consider the rather complex evolving views of number from the sixteenth to the eighteenth century. Furthermore, it skips over the tremendous advances in mathematics and mathematical physics from the seventeenth to the early eighteenth century. Despite these advances, however, the background for thinking about continuous and discrete magnitudes, number, and the basis of a universal mathematics founded on the theory of proportions remained influential into the eighteenth century and was only fully overthrown through the arithmetization of mathematics at the close of the nineteenth century. What we have seen is therefore sufficient to provide enough historical context to help explain the way in which mathematicians thought and wrote about the role of magnitudes in mathematics during Kant's time.

6.8 Magnitudes and Mathematics in Kant's Immediate Predecessors

To gain a sense of the way mathematics was understood in the second half of the eighteenth century in Germany, I will examine the work of a few

[31] De Risi (2016, p. 17).

influential figures with whose work Kant was familiar: Christian Wolff, Abraham Kästner, and Leonhard Euler. A close consideration of their views of magnitude is helpful, because it reveals the immediate background for Kant's theory of magnitudes and helps to explain his own thinking. I will focus on their views of magnitudes, including both continuous and discrete magnitudes and the relationship between magnitudes and number.

Wolff's work reflects the understanding of number and magnitude that derived from the Euclidean tradition and persisted into the eighteenth century. In his *Mathematical Lexicon* (1716), for example, he defines mathematics as "a science of measuring all that can be measured," and continues:

> one commonly describes it *per scientiam quantitatum* [as a science of quantity], as a science of magnitudes [*Wissenschaft der Größen*]; that is, all those things that allow of increase or decrease. Since all finite things allow of being measured in all that they have that is finite, that is what they are [i.e., magnitudes]; hence, there is nothing in the world to which mathematics cannot be brought to bear. (Wolff (1965, p. 863))

Wolff adopts a common characterization of magnitude as that which allows of increase and decrease and asserts that magnitudes are those things that can be measured. He adds that all that is finite in the world can be measured – a reflection of the mathematical character of the world. His definition of pure, genuine mathematics also places magnitude at the foundation of mathematics, and makes explicit that numbers fall under the rubric of magnitudes:

> [Pure or simple, genuine mathematics] is what is called that which merely views magnitudes as magnitudes, e.g., a straight line as a straight line, the number 6 as 6. And hence arithmetic, geometry, along with trigonometry and algebra, belong to mathematics. (Wolff (1965, p. 868))

Wolff apparently assimilates numbers to magnitudes by considering them to be discrete magnitudes, for he explicitly endorses the Euclidean conception of number as a collection of units. In his *Foundations of All Mathematical Sciences* (1709/10), he states that when one takes together several single things of the same kind, a number arises, and that for this reason Euclid defines number as a multitude of units.[32] His very popular abridged version of this work uses the same definition.[33]

Wolff's treatment of numbers in the Euclidean sense as magnitudes and his characterization of mathematics as the science of measuring magnitudes show

[32] Wolff (1963, p. 38). This work went through six editions in Wolff's lifetime. My reference is to the 1750 edition reprinted by Georg Olms Verlag in *Christian Wolff: Gesammelte Werke*.

[33] *Auszug aus den Anfangs-Gründen aller Mathematischen Wissenschaften*, Wolff (2009, p. 11). This also appeared in multiple editions. My reference is to the 1728 edition reprinted in in *Christian Wolff: Gesammelte Werke*.

that he employs a sense of measurement that would regard determining the size of a collection as a kind of measurement. This was a common view of measurement that parallels Euclid's appeal to the notion of measure in his account of number in Book VII. We will see this broad conception of measurement in Euler below as well, and will return to this notion of measurement and its role in the theory of magnitudes in Chapter 9.

There is an important caveat to Wolff's endorsement of the Euclidian conception of number as a collection of units and its assimilation into the theory of magnitude. Wolff also published a mathematics text in Latin, *Elementa Matheseos Universae* (1713/15), a work intended to be more rigorous. There, Wolff defines a number as that which relates to unity in the way that one straight line relates to another straight line. He explains in his *Mathematical Lexicon* (1716) that he used Euclid's definition in the German *Foundations* only because it is very clear to beginners, and that he changed the definition in the Latin *Elementa* so that it would include irrationals (Wolff (1965, pp. 944–5)). Wolff's views here reflect the changing understanding of number in the early modern period; I will undertake a fuller discussion of narrower and broader senses of number in the eighteenth century and in Kant in a future work focused more specifically on Kant's philosophy of arithmetic. For now, it is sufficient to note that Wolff and others endorsed narrower and broader notions of number, and that the Euclidean notion of number was considered more basic.[34]

Wolff's mathematical texts were widely published and used in Germany from the beginning of the eighteenth century. By mid-century, however, some came to think that they needed to be replaced with a treatment that was more rigorous and thorough.[35] Abraham Kästner, Andreas Segner, and Wenceslaw Kartens all published their own versions of foundations of mathematics. All three begin with arithmetic rather than geometry, and all three include an account of ratios in the section on arithmetic, which may indicate an awareness that the Eudoxian theory of proportions presupposes number in the sense of collections of units.

Of these three post-Wolffian works, Kästner's was the most well-known and influential. Kästner was a highly regarded professor of mathematics in Kant's

[34] There is a further feature of Wolff's foundations of mathematics worth noting. Both his German and Latin versions begin with arithmetic and only then turn to geometry. The more rigorous and extensive Latin *Foundations* includes a much larger arithmetic. The third chapter is devoted to the theory of ratios and proportions, and only after that does Wolff turn to the arithmetic of fractions and to square and cubic roots; those chapters are followed by another on the principles of the rules of proportion. Only then does geometry appear. This reordering may reflect the belief that arithmetic is more basic and that the theory of proportions is more general than geometry; as mentioned shortly below, it may also reflect the view that the theory of proportions presupposes arithmetic.

[35] See Kästner's second edition Preface to his *Anfangsgründe*, p. **1.

time, particularly famous for his epigrammatic poetry. In mathematics, he was known more for his textbooks than original contributions, although he did seminal work on the grounding of the parallel postulate. In 1756, he published *Foundations of Arithmetic, Geometry, Plane and Spherical Geometry and Perspective*, which went through at least five editions.[36]

Kästner starts his work with a preface on mathematics in general and its teaching, and he begins with magnitude. He describes magnitude as that which allows of increase and decrease. He then adds by way of clarification:

> This property thus distinguishes things in which one otherwise views nothing as differing, e.g., a collection of ducats from each other, which all have the same imprint and weight. (Kästner (1800, p. 1))

Thus, the second sentence of his influential book shows that he views discrete individuals as magnitudes.[37] He also asserts that pure mathematics abstracts magnitude from things, tying mathematics directly to the world:

> Magnitudes can be viewed as abstracted from all other properties of a thing, and this is what **pure and abstract** mathematics [*Mathesis pura vel abstracta*] does . . . (p. 3)

He also ties the representation of magnitudes to part–whole relations, and in particular, to the representation of magnitudes as collections of discrete parts:

> One can regard magnitudes merely as a collection of parts, as a **whole** (*Totum*); or one can at the same time view the connection or order of these parts that constitute a certain **composed thing** (*compositum*). (p. 3)

In summary, Kästner thinks of mathematics as the science of magnitudes. Since numbers are a part of mathematics abstracted from the world, and mathematics is the science of magnitudes, he implicitly holds that numbers are a species of magnitude. Moreover, he explicitly endorses the Euclidean conception of number as a collection of units, so he at least implicitly holds that numbers are discrete magnitudes.

[36] Kant owned a copy of this work and expressed admiration for Kästner, especially his treatment of magnitudes: in *Negative Magnitudes* Kant states that "No one, perhaps, has indicated with greater distinctness and precision what is to be understood by negative magnitudes than the celebrated *Kästner* in whose hands everything becomes exact, intelligible and agreeable" (2:170). Kästner was an adherent of the Wolffian philosophy, and he later clashed with Kant on the proper account of mathematical cognition after the *Critique of Pure Reason* was published, but on matters of mathematics itself, Kant appears to have held Kästner in high esteem.

[37] Note as well that he refers to not distinguishing the individuals by any other property than being a magnitude. This is a point about the representation of magnitudes, continuous and discrete, which will acquire significance in Kant's views, as we will see in the next chapter.

Kästner gives the following account of number:

4. Explication: That which the **things** that are to be counted have in common is called the **unit** [*Einheit*], and **number** is hence a collection of units. (p. 25)

He goes on to make clear that the sense of number he defined as a collection of units is whole number (*ganze Zahl*), and at the beginning of the arithmetic chapter, declares it the default understanding of number. Addition, subtraction, multiplication, and division (with a remainder) are defined for whole numbers.[38] Nevertheless, like Wolff, Kästner also allows a broader notion of number that includes rationals and irrationals. Again, we will postpone deeper consideration of the different senses of number for an investigation focused on Kant's philosophy of arithmetic. What is most important now is that he held that the whole numbers, understood as collections of units, are foundational. In the preface to the *Anfangsgründe*, he states:

> All concepts of arithmetic are founded, in my opinion, on the whole numbers; fractions are whole numbers whose unit is a part of what was the original unit, and one must represent irrational magnitudes as fractions . . . (Kästner (1800, pp. 3*–4*))

In Kästner's view, the most basic sense of number is the Euclidean.

Let us now turn to the views of Euler, one of the most distinguished mathematicians of the eighteenth century, or indeed any century. I will consider two of his texts, his two-part *Introduction to Arithmetic* published in 1738 and 1740, and his *Elements of Algebra*, first published in 1770. The latter includes a general description of mathematics, so I will begin with it before looking at his conception of number in particular in both works.

The first part of *Elements of Algebra* lays the groundwork for algebra, and is divided into three sections. Significantly, the first two concern magnitudes, while the third considers ratios and proportions. As noted already in Section 1.5, Euler opens his work with a discussion of magnitudes. He begins "Chapter 1: Of Mathematics" by explaining that "Whatever is capable of increase or decrease, or to which something can be added or from which something can be taken away, is called magnitude" (Euler (1802, p. 3)), adding, "Mathematics in general is nothing but a science of magnitudes, and seeks the means of measuring them."[39] Euler often uses "magnitude" for

[38] In the case of multiplication and division, the operations are made visually evident (*augenscheinlich*) using arrays of asterisks (pp. 25, 27, 28).

[39] Note that Euler (1984), a reprint of the 1890 translation of this work by John Hewlett, is a helpful resource. It takes liberties in translation, however, that obscure the exact way in which Euler expresses himself, which matter for our purposes. For example, Hewlett translates "magnitude" as "magnitude, or quantity."

something concrete; his first example of a magnitude is a sum of money. Euler continues by describing the conditions for measuring, stating:

> We cannot determine or measure any magnitude, except by considering some other magnitude of the same kind as known, and indicating the ratio [*Verhältniß*] in which it stands to any magnitude of the same kind. (Euler (1802, p. 4))

Euler glosses measurement as "determining a magnitude," and also uses "magnitude" in the same sentence to refer to "some other magnitude of the same kind" as that which is taken as known. His example is of determining the magnitude of a sum of money, taking as a unit a thaler or ducat or some other coin. This makes it clear that he considers collections of individuals to be magnitudes (a point confirmed in the next paragraph below). Whether or not he uses the term "discrete," he endorses the notion of discrete magnitudes. Euler goes on to make a parallel claim about measuring a continuous magnitude, namely, choosing a unit of a pound to determine the quantity of weight of something. Both discrete and continuous magnitudes are brought under a common theory of measurement according to a unit.

Focusing now on Euler's understanding of number, we find that he, like Wolff, also expounds two conceptions of number, the first in relation to a collection of units, construed as magnitudes, and the second as a ratio between magnitudes. In his *Introduction to Arithmetic*, designed for use in the schools, he states that numbers are a particular species of magnitude (Euler (1738, p. 2)). Like many of the authors of mathematical texts of the time, he took a rather practical approach, but he still insisted on acquainting his readers with the grounds of the rules of arithmetic. He holds that, above all, we need a clear concept of number, and states as clarification: "When many pieces [*Stücke*] of one kind [*Art*] are at hand, this plurality [*Vielheit*] is indicated through a number. And for this reason, one understands through a number how many pieces are spoken of" (Euler (1738, p. 4)). Euler here indicates that he understands number in Kant's sense of *quantitas*, that which represents the size of a collection, rather than the collection numbered. Euler gives rubles as an example of the kind of "pieces" (*Stücke*) that are numbered. His arithmetic includes fractions, but does not discuss irrational quantities.[40]

His *Elements of Algebra* of 1770, on the other hand, shifts to a more abstract and broader conception of number. He states in the first chapter that the proportion of a magnitude to a unit of measure is "always expressed by numbers, so that a number is nothing but the proportion of one magnitude to another" (Euler (1802, p. 5)). (As noted earlier, at least by the early modern

[40] Euler, *Einleitung zur Rechen-Kunst zum Gebrauch des Gymnasii bei der Kaiserlichen Academie der Wissenschaft in St. Petersburg*, reprinted in *Euler Opera Omnia*, ser. 3, v. 2, pp. 1–303.

period, the term "proportion" was sometimes used to refer to ratio and was no longer used to refer only to the sameness of two ratios; Euler is using "proportion" here to mean ratio.)

In summary, Wolff, Kästner, and Euler all express the view that mathematics is the science of magnitudes and their measurement, a notion of magnitude and measurement that includes collections of units, whether or not they explicitly refer to those collections as "discrete" magnitudes. On the other hand, in their more scholarly work, Wolff and Euler also define number as a ratio between magnitudes. The term "number" was used for both the narrower and the broader notion of number into the eighteenth century, which was part of the complex development of the notion of number from the Renaissance to the late nineteenth century.

6.9 Conclusion

I have attempted to provide a selective overview of Euclid's *Elements* and the very long Euclidean tradition that followed up to Kant's immediate predecessors, with a focus on those features that help explain the ways in which mathematicians and philosophers thought about mathematics as a science of magnitudes and their measurement. I have highlighted the theory of proportions and its central role in the pursuit of a universal mathematics, and the pressure that put on extending the notion of magnitudes to include nonspatial magnitudes such as time and motion, as well as numbers, conceived as collections of units. Along the way, I have explained the foundations of the theory of proportions and the long tradition of engaging with Euclid by adding principles and reforming and rearranging the *Elements*. All of this was rooted in Euclid's understanding of magnitude, and we have seen that the idea that mathematics is a science of magnitudes was still very much alive in the eighteenth century. We are now in a position to return to Kant's understanding of mathematics and the mathematical character of the world. We saw in the previous chapters that Kant adopts the view that mathematics is about magnitudes and views the objects of experience as extended magnitudes whose real properties have intensive magnitudes. In the following chapters, we will see that Kant rethinks the nature of magnitudes – what makes them mathematical and how we cognize them – in light of his account of human cognition.

PART II

Kant's Theory of Magnitudes, Intuition, and Measurement

7

Kant's Reworking of the Theory of Magnitudes

Homogeneity and the Role of Intuition

7.1 Introduction

Chapters 2–4 revealed that Kant's philosophy of mathematics and his account of the mathematical character of the world rests on a theory of magnitudes. Kant was influenced by the long Greek mathematical tradition on both points, so Chapter 6 sketched that background and traced its influence into the eighteenth century. But Kant did not simply take on the conception of magnitude found in the Greek mathematical tradition; he transformed it in light of his understanding of mathematics and mathematical cognition. Chapter 5 began an investigation of Kant's view of the role of intuition in representing magnitudes, with a focus on the singularity and concreteness of intuition. This chapter will delve more deeply into the role of intuition, Kant's understanding of magnitudes, and how it relates to the Greek mathematical tradition. I will argue that Kant reworks the traditional notions of magnitude and homogeneity in an attempt to give a deeper analysis of magnitudes. More specifically, Kant appeals to what I call "strict homogeneity" in an attempt to capture what is distinctive about magnitudes. Kant also holds that concepts without the aid of intuition cannot represent strict homogeneity, but that intuition makes it possible for us to do so. Hence, intuition plays an important role in mathematical cognition by making it possible for us to represent homogeneous magnitudes. The chapter then concludes with some clarifications about the role of intuition in representing magnitudes.

7.2 Kant's Rethinking of Magnitude and Homogeneity

As we saw in Chapter 6, Euclid did not define what it is to be a magnitude, or explain what is special about magnitudes, or what gives them the mathematical properties they have. Our understanding of Euclid's theory of magnitude derives from the examples Euclid provides and the claims he makes about them, including the claims that homogeneity is a necessary condition of their standing in ratios and that satisfying the Archimedean property is a sufficient condition of standing in a ratio. More generally, the Euclidean understanding of magnitude is revealed by their role in the Eudoxian theory of proportions;

magnitudes are the sorts of things that can stand in ratios and proportions and hence have the mathematical properties articulated in the propositions of Book V (see Section 6.4 above). It is homogeneous magnitudes that can be combined and composed and can stand in ratios with one another, although Euclid does not explain what being homogeneous consists in. Those who followed Euclid, most notably Proclus, broadened the notion of magnitude in light of the power and generalizability of the Eudoxian theory of proportions found in Book V. Still, neither Proclus nor others gave an analysis of what it is to be a magnitude, much less what it is for them to be homogeneous with each other and hence to be such that they can be combined and composed and stand in ratios.

Kant attempted to fill this lacuna by giving a deeper explanation without, however, overthrowing Euclidean geometry or beginning from a different foundation. He differed in this regard from others who pushed for stronger reforms, notably Leibniz and Wolff, whose views we will examine more closely in Chapter 8. Kant by and large endorsed the framework of Euclidean geometry, including the Eudoxian theory of proportions it contains. Kant's aim is to push deeper and think through the foundational concepts and principles with an eye to explaining the possibility of the cognition required for the theory of magnitudes more generally. In a manner noted in Chapter 1 as characteristic of Kant, he reworks the foundations of the theory while preserving most of its consequences. Kant does not believe in throwing the baby out with the bathwater, and in this case, the baby is the theory of magnitudes on which the Greek geometrical tradition rests.

How do Kant's notions of magnitude and homogeneity differ from Euclid's? The Greek notion of homogeneity divides magnitudes into kinds, and the kinds Euclid has in mind are generally differentiated by the number of spatial dimensions involved – lines are homogeneous only with lines, areas with areas, volumes with volumes, and angles with angles. As noted in Chapter 3, Kant defines a magnitude as a homogeneous manifold in intuition in general, insofar as it makes the representation of an object first possible. Kant thereby shifts away from the Greek account, for Kant makes homogeneity not simply a requirement on two or more magnitudes that stand in a ratio to each other, but essential to what any magnitude is; homogeneity is now a property of the manifold of anything that counts as a magnitude at all. That is, the manifold parts of a magnitude are all homogeneous with each other.[1] Despite Kant's shift, there is a close connection between his and the Greek understanding of homogeneity and magnitude, which will be seen once we've explained Kant's understanding of homogeneity.

[1] I would like to thank Lisa Shabel and Bill Tait for prompting me to directly address this difference in the way Kant employs the notion of homogeneity. (I discussed the sense of "part" when applied to a continuous homogeneous manifold in Chapters 4 and 5.)

7.3 Strict Logical Homogeneity and Magnitude Homogeneity

Kant's notion of homogeneity ultimately rests on a characterization that belonged in Kant's time to logic and the study of genus–species relations (as described in Section 5.3.2). This logical notion of homogeneity reflects its etymology, as it derives from the Latin *homogen*, and ultimately from the Greek for "same genus," while the German translation for this notion is *gleichartig*, which derives from "same species." Two concepts or things are homogeneous with respect to a concept if they both fall under that concept. For example, the concept <Clydesdale> and the concept <Shetland> are homogeneous with respect to the concept <horse>.[2] I will call this "logical homogeneity" because it concerns the relations among concepts. In the logic of Wolff and his followers, as well as in Kant, concepts can be ordered into genus–species hierarchies.[3] Concepts or things can be more or less logically homogeneous with each other depending on how general the lowest common concept is under which they fall, and hence logical homogeneity comes in degrees. For example, the concept <Clydesdale> is more homogeneous with the concept <Shetland> than with the concept < jaguar>, since the lowest common concept under which both the concepts <Clydesdale> and <jaguar> fall, the concept <mammal>, is more general than the concept <horse>. It was also commonly held that what I am calling "logical homogeneity" is a requirement of counting; that is, objects that are counted must be counted under a common concept.[4] Kant also uses the notion of logical homogeneity in his defense of a principle of reason concerning the systematic unity of experience, a principle of the homogeneity of forms (A651–63/B679–91). More importantly for our interests, however, Kant uses logical homogeneity to explain the special sort of homogeneity required to be a magnitude.

As we saw in Chapter 3, Kant distinguishes between two sorts of magnitude, *quanta* and *quantitas*, and as described in Chapter 5, the former is a concrete magnitude, while the latter is a more abstract counterpart. The definition of magnitude Kant gives in the Axioms is of *quanta*, which is our current focus. There is important evidence for Kant's understanding of the nature of *quanta* in his lectures on metaphysics and in his annotations of Baumgarten's *Metaphysica*. In his lectures, Kant uses logical homogeneity to contrast a

[2] I am referring here to the concept <*equus*>, not to a sense of "horse" that contrasts with "pony."

[3] For a thorough and careful account of the importance of the genus–species hierarchy of concepts in both Wolff and Kant, see Lanier Anderson (2015).

[4] Some have thought that this is Kant's primary notion of homogeneity in mathematical contexts. Longuenesse (1998, pp. 249–50, esp. note 16), for example, holds that this is the notion of homogeneity Kant has in mind for the homogeneity of magnitudes. As will become apparent, I think Kant has a narrower notion in mind.

quantum with a *compositum*. He states that both a *quantum* and a *compositum* contain a plurality, but a *compositum* allows for an aggregate of heterogeneous parts, while a *quantum* requires homogeneity among the parts. Kant articulates the homogeneity requirement of *quanta* as follows:

> *Homogeneitatem*, i.e. things from one and the same genus (*genus*) [*Gattung* (*genus*)], hence *compositum* differs from *quantum*, and the many would in that case be able to be a variety [*varietaet*], every *quantum* contains a multitude [*Menge*] but not every multitude is a *quantum*; rather, only when the parts are homogeneous. (*Metaphysik Vigilantius* 1794–5, 29:991)

Thus, a *quantum* requires that the manifold be at least logically homogeneous, that is, must belong to one and the same genus.

Kant next explains his use of genus and species in this context. Kant introduces the terms "quiddity," "quality," and "quantity," corresponding to a traditional trio of metaphysical questions one can ask about a thing: What? What sort? How much? He explains:

> *Quidditas* if one should speak that way, would be distinguished from quality as the determination of genus and the specific difference; e.g., quiddity the genus of which is essence: but whether it is hard or soft belongs to quality; therefore, in regard to species conceived under the genus. (*Metaphysik Vigilantius* 1794–5, 29:991)

Quiddity consists of the genus and the specific difference that together define the essence, while quality concerns further specific differences. But Kant is not fond of the notion of *quidditas* as a separate type of determination, calling it a "barbarous" way of speaking. He is instead focused on the more fundamental distinction between *quantitas* and *qualitas*. Kant goes on to contrast quality and quantity:

> quality differs from quantity [*Quantität*] in that, and to the extent that, the [*former*][5] indicates something in the same object which is inhomogeneous [*ungleichartiges*] with regard to other determinations found in it. Hence, quality is that determination of a thing according to which whatever is specifically different finds itself under the same genus, and can be distinguished from it. This is heterogeneous [*heterogen*] in opposition to that which is not specifically different, or to the homogeneous [*homogen*]. (*Metaphysik Vigilantius* 1794–5, 29:992)[6]

[5] I follow the editors of Kant's Lectures on Metaphysics in thinking that the term "latter" had been mistakenly inserted for "former" either by Kant, by the student taking the notes, or by a transcriber (students very often hired someone to rewrite their notes). (See Kant (1997, p. 460, note b.) The last line of the quote and other passages substantiate the claim that this mistake had been made.

[6] In an extensive survey of passages, I have not been able to detect any distinction between Kant's use of *gleichartig* and *homogen*, and I take them to be synonymous.

Qualities are specifically different characteristics, and these specific differences are heterogeneous to each other. Quantity, in contrast, does not even allow specific differences, and the lack of specific difference is called homogeneity (see also *Metaphysik Mongrovius*, 1782–3, 29:839). In short, qualitative differences are heterogeneous, and the homogeneity of quantity excludes any qualitative difference at all.

I will call the logical homogeneity that excludes all qualitative difference "maximum" or "strict" logical homogeneity. Kant appeals to strict logical homogeneity to explain quantity, and it is the same notion that explains the homogeneity of magnitudes.

It is important to note that logical relations among concepts, in particular, genus and species relations, are used to define strict logical homogeneity, and hence to define the homogeneity of magnitudes.[7] This does not mean that logic has the resources to represent the homogeneity of magnitudes. That is, one cannot represent the difference of the parts of a homogeneous manifold by means of the genus–species relations among concepts, or any other relations among concepts. In fact, the contrary is true: the representation of strict logical homogeneity requires something extra-logical in the sense that it requires something beyond the reach of conceptual representation on its own, as we shall see when we take a closer look at Kant's views on specific and numerical difference. But to avoid the misleading suggestion that a strict logical homogeneity is something that can not only be characterized but also represented by means of resources available in logic, I will henceforth refer to it simply as "strict homogeneity."

7.4 Strict Homogeneity and the Limits of Conceptual Representation

In further notes from his lectures, Kant explains the notion of strict homogeneity in terms of numerical difference. We have seen that the distinction between quiddity, quality, and quantity corresponds to generic and essential specific difference, mere specific difference, and the not-even-specifically different. Kant claims in other lectures on metaphysics that all difference is generic, specific, or numerical (*Metaphysik Volckmann* 1784–5, 28:422; *Metaphysik Pölitz*, 1790–1?, 28:561). He adds in one of these lectures that "two drops of water on 2 needle points are numerically different and

[7] I mean "define" in a broad sense of characterize or explicate, not the narrow Kantian sense of define as "exhibit originally the *exhaustive* concept of a thing within its boundaries." See Section 3.3 above.

specifically identical" (28:422). Kant elaborates on the notion of numerical difference in a lecture on metaphysics most likely delivered in the decade of the *Critique*:

> The concept of a man [*Mann*] is already more closely determined than the concept of a human [*Menschen*]; this is in relation to that genus and that species. Meanwhile, this species can again become a genus with respect to another species. The genus differentiates itself from a species, insofar as different species can be contained under a genus. Under a genus [*genera*] these are called inferior concepts [*conceptus inferiores*] and their difference is specific difference [*differentia specifica*]. If this species cannot itself again be regarded as a genus, then it is a lowest species [*species infima*] and the difference of multiple [*mehrerer*] lowest species is numerical difference [*Differentia numerica*]. (*Metaphysik von Schön*, 28:504, late 1780s)

There are two points here. First, a concept's status as either a genus or species depends on whether it is viewed in relation to a concept that falls under it or in relation to a concept under which it falls. Thus, concepts represent both the generic differences of quiddity and the specific differences of quality, and both quiddity and quality can be described more broadly as specific differences in the sense that concepts represent qualitative differences. Second, an *infima species* is a concept under which no further concepts can fall, and hence is a species but not a genus. Numerical difference is difference even where no further specific difference is possible.

There is an apparent difficulty with Kant's account of bare numerical difference that we need to address. He ties it to the notion of *infimae species*. As we saw in Section 5.3.2, however, Kant claims in the *Critique* that it is a presupposition of the faculty of reason that there are no *infimae species*. This presupposition is based on the fact that concepts are by their nature general representations; that is, they are capable of referring to more than one thing by means of a common characteristic (A655–6/B683–4). This might suggest, however, that there could be no representation of bare numerical difference.

The solution of this difficulty is found in two ways of considering concepts. Kant claims that if we consider a concept apart from all relation to an object – that is, consider it from the point of view of general logic – we see that because it is a general representation, it is always possible in principle to add further specifications to it. For that reason, a concept is by its nature not an *infima species*. Hence, from the point of view of general logic, there are no *infimae species*.

According to Kant, however, we can also consider concepts as they are employed in the cognition of objects. Concepts so employed are not determined or qualified *ad infinitum*. Kant states in the *Jäsche Logik* that when we employ a concept in the cognition of an individual, we either do not notice or ignore the further possible specifications of the concept: "Only comparatively for use are there lowest concepts, which have attained this significance [*Bedeutung*] through convention, as it were, insofar as one has agreed not to go deeper here" (9:97). And in a lecture on Logic, he states:

> A *species infima* is only comparatively *infima*, and in use the last. It must always be possible to find another species, whereby this latter would in turn become a genus. But applied immediately to *individua*, a species can be called a *species infima*. (*Wiener Logik*, c. 1780, 24:911)

Returning to the view we have been fleshing out, and drawing together several points discussed, we see that Kant's position implies that a *quantum* differs from a *compositum* in being homogeneous, that being homogeneous corresponds to quantity, and that homogeneity requires specific identity with numerical diversity. Kant is explicit about this result:

> Homogeneity is specific identity with numerical diversity [*numerischen Diversitaet*], and a *quantum* consists of homogeneous parts [*partibus homogeneis*] . . . (*Metaphysik von Schön*, 28:504, late 1780s)

Quanta, that is, concrete magnitudes, exhibit strict homogeneity; that is, the parts are specifically identical yet numerically diverse.

Kant's understanding of conceptual representation precludes concepts representing numerical difference with specific identity. That is because concepts distinguish what they represent through specific differences. This is a fundamental limitation on what concepts can represent. In contrast, intuition can and does represent bare numerical difference. Both space and time contain a manifold of qualitatively indistinguishable and specifically identical and yet numerically distinct potential parts. Putting this together with the results of Chapter 5, intuition allows us to singularly represent an indeterminate homogeneous manifold, that is, to singularly represent a manifold the indeterminately represented parts of which are qualitatively identical yet numerically diverse.

The limitation of conceptual representation I've described requires two important clarifications. First, Kant thinks that we possess concepts that are derived from intuitions; the concept of a region of space, for example, is what Kant calls a "sensible concept" that is based on an intuition of space. Such concepts draw on the properties of intuition for their content and are not subject to the same representational limitation. Thus, we may derive concepts that depend on bare numerical difference from intuition, but we cannot represent that difference by means of concepts on their own, without relying on intuition.

A second closely related point is that concepts on their own could allow us to provide a negative characterization of strict homogeneity by means of the concepts of species and difference: homogeneity is a difference where there is no specific difference. This is what we find in Kant's characterization of strict logical homogeneity. This negative characterization, however, has no positive content; it doesn't reveal what such a difference would be. On the other hand, intuition allows us to represent mere numerical difference, and hence allows us to give content to and establish the objective validity of what would otherwise be an empty characterization of strict homogeneity. Since Kant employs the notion of strict homogeneity in his account of the homogeneity of magnitudes, intuition makes it possible for us to represent concrete magnitudes, that is, *quanta*.[8]

Since the only forms of intuition for humans are space and time, the argument establishes that the intuitions of space and time make it possible to represent *quanta*. It is important to note that the argument does not establish that only spatial and temporal manifolds are magnitudes. As we saw in Section 4.1, Kant is quite clear in the Anticipations of Perception that qualities themselves have a magnitude in virtue of their intensity. Kant holds that every quality has homogeneous parts that "coalesce" in giving the quality its intensity. A light, for example, has an intensity that can in principle be increased or decreased without changing the quality itself, only the amount of that quality. Nevertheless, the representation of the quantity of an intensive magnitude, and in fact the representation of it as a magnitude at all, requires the aid of the representations of space and time as extensive magnitudes.

Kant wrote notes in his copy of the first edition of the *Critique*, and in the margins of the Axioms of Intuition, he emphasizes the importance of the homogeneity of space and time in representing a homogeneous manifold. He states:

> We can never take up a manifold as such in perception without doing so in space and time. But since we do not intuit these for themselves, we must take up the homogeneous manifold in general in accordance with the concepts of magnitude. (23:29)

Kant claims that space and time are required to take up a manifold as such in perception. In this context, I take him to use this locution to leave room for a distinction between perceiving a homogeneous manifold "as such," and perceiving something whose manifold is not perceived "as such," namely, the real corresponding to a sensation which has intensive magnitude, and is perceived not as a manifold but as a unity. There are manifolds of qualities in perception, but to take up a manifold as such in perception requires the aid of the extensive magnitude of time. I will return to intensive magnitudes below.[9]

[8] I would like to thank Tyler Burge for assistance in clarifying several points in this argument.

[9] Kant's contrast between quality and quantity and its relation to the representation of homogeneous manifolds in mathematics raises further questions, such as how to understand the qualitative properties of geometrical figures. I examine Kant's understanding of the metaphysics of quality and quantity and its relation to geometry in Chapter 8.

7.5 The Role of Intuition: Kant contra Leibniz on the Identity of Indiscernibles

We have seen that in Kant's view, concepts alone cannot represent bare numerical difference. On the assumption that we have, and only have, two forms of representation, concepts and intuitions, it falls to intuition to play that role. This is only an argument from elimination, however. Confirmation that Kant holds that intuition can represent bare numerical difference would be welcome. Furthermore, I have relied heavily on notes of Kant's lectures and unpublished writings, and evidence for Kant's views from published work would be helpful.

The Amphiboly of Pure Reason in the *Critique* provides that evidence. Kant uses the Amphiboly to attack Leibniz's principle of the identity of indiscernibles, which claims that two completely indiscernible individuals are identical. According to Leibniz:

> There is no such thing as two individuals indiscernible from each other.... Two drops of water, or milk, viewed with a microscope, will appear distinguishable from each other. (Leibniz (1989, p. 687))

Despite Leibniz's claim that two drops of water or milk can be distinguished through a microscope, his point is not that there will always be differences among individuals that as a matter of fact we can empirically detect. Leibniz is making a different point: any two drops can in principle be distinguished. Leibniz believes in an underlying rational order of the world that reflects God's intellect. He also believes that there must be a sufficient reason for every fact, and that all reasons correspond to subject–predicate relations between concepts. As Kant understands him, Leibniz's point in this passage is that God, if no one else, can always distinguish any two individuals by their qualities, no matter how similar they may appear to be, and that means that in principle, distinct individuals are always conceptually distinguishable. As Benson Mates puts it, Leibniz's position is that

> God's concepts . . . are fine-grained enough to distinguish each individual from all of the others. It is obvious that by virtue of their accidents, any two individuals will fall together under a very large number of concepts, that is, will have a large number of attributes in common. But the principle assures us that however similar they may be, there will always be some concept under which one of them falls and the other does not. (Mates (1989, p. 135))

This is the way Kant understands Leibniz, and Leibniz's position can be recast in Kantian terms: between distinct individuals there are always specific differences that can be represented conceptually. It is this claim that Kant attacks in the Amphiboly. Kant claims that the role of intuition in our cognition allows for numerical difference even when two objects are conceptually indistinguishable:

> However identical everything may be in regard to [the comparison of two objects in respect of their concepts], the difference of the places of these appearances at the same time is still an adequate ground for the **numerical difference** of the object (of the senses) itself. Thus, in the case of two drops of water one can completely abstract from all inner difference (of quality and quantity), and it is enough that they be intuited in different places at the same time in order for them to be held numerically different. (A263–4/B319–20)

Kant reverses Leibniz's claim concerning two drops of water, and argues that if we abstract from all inner qualitative and quantitative differences, their simultaneous location in different regions of space is sufficient to ground cognition of their (discrete) numerical diversity.

Kant diagnoses the source of Leibniz's error as a misunderstanding of the nature of human cognition. Leibniz, he claims, fails to recognize that we have a faculty of sensibility with its own distinctive kind of representation, namely, intuition. According to Kant, Leibniz believes we have only conceptual representations, the sort that belong to the intellectual faculty, and that sensation is but a confused form of conceptual representation. Thus, in Kant's view, Leibniz in effect assimilates intuitive representation into his model of conceptual representation. Since Leibniz thinks that objects of experience are cognized only through the understanding, all representation is of a fundamentally conceptual nature. Hence, the difference between any two individuals can be represented conceptually only as a specific difference, a difference in quality. As Kant states it:

> Leibniz took the appearances [i.e., objects of sensibility] ... for *intelligibilia*, i.e., objects of the pure understanding ... and on that assumption his principle of **nondiscernibility** (*principium identitatis indiscernibilium*) certainly could not be disputed, but since they are objects of sensibility ... plurality [*Vielheit*] and numerical difference are already given us by space itself.... For a part of space, even if it may be completely similar and equal to another [part], is nevertheless outside of it, and for this very reason is a different part from that which is added to it in order to constitute a larger space. (A264/B320)

Space, the form of the faculty of sensibility, is a source of plurality and numerical difference, and can give numerical diversity to specifically identical individuals.[10] Our interest here is not in whether his argument against Leibniz

[10] Note that since Kant holds that space and time are not discrete but continuous, his reference to plurality (*Vielheit*) confirms that the category of plurality applies to continuous as well as discrete magnitudes.

is fair or his diagnosis correct, but in Kant's claim that if objects of experience were objects of understanding alone, then the principle of the identity of indiscernibles would hold (see also A272/B328). In other words, if the only means for us to represent objects were conceptual, then we could not represent specific identity with numerical diversity. Furthermore, it is intuition that allows us to overcome this limitation of conceptual representation. As a consequence, intuition allows us to represent strict homogeneity and hence magnitudes.

It is important to note that there are two notions of numerical diversity in play in the Amphiboly: the numerical diversity of discrete objects and the continuous numerical diversity of regions of space, which Kant describes as the "plurality and numerical difference given to us by space itself." Kant argues that intuition allows us to represent the numerical diversity of two water drops, that is, the numerical difference of discrete objects. He argues that if we abstract from any difference of quantity, and hence any difference between the amount of space each object occupies, their simultaneous position in different locations in space would still distinguish them. In other words, the discrete numerical diversity of indistinguishable objects is grounded in the continuous numerical diversity given by intuition. This supports the claim made in Chapter 5 concerning singularity: the cognition of singular spatial objects is grounded in the singularity of the homogeneous manifold of space in which they appear.[11]

7.6 The Relation between Kantian Homogeneity and Euclidean Homogeneity

We have seen that Kant refashions the notions of homogeneity and magnitude so that they are founded on the representation of numerical difference without specific difference. Kant is thereby attempting to capture something he believes is essential to the mathematical character of the magnitudes treated in the Euclidean theory of magnitudes. But if Kant wishes to retain the essence of the Euclidean theory, what is the relation between his notions of homogeneity and magnitude and those of Euclid?

As Kant defines homogeneity, a line counts as a magnitude because the manifold parts of the line are homogeneous with each other. Similarly, an area

[11] Kant's treatment of discrete numerical diversity rests on his understanding of discrete *quanta*. Kant thinks that continuous *quanta*, and in particular, the continuous *quanta* of space and time, are more fundamental, and that discrete *quanta* are *quanta* in a secondary sense that reflects their dependence on the representation of continuous *quanta* that are distinct from one another. I will examine the relationship between Kant's understanding of discrete *quanta* and number in more detail in work focused on Kant's philosophy of arithmetic.

contains a manifold of smaller areas each homogeneous with each other in Kant's sense, and a volume contains a manifold of smaller volumes that are homogeneous with each other. Moreover, two distinct lines are homogeneous with each other in virtue of the fact that the parts contained in one are qualitatively indistinguishable from the parts contained in the other – they and all lines contain a one-dimensional qualitatively indistinguishable manifold. The same point holds for areas and for volumes. Furthermore, in Kant's view, a difference in dimensionality is a difference in quality, so that there is a qualitative difference among the manifolds contained in a line, an area, and a volume. Hence, lines are inhomogeneous with areas and volumes, and so on. The result is that Kant's notion of homogeneity and magnitude yield the same distinctions among spatial magnitude kinds as found in the Eudoxian theory.

Kant's understanding of homogeneity will not align with the Euclidean if we interpret the latter as requiring that two magnitudes are homogeneous only if they satisfy the Archimedean property, so that magnitudes that cannot be multiplied to exceed one another will not count as homogeneous. Kant has no such requirement on magnitudes being homogeneous. Recall from Section 6.4 above that some have taken V.D4, which expresses the Archimedean property, as implicitly defining Euclid's notion of homogeneity, but that a careful reading of the definitions reveals that the Archimedean property is not a necessary condition of homogeneity, but a sufficient condition of two homogeneous magnitudes standing in a ratio, and hence as being subject to those propositions concerning ratios covered by the Eudoxian theory of proportions. Thus, Kant's notion of homogeneity does not leave out a feature that, according to some, Euclid held to be essential to being homogeneous. On the other hand, this is not to say that Kant endorses the idea that there are non-Archimedean ratios, which would be a substantive position to take with respect to the foundations of analysis.[12] It is only a point about how much is included in Kant's characterization of the homogeneity of magnitudes, and the extent to which his understanding of homogeneity corresponds to the Euclidean.

There is an important way, however, in which Kant's notions of homogeneity and magnitude do not coincide with Euclid's. As discussed in Section 6.7, the tradition following Euclid expanded the notion of magnitude beyond spatial magnitudes, and Kant did as well. In particular, he maintains that time is a homogeneous manifold in intuition, and hence is a magnitude. We have also seen that phenomena such as a light or a force have intensive magnitude.

[12] I consider Kant's account of the foundations of infinite analysis and infinitesimals in Sutherland (2014) and Sutherland (2020b).

Finally, Kant treats numbers, understood as collections of units, as discrete magnitudes.

We have seen that spatial magnitudes of different kinds, such as lines and areas, are not homogeneous. It is their dimensionality that distinguished them; hence, dimension is treated as a qualitative property distinguishing these kinds of spaces. On the other hand, time is one-dimensional, a property it shares with a line, yet times are not homogeneous with lines. In this case, dimensionality is not the qualitative difference that distinguishes them. In Kant's view, there is a qualitative difference between time and space that distinguishes the one-dimensional manifold of time from the one-dimensional manifold of a line.

As discussed in Section 4.1, Kant also follows the Euclidean tradition by expanding the notion of magnitude to include intensive magnitudes, such as the intensity of a light or speed. Those features of the world that count as magnitudes will also be individuated on the basis of qualitative difference in contrast to qualitative identity with numerical difference. A light varied in intensity alone, for example, is identical in quality, while a light varied in hue will have a different quality and hence be inhomogeneous with the starting light.[13]

In summary, we have seen that Kant employs different notions of homogeneity and magnitude from the Euclidean, that his notions are grounded in what I have called strict homogeneity, that concepts on their own cannot represent strict homogeneity, and that intuition makes it possible for us to represent strict homogeneity. Despite Kant's fundamental reworking of these notions, however, and his broader notion of magnitudes that goes beyond continuous spatial magnitudes, his notions comport with the Euclidean. At the same time, the larger aim of Kant's reformulation was to give a deeper analysis of homogeneity and magnitudes that helped explain the specifically mathematical properties of Euclidean magnitudes. But what is the relation between the representation of strict homogeneity and the mathematical properties of magnitudes articulated in the Eudoxian theory? Answering that question will occupy the remainder of this chapter and all of Chapter 9. The remainder of this chapter takes the first step by returning to the categories of quantity and how they are employed in the cognition of magnitudes.

[13] As noted in Section 4.1, Kant thinks that objects of experience are extensive magnitudes all of whose real properties have an intensive magnitude, which grounds his view of the fundamentally mathematical character of the world. Nevertheless, his understanding of intensive magnitudes is one-dimensional; he does not seem to accommodate multidimensional features, e.g., hue and saturation as well as the intensity of color.

7.7 The Categories, Intuition, and the Part–Whole Relation of Magnitudes

In a letter to Johann Teiftrunk in 1797, Kant explicitly connects the mathematical employment of the categories of quantity and quality to homogeneity:

> All the categories are directed upon something composed *a priori*, and, if it is homogeneous, they contain mathematical functions, but if it is not homogeneous, they contain dynamic functions. E.g., what the first concerns: that concerns the category of extensive magnitude: one in many; what concerns quality or intensive magnitude: many in one. The former the **multiplicity** [***Menge***] of the homogeneous (e.g., the square inches in an area), the latter the **degree** (e.g., the illumination of a room). (12:223)

As we saw in Section 4.2, the categorial foundation of mathematical cognition and the mathematical features of the objects of experience is mereological; that is, the primary role of the categories of quantity is to make it possible to cognize the part–whole relations of magnitudes. Moreover, Kant uses the part–whole relation to distinguish extensive and intensive magnitudes and Kant defines extensive magnitude as that magnitude "in which the representation of the parts makes possible the representation of the whole (and thus necessarily precedes it)" (A162/B203). Extensive magnitudes manifest their part–whole relational structure in a way that intensive magnitudes do not, and extensive magnitudes play the leading role in Kant's philosophy of mathematics. We are now in a position to look more closely at the mereological role of the categories in light of the role of intuition in representing strict homogeneity.

I have argued that in Kant's view, intuition allows us to represent strict homogeneity, but Kant also holds that intuition is required for the representation of the part–whole composition of homogeneous manifolds. Many of the notes in which Kant discusses the one–many–one and unity–plurality–totality concepts also concern the role of intuition. For example, lecture notes from a course on metaphysics state the following:

> The concept of magnitude is properly characteristic of understanding [*zu dem Verstand gerade zu eigen*] because it concerns itself with the connection of the manifold homogeneous [*mannigfaltigen gleichartigen*]. Magnitude [*Größe*] is employed in mathematics through the help of a pure intuition in sensibility, i.e., through the form of space and time in the determination of each figure or number. But in philosophy it cannot be determined from the concept alone whether the category of magnitude [*Größe*] has objective reality. I.e., it cannot be cognized that many together constitute a one [*daß Vieles zusammen Eins ausmache*]. (*Metaphysik Vigilantius*, 1794–5, 29:992)

This passage, which concerns the role of magnitudes in mathematics, states that intuition allows us to establish the objective reality of the category of magnitude, where that amounts to allowing us to cognize a homogeneous many constituting a one, that is, the part–whole structure of magnitudes.

For humans, the two forms of intuition are space and time, so it is not surprising that Kant considered the role of each in representing the relations between parts and wholes. For example, Kant expressed the view that the temporal act of drawing a line successively represents the parts of a line, while the spatial figure of the line simultaneously represents all those parts, and hence represents them as coexisting in a whole (*Metaphysik Vigilantius*, 1794–5, 29:994). In several important passages of the *Critique*, Kant refers to drawing a line in thought through a figurative synthesis, which I argued in Chapter 4 is a continuous synthesis that avoids the extensive magnitude regress. Even setting aside these particular roles for space in time in representing part–whole relations, he often mentions the role of space and time in representing homogeneous parts:

> the category of magnitude [*Größe*], as a homogeneous many that together constitutes [*ausmacht*] one: this cannot be grasped without space and time. (29:979)

Space and time allow us to grasp the category of magnitude by allowing us to grasp that a homogeneous many together constitute a one, that is, that the strictly homogeneous parts constitute a whole.

7.8 The Composition of Magnitudes and Intuition

So far, we have seen that in Kant's view, strict homogeneity, and hence intuition, plays an important role in the cognition of the part–whole relations of magnitudes; intuition allows us to grasp that a strictly homogenous many "constitutes" one. Cognition of this constituting relation requires the synthesis of composition, which is also central to his account of mathematical cognition.

As discussed in Chapters 2 and 3, Kant calls the Axioms of Intuition and the Anticipations of Perception mathematical principles. In the second edition of the *Critique*, Kant added a footnote to further explain what makes these principles relevant to mathematics. He delineates various sorts of synthesis or combination (*Verbindung*), and the combination at the root of the mathematical principles is composition (*Zusammensetzung*). This special act of synthesis is employed only in the representation of magnitudes (B201n; see also 4:343).[14] In contrast to the synthesis underlying the dynamical principles,

[14] Kant sometimes uses "composition" in a broader sense for any sort of combination; it is the more narrowly defined sense of composition that interests us here.

it involves a synthesis of "what does not necessarily belong to each other"; as a result, the synthesis of composition is exclusively a synthesis of a homogeneous manifold:

> All combination (*conjunctio*) is either **composition** (*compositio*) or **connection** (*nexus*). The former is the synthesis of a manifold of what **does not necessarily** belong **to each other**, as e.g., the two triangles into which a square is divided by the diagonal do not of themselves necessarily belong to each other, and of such a sort is the synthesis of the **homogeneous** in everything that can be considered **mathematically**. (B201n)[15]

The synthesis of composition of a strictly homogeneous manifold makes our representations of magnitudes possible. This and previous passages show that by representing a strictly homogeneous manifold, intuition makes possible a special synthesis of composition underlying our cognition of "everything that can be considered mathematically." But why does Kant think that the composition of a strictly homogeneous manifold is special, and what does it have to do with mathematics?

Kant's views of composition are best brought out by contrasting it to his understanding of the synthesis of concepts and its implications for conceptual representation. As discussed in Section 5.3, the intension of a concept consists of any constituent concepts it has; the concept <horse>, for example, might contain the concepts <mammal> and <ungulate>. Kant calls constituent concepts "partial" concepts (*Teilbegriffe*), and the relation between a concept and its partial concepts is that of whole to part. And to repeat a point made in Chapter 5, Kant's notion of intension is familiar, but his notion of extension differs from our contemporary concept. On our understanding, an extension consists of the objects which fall under a concept. Kant holds that objects fall under concepts, but he also holds that concepts fall under concepts, and in his lectures on logic, Kant describes the logical extension of a concept as the concepts which fall under a concept (*Jäsche Logik*, 1800, 9:98). The concept <horse>, for example, is part of the extension of the concept <mammal>. This is the primary notion of extension, and an object falls under a concept in virtue of the object possessing the marks contained in the concept employed in its cognition. The part–whole relation between concepts and their intensions is reciprocal to that between concepts and their extensions; the concept <mammal> is part of the intention of the concept <horse> if and only if the concept <horse> is part of the extension of the concept <mammal>.

We have seen in this chapter that Kant holds that concepts represent qualitative differences, that is, specific differences. If we combine the concept <ungulate> with the concept <mammal>, we generate a more specific concept and restrict the extension of the concept <mammal> to a smaller

[15] See also *The Critique of Practical Reason*, 5:104.

extension – smaller in the sense that the extension now excludes any concepts that do not include <ungulate> in their intension, and hence exclude non-ungulates. We also saw that in Kant's view, concepts can only represent qualitative differences; hence, the combination of unique concepts always results in ever more specific concepts and ever smaller extensions. There is no room in this theory for the combination of concepts that would yield the representation of more than what the constituent concepts represent.

In contrast, the composition of a strictly homogeneous manifold, of bare numerical difference without specific difference, yields more of the same. Kant makes the importance of this property of intuition clear in a passage from the Amphiboly already cited:

> plurality [*Vielheit*] and numerical difference are already given us by space itself . . . For a part of space, even if it may be completely similar and equal to another [part], is nevertheless outside of it, and for this very reason is a different part *from that which is added to it in order to constitute a larger space.* (A264/B320, my emphasis)

Kant states that parts of space are numerically different while being similar and equal, and then adds that adjoining parts of space together constitute a greater space. This latter property is not directly relevant to the point of his argument, which is that space allows the representation of bare numerical difference. Kant mentions it because he thinks that this important property of space is intimately tied to the representation of bare numerical difference, that is, the representation of a homogeneous manifold in intuition.

Kant holds that the composition of parts that have strict homogeneity results in a whole that is greater than those parts. Concepts on their own, no matter how they are combined, do not allow the representation of this kind of part–whole composition. And it is exactly this sort of composition that Kant thinks is characteristic of "that which can be considered mathematically." Without the representation of this sort of combination, mathematical cognition would not be possible at all.[16]

The inability of concepts to represent mathematical composition can be brought out in another way. As discussed in Section 7.5 above, Kant claims that if objects were objects of understanding alone and hence represented only by means of concepts, then Leibniz's principle of the identity of indiscernibles

[16] Longuenesse draws a similar contrast between the combination of marks in a concept and the generation of a multiplicity from two given multiplicities "by means of an operation which has nothing in common with the combination of marks making up the content of a concept" (Longuenesse (1998, p. 277)). The conclusion she draws is quite different, however; in her view, this shows that Kant's concept of number is implicitly a second-order concept, one that reflects a rule for constituting the extensions of concepts. In my view, the contrast shows that mathematics requires a special synthesis of combination of a strictly homogeneous manifold, that is, of intuition.

would hold. Kant uses this hypothetical to bring out the limitation of conceptual representation. If we attempted to use only concepts to represent the composition of strictly homogeneous units of space into a larger space, we would have to represent each of the spaces as an instance of one and the same concept, such as the concept <cubic inch>. In that case, however, each instance of the concept would be indistinguishable and hence identical, which means we could at best succeed in repeatedly picking out one and the same object. Kant makes this point in *What Real Progress Has Metaphysics Made since the Time of Leibniz and Wolff?*, published in 1791:

> According to mere concepts of the understanding, it is a contradiction to think of two things outside of each other that are nevertheless fully identical in respect of all their inner determinations (of quantity and quality); *it is always one and the same thing thought twice (numerically one).* (20:280, my emphasis; cf. A263/B319 and A282/B338)

In other words, concepts alone cannot represent pure numerical diversity, and our attempt at representing distinct spaces collapses into representing just one. As Kant very strikingly puts it, if we had only conceptual representation alone, that would require us to say that different similar and equal spaces

> are one and the same space (since in that way we could bring the whole of infinite space into a cubic inch, and even less) . . . (20:282)

Since concepts alone cannot represent the bare numerical diversity of mathematical composition, we must rely on intuition.

In summary, the cognition of part–whole relations of magnitudes in intuition, and in particular, cognition of the composition of those parts into a whole, plays a crucial role in making mathematical cognition possible. Kant's view is that only the representation of a strictly homogeneous manifold allows the representation of the sort of composition required for the Eudoxian theory of magnitudes and characteristic of mathematical cognition.

7.9 The Role of Strict Homogeneity in Representing Magnitudes: Clarifications

We have seen that in Kant's view, the representation of strict homogeneity, and hence intuition, plays an important role in making mathematical cognition possible. One might think, however, that the representation of strict homogeneity is too strong a condition on the representation of mathematical composition, a worry that might arise in three distinct ways.[17] Responding to each helps to clarify Kant's account.

[17] I would like to thank Michael Friedman for pressing me to be more perspicuous about Kant's position on these issues.

One might think Kant's account entails that there is only mathematical combination when there are no qualitative differences at all among the things combined. For example, if I represent the combination of blue and red blocks of the same size into larger groupings, that would not count as mathematical composition because the blocks are qualitatively distinct. But this misconstrues the things to which Kant's claim applies. Kant's account requires that only those features that are directly represented as mathematically composed be represented as strictly homogeneous. In the case of blocks of the same size, what is represented as mathematically composed are their volumes, that is, their spatial three-dimensional manifolds, which are homogeneous. While there are different features of the blocks, the feature that is mathematically composed, their spatial extension, is strictly homogeneous. In Kant's view, color qualities can be mathematically combined under the right circumstances, that is, when the colors are strictly homogeneous and hence qualitatively identical and they are combined in a way that leads to an increase in intensity, but that is another matter. In the present case, the qualitative differences among the blocks do not prevent the representation of their mathematical combination with respect to their spatial properties.

A different worry is that requiring qualitative identity of the manifold is too stringent even in the paradigm case of spaces, since we can mathematically combine spaces of different sizes. Lines of different lengths, for example, differ in quantity, yet they can be mathematically composed into a longer line. This worry is reinforced by the fact that Kant's discussion in the Amphiboly refers to parts of space that are identical in both quality and quantity. Here it must be said that although the Amphiboly has important implications for mathematical cognition, it is not the focus. Kant is arguing against Leibniz's assumption that all representation is fundamentally conceptual in nature, and he does so by pressing on Leibniz's principle of the identity of indiscernibles. His argument focuses on the case in which the spaces or the raindrops are identical in both qualitative properties and quantitative properties such as size, but this is not a requirement for strict homogeneity and hence mathematical composition. He includes these quantitative properties for the sake of his argument against Leibniz: even if we grant that the spaces or raindrops are the same size, the difference in their locations will still distinguish them. But to distinguish two raindrops or spaces by their size already presupposes the sort of mathematical composition underlying the representation of the spatial extent of the raindrops that Kant's appeal to strict homogeneity is meant to explain. Moreover, the relative sizes of the raindrops or spaces is a quantitative determination, that is, a determination of the *quantitas* of the raindrops or spaces as *quanta*. Kant's requirement of strict homogeneity concerns the representation of *quanta* and their combination; the conditions for the determination of their *quantitas* is a further matter. What is required is that there not be any qualitative difference in the manifold making up the *quantum*, and magnitudes of different sizes meet this requirement.

Finally, one might think that if only the qualitatively indistinguishable manifolds of space and time are strictly homogeneous, then only spaces and times are magnitudes. Kant, however, clearly thinks of intensive magnitudes, such as a light or a force, as magnitudes. One might then conclude that the strict homogeneity requirement as I have described it contradicts Kant's own understanding of magnitudes. The antecedent of this conditional is false, however; Kant holds that intensive magnitudes, and not just space and time, contain a strictly homogeneous manifold. A particular color of light, for example, will differ in quality from other colors, and hence be specifically distinguished from them, but the different intensities of that light will be homogeneous with each other. Kant states this in the *Prolegomena*:

> the transition to sensation from empty time and space is possible only in time, with the consequence that although sensation, as the quality of empirical intuition with respect to that by which a sensation is distinguished specifically from other sensations, can never be cognized *a priori*, it nonetheless can, in a possible experience in general, as the magnitude of perception, be distinguished intensively from every other homogeneous sensation. (4:309)

He makes the same point in the First Introduction to the *Critique* of the Power of Judgment:

> One can definitely say: that things must never be held to be specifically different through a quality that passes into some other through the mere diminution or augmentation of its degree. (20:226n)

The fact that sensations can represent a strictly homogeneous manifold, however, might seem to lead to the conclusion that intuition is not required to represent magnitudes after all, contrary to the argument outlined in Sections 7.3 and 7.4 above. Kant nevertheless makes it quite clear that sensations are homogeneous manifolds, yet apart from space and time, they do not and cannot allow us to represent a strictly homogeneous manifold. Recall that because they are intensive, their intensity is represented as a unity. That they have a manifold at all, much less what their intensity is, can be represented only with the aid of intuition. The Anticipations of Perception makes it clear that the representation of the variation of a sensation and the real corresponding to it over time is required to reveal and represent the manifold of the quality (see Section 4.1). This dependence of the representation of intensive magnitudes on the extensive magnitudes of space and time is what Kant has in mind in the *Prolegomena* when he states that the principle of the Anticipations

> does not subsume ... sensation ... directly under the concept of magnitude, since sensation is no intuition **containing** space or time, although it places the object corresponding to it in both. (4:306)

Thus, when Kant defines a magnitude as a homogeneous manifold in intuition, the sense of "in" comprises both manifolds of space and time themselves, and sensations whose objects are placed in space and time (see Section 4.1). Space and time make it possible to represent strict homogeneity, and the representation of spatial and temporal manifolds makes it possible to represent intensive magnitudes as magnitudes.

7.10 Conclusion

We have seen that Kant provides a deeper analysis of the notions of homogeneity and magnitude found in Euclid and the Euclidean tradition. Kant appeals to the notions of quality and quantity, and the latter is explicitly characterized as numerical difference without specific difference, which is the defining feature of what I have called "strict homogeneity," the homogeneity of magnitudes. We have also seen that in Kant's view, concepts alone cannot represent a strict homogeneity, while intuition makes it possible, which reveals an important role for intuition in the representation of magnitudes. Despite Kant's fundamental reworking of the notions of homogeneity and magnitude, Kant's views align with the Euclidean, for they accommodate the fact that lines are homogeneous with lines, areas with areas, and volumes with volumes, while these kinds are inhomogeneous with each other.

We then stepped back to reconsider the roles of the categories of quantity and intuition in the cognition of magnitudes. The categories of unity, plurality, and totality are employed in the representation of their part–whole relations, and we found that intuition allows us to grasp the part–whole relations that constitute a magnitude. More specifically, intuition underlies our cognition of magnitudes through a special synthesis, composition, that is particular to magnitudes. In Kant's view, this special synthesis has properties required for mathematical cognition, for it allows the representation of a composition of a manifold the result of which is a magnitude that is homogeneous with that which one started and yet is greater than it. Since the part–whole composition relations are fundamental to Euclid's *Elements*, including the Eudoxian theory, Kant has given a deeper analysis of important features of magnitudes that is a key part of explaining the mathematics of magnitudes developed in the Eudoxian theory of proportions.

As it stands, however, this explication of Kant's understanding of magnitude is incomplete. We have seen that Kant defines the homogeneity of magnitudes by appealing to a fundamental metaphysical distinction between quality and quantity, which has ancient roots, but Kant departs from that tradition by arguing that intuition is required for the representation of the latter. How was the metaphysics of quality and quantity understood in Kant's time, and how did he alter it to make room for the role of intuition? The next chapter takes up this topic.

8

Kant's Revision of the Metaphysics of Quantity

8.1 Introduction

We saw in the previous chapter that Kant's new account of magnitude appeals to the notion of strict homogeneity, by which Kant means specific identity with numerical diversity, and Kant further characterizes specific identity and numerical diversity by appealing to the quality/quantity distinction. It is clear that the quality/quantity distinction is fundamental to Kant's account of magnitudes. Moreover, the relation of equality is elemental in all mathematical sciences and the relation of similarity plays a key role in geometry; and Kant, following a long philosophical tradition, characterizes similarity as identity of quality and equality as identity of quantity. At the same time, Kant departs from the philosophical tradition in his claim that quantity, more specifically *quanta*, requires intuition for its representation. This chapter will examine the metaphysics of quantity that Kant inherited, how that metaphysics was brought to bear on mathematics, and how Kant revises it.

Sections 8.2 and 8.3 start with a survey of the accounts of quality and quantity and their relation to geometry found in the rationalist metaphysics preceding Kant, and more specifically, in the work of Leibniz, Wolff, and Baumgarten. Sections 8.4 and 8.5 then examine how Kant understood the distinction between quality and quantity, and how it contrasts with his predecessors. Just as Kant preserves the Eudoxian theory of proportions while reworking the Euclidean theory of magnitudes to take into account the conditions of the cognition of magnitudes, Kant keeps the framework of the metaphysics of quantity while reforming it at its foundation. Understanding the way in which he accomplishes this task places his innovative reaction to prior metaphysics in context and fleshes out his understanding of the mathematical nature of magnitude and quantity. It also clarifies the distinction between qualitative and quantitative properties in mathematics.

8.2 Leibniz on Identity, Quality, and Quantity

Leibniz famously envisioned the possibility of a unified theory of all knowledge, in which representations of primitive ideas could be combined to reveal the relationships between ideas and the truths based on them. He referred to different aspects of that theory as an *ars combinatora*, a universal characteristic, and a *lingua philosophica*.

An important part of his vision required identifying primitive ideas that correspond to reality and are the building blocks of other ideas. Leibniz's search for primitive ideas was influenced by Aristotle's Categories and its long history in Western philosophy from Porphyry to Melanchthon.[1] While Leibniz's ideas for an *ars combinatora* are familiar, his search for primitive ideas and his attempt to reform the categories has received less attention.[2] By late scholasticism, a treatment of categories was not limited to the particular Aristotelian categories such as substance, quality, and quantity, but encompassed the ordering of fundamental concepts under a highest genus, and Leibniz's own views were influenced by this tradition.[3] Leibniz sometimes believed that absolutely primitive ideas could not be discovered, but that determining even relatively primitive ideas would provide insight into what is true. Leibniz's own efforts to identify at least relatively primitive ideas included the ideas of similarity, equality, quality, and quantity. In notes taken around 1680, for example, Leibniz explored the relations among a variety of fundamental concepts, including coincidence, similarity, quality, congruence, and quantity. Equality, similarity, and congruence are obviously fundamental to mathematics and geometry.[4]

The concept of identity is central to Leibniz's logic and metaphysics: Leibniz took the logical statements that reveal the containment relations among ideas to be identity statements. It is therefore unsurprising that the concept of identity plays a lead role in Leibniz's attempts to reform the Aristotelian categories. In particular, he defined similarity as identity of quality and equality as identity of quantity, thereby making similarity and equality dependent on his account of quality and quantity.[5]

[1] Schepers (1969, p. 38). For a broad understanding of Leibniz's metaphysics, I'm greatly indebted to Adams (1994).

[2] See Rutherford (1995, p. 233). As he notes, chapter V of Couterat's *La Logique de Leibniz* discusses Leibniz's larger aim of an encyclopedia of human knowledge, but it is the work of Heinrich Schepers on Leibniz's attempts to reform the categories that has most helped to bring this side of Leibniz's work to the fore; see Schepers (1966) and (1969).

[3] See Schepers (1969, p. 38). I am simplifying the range of Leibniz's thought on this issue; Leibniz entertained various approaches to thinking about the primary ideas, including for a time that there were only two absolutely primitive ideas, that of God and nothing. It isn't clear that this last possibility is most usefully considered on a genus–species model.

[4] See Schepers (1969, p. 45) for this and other examples of Leibniz's attempts to reform Aristotle's categories.

[5] See, for example, "On *Analysis Situs*," Leibniz (1971), V 179, translated in Leibniz (1956, I, p. 392).

The definition of similarity as identity of quality and quality as identity of quantity had a very old pedigree; it has roots in ancient Greek philosophy and appears in Aristotle's metaphysics, where he discusses "relatives," that is, ways in which things can be related to one another. In this context, unity plays the role of being one and the same, and hence identical. Aristotle states:[6]

> So again, in another sense [i.e., from numerical relatives], are the equal and similar and the same. For they are all so called in respect of one; for things are the same whose substance is one, similarity whose quality is one, equal whose quantity is one.

Leibniz was undoubtedly influenced by these traditional definitions. Leibniz, however, also characterizes the distinction between quantity and quality in terms of how we grasp them or know them:

> Quantity can be grasped only when the things are actually present together or when some intervening thing can be applied to both. Quality represents something to the mind which can be known in a thing separately and can then be applied to the comparison of two things without actually bringing the two together either immediately or through the mediation of a third object as measure.[7]

Thus, Leibniz, characterizes the quantitative and qualitative properties of a thing according to whether they can be grasped or known on their own or require compresence of something else for that grasp or knowledge. Leibniz's characterization of the quality and quantity of things is intended to have the broadest metaphysical application and to support the definitions of similarity and equality. Since similarity is defined as identity of quality, if two completely similar things are each viewed in isolation, they cannot be distinguished. Distinguishing them requires comparing them with each other or to some other object used as a standard. His view is that a thing's being red, for example, is a property of a thing that can be grasped and known in the thing on its own, and "then be applied in the comparison of two things," and hence is a quality, while the size of something cannot be grasped or known in the thing on its own – our knowledge of its size is entirely grounded in its relation to things outside it – and hence is a quantity.

The most important feature of Leibniz's account is that his metaphysical definitions of similarity and equality include geometrical similarity and equality as a special case, and hence account for their metaphysical grounding. Two geometrical figures are similar if and only if all the properties internal to one figure are the same as the other figure so that the two figures, considered only with respect to those internal properties, are indistinguishable. Two equilateral

[6] See Aristotle (1984), Met. V, 1021^{a}10.
[7] Leibniz (1971), V 180. Translated in Leibniz (1956), I 392.

triangles in a Euclidean space, for example, are similar; considered each on its own, they are completely indistinguishable. This is true even if one is twice as large as the other, since that is something that could be grasped or known only by comparing the two with each other or some third figure taken as a measure.

As we saw in Chapter 6, Euclid gives a quite different kind of definition of similarity for rectilineal plane figures, one that is entirely geometrical. Recall that the *Elements* develops plane geometry in Books I–IV until the relation of similarity is required, and then inserts Book V presenting the theory of proportions (see Section 6.5). Euclid then appeals to proportions in Book VI to define similarity:

VI, Def. 1: Similar rectilineal figures are such as have their angles equal one by one and the sides around the equal angles proportional.

Although Euclid's definition is geometrical, Leibniz's characterization coincides with it. If the conditions articulated in Euclid's geometrical definition are not satisfied, then the rectilineal figures will be distinguishable from each other when each is regarded on its own, and hence they will not be similar according to Leibniz's metaphysical definition. And if the conditions are satisfied, then the figures will not be distinguishable from each other when each is regarded on its own.[8]

As will be discussed in the next chapter, Euclid does not define equality, and simply assumes that geometrical objects – lines, areas, volumes, and angles – stand in the relation of equality, and that a circle, which is defined as a plane figure contained by one line and whose radii are equal, can be constructed. Nevertheless, the geometrical notion of equality he employs fits with Leibniz's characterization of equality, since figures that are regarded each on their own cannot be distinguished merely by their size; they must be directly compared with each other or both must be compared with some third thing in order to

[8] Incongruent counterparts provide a problem, however. For example, consider two 3-4-5 right triangles (i.e., each of which has a hypotenuse that stands in the ratio 5:4 and 5:3 to the remaining sides) whose bases are parallel, but which "point" in opposite directions, one to the left and one to the right. The triangles meet Euclid's definition of similarity. Hence, if Leibniz's characterization of similarity is to coincide with Euclid's, then he must claim that handedness cannot be used to distinguish two rectilineal plane figures, each taken on their own. Leibniz's characterization of similarity will be coextensive with Euclid's only if our ability to distinguish them when viewed one at a time implicitly appeals to something beyond either triangle that serves to distinguish the handedness of each – for example, in relation to the left or right side of my body. This would further entail that handedness is a quantitative property. An alternative might be to deny that we can distinguish two figures by appeal to handedness. We can make this claim plausible if we regard the triangles as not constrained to a plane. In that case, free rigid motion in three-space would render them indistinguishable. I consider the views of Euclid, Leibniz, and Kant on similarity and handedness more carefully in Sutherland (2005a).

grasp or know their relative size. Leibniz's metaphysical definitions of quality and quantity therefore allow his metaphysical definitions of similarity and equality to subsume Euclid's geometrical notions as a special case. This is crucial for his project of bringing all knowledge under the umbrella of one system of primitive ideas.

The success of Leibniz's account depends on a very important detail. Geometrical similarity will be a special case of metaphysical similarity only if Leibniz's notions of quantity and quality are understood properly: whether a property is quantitative or qualitative is relative to what one is taking as an object. In particular, the quantitative properties of the parts of an object might be qualitative properties of the object as a whole.[9] Suppose, for example, that a triangle has a side three times the length of another. This ratio is a quantitative property of each of the two sides, since it can be grasped or known only with reference to something outside each, that is, with reference to each other or some measure common to both. At the same time, it is a qualitative property of the triangle as a whole since it can be known without reference to anything outside the triangle. Geometrical similarity requires that the ratios among the sides of a triangle be quantitative properties of the sides. Nevertheless, these ratios can be known without reference to something outside the triangle, and hence count as qualities of the triangle as a whole. Leibniz's metaphysical definition of similarity has the consequence that two triangles are similar if they have the same qualities, even if those qualities themselves are quantitative properties of the parts of the triangle.

Leibniz's characterization of the qualities and quantities of a thing in terms of whether they can be grasped or known to the mind in a thing separately raises two issues. First, in claiming that a property of a thing is a quantity if it can be grasped or known only when it is compresent with something else, in what sense are two things actually compresent? Leibniz's reference to being known "to the mind" might seem to undercut the idea that they must be actually present to the senses. Regardless of whether the senses are involved, what is required for compresence?

The second issue concerns the fact that the quantity of a thing is relative to something beyond it. A meaningful specification of the quantity of something, in the sense of how big it is, requires relating it to something else or to a third thing as a unit of measure, and hence is relative, which would make it an external or relational property of a thing. Yet the relation itself, that is, whether one thing is greater in size than another, seems to be determined by some

[9] I came to this conclusion in an analysis of quality and quantity published in Sutherland (2010); I discovered as that article was prepared for printing that Rusnock and George (1995) had already made the same point fifteen years earlier in their work on Kant on incongruent counterparts. The fact that I arrived at the same conclusion through a different route provides confirming evidence for it.

property each of the things independently possesses. For example, if we fix a standard of measure – inches, say – whether a thing is four inches long is determined by a property of the thing, even if the specification of the size is relative to a measure outside it, such as inches.

While Leibniz's views warrant closer scrutiny in their own right, we will not undertake an analysis of how he might respond to these worries; we are most interested in how Leibniz's views were taken up by his German rationalist successors, Wolff and Baumgarten, in order to examine Kant's reaction to them, and through them to Leibniz.

8.3 Quantity and Quality in the Metaphysics of Wolff and Baumgarten

As noted, the definitions of equality and similarity as identity of quantity and quality had a venerable tradition. Wolff was directly influenced by his study of late scholastic philosophy at least as much as by Leibniz, and he endorsed these definitions. Significantly, Leibniz also communicated to Wolff that he thought not only that these definitions were correct, but that, employed properly, they would allow a radically new foundation for geometry, one that Wolff attempted to carry out.[10] In overall approach, Wolff adheres to tradition and Leibniz's account of identity, similarity, and equality; he defines identity in terms of substitutivity and then defines equality as a property of those things which can be substituted for each other without affecting their quantity.[11] And at least in his early writings, he followed Leibniz's definition of similarity as well.[12] As in Leibniz, these definitions rely on the definitions of quantity and quality, and like Leibniz, Wolff characterizes them in relation to their ability to be grasped or known; qualities are those properties of a thing that can be intelligible without assuming anything else, whereas quantities are those properties of a thing that are not intelligible in themselves.[13] Finally, Wolff stresses that his metaphysical definition of similarity conforms to Euclid's, a requirement of any successful account.[14]

[10] I examine Leibniz's suggested approach to geometry and Wolff's failed endeavor to fulfill its promise in Sutherland (2010). I argue there that Wolff's inability to provide a foundation for geometry based on definitions provided Kant a shining example of the inadequacy of mere analytic truths to account for mathematical knowledge. Here, I focus primarily on the relation of similarity and equality to quantity and quality.

[11] Wolff, *Ontologia* (1736, pp. 86, 160–1).

[12] Wolff, *Lexikon* (1965, p. 1278).

[13] Wolff, *Ontologia* (1736, pp. 93, 203). Wolff's declared definition of quantity appeals to similarity: it is that which can be different when two things are similar. But the definition refers back to his earlier claim that the quantity of a thing is that which is not intelligible in itself, adding that similar things can differ in quantity.

[14] Wolff (1736, pp. 106–8).

On the other hand, by 1715 Wolff came to find Leibniz's characterization of similarity problematic.[15] Wolff states that Leibniz defines things being similar as being distinguishable only through compresence (*per compræsentium*). His objection concerned the first issue mentioned above: he was troubled by the sense in which two things must be compresent for comparison, stating in a note inserted to his *Ontologia* that the notion of compresence is obscure.[16] As a consequence, Wolff adopted a somewhat different definition of similarity in later editions of his works.[17] He does not mention the second issue raised above, that Leibniz's account makes quantity an external relation of things, but his follower Baumgarten does.

Baumgarten's *Metaphysica* begins with ontology, which first treats internal predicates of a thing and then relative or external predicates of a thing. His characterizations of quality (*qualitas*) and quantity (*quantitas*) fall under the former and are internal predicates of a thing. They appear in §69, which is immediately followed by characterizations of similarity, equality, and congruence in §70. These three relations count as outer determinations, since they are relations between a thing and other things, and hence properly belong to the chapter on relative or external predicates of a thing, but Baumgarten presumably includes them and their relation to quality and quantity early because of their significance. Their importance is also apparent when Baumgarten turns to the relative predicates of a thing in chapter 3, where he immediately returns to similarity, equality, and congruence. In his earlier introduction of them, he offers the following characterizations of these notions:

> §69
>
> The inner determinations can be represented in a being viewed in itself (§67, §37), and hence can be in some way cognized, or GIVEN. We can either CONCEIVE or understand (i.e., distinctly cognize) the given [inner determinations of a thing] without assuming or relating them to anything else (without the compresence of anything else), or we cannot. The former are QUALITIES, and the latter are QUANTITIES.
>
> §70
>
> Things that are the same according to quality are SIMILAR (~); according to quantity, EQUAL (=); according to both, CONGRUENT (≅). Things that are different according to quality are DISSIMILAR (Λ); according to quantity, UNEQUAL (≠); according to both INCONGRUENT (≇).[18]

[15] Wolff, *Lexikon* (1965, p. 1278); Wolff, *Anfangsgründe* (1973, pp. 118f.).

[16] Wolff, *Lexikon* (1965, p. 1278); Wolff, *Ontologia* (1736, p. 96).

[17] I will pass over Wolff's new definition of similarity here, as it does not bear directly on Kant's views.

[18] See (17:41); I depart from the translation of Fugate and Hymers in substituting "inner determination" for "internal distinguishing mark" and "cognize" for "know"; see Baumgarten (2014, p. 113).

Baumgarten characterizes an internal or inner determination of a thing as that which can be *represented* in a thing when it is regarded on its own and independently of other objects. An outer relation is one in which the determination cannot be so represented. In contrast, whether something is a quality or a quantity depends on whether it can be *distinctly cognized* in an object. Both qualities and quantities of a thing count as inner determinations since they can be represented in it.[19] But qualities differ from quantities in having a closer relation to a thing: qualities can be distinctly cognized in a thing without reference to any other thing, while quantitative properties cannot be distinctly cognized when regarding just the thing.

Baumgarten's distinction between representing a determination and distinctly cognizing it in a thing taken on its own allows him to count the quantitative properties of a thing as inner determinations, while maintaining the thrust of the Leibnizian distinction between quality and quantity in reference to how they can be cognized. The important difference from Leibniz's view is that the quantitative properties of a thing are inner determinations, and hence properly belong to the thing, despite the fact that they require a relation to something outside the thing for their distinct cognition. Baumgarten's innovation thereby resolves the second issue we raised concerning Leibniz's characterization of the distinction between *qualitas* and *quantitas*. Figure 8.1 summarizes Baumgarten's position in a diagram.

It is worth noting that the significance of Baumgarten's distinction between representing and distinctly cognizing was not lost on others. Georg Friedrich Meier was another follower of Wolff who wrote a preface to the first edition of Baumgarten's *Metaphysica* and translated it into German in 1766. He added notes in his copy of the work. After the first sentence of §69, in which Baumgarten claims that the inner determinations of a thing can be represented in a being considered in itself, he jotted:

> However, it does not follow from this that they can be entirely known with distinctness when they are represented in the thing merely considered in itself.[20]

Meier highlights the key innovation that Baumgarten introduces.

There is a final important point about Wolff's and Baumgarten's understanding of how we cognize quantity. Both understand distinct and indistinct cognition in a way that ties quantity to sensible cognition. In their view, a concept is any representation in the understanding that represents a thing.[21] Any representation of a thing that is not distinct is confused, and confused representations are sensible. Since quantitative properties of an object viewed

[19] Kant 17:35; Baumgarten (2014, p. 108).
[20] See Baumgarten (2014, p. 113).
[21] Wolff (1973, p. 6).

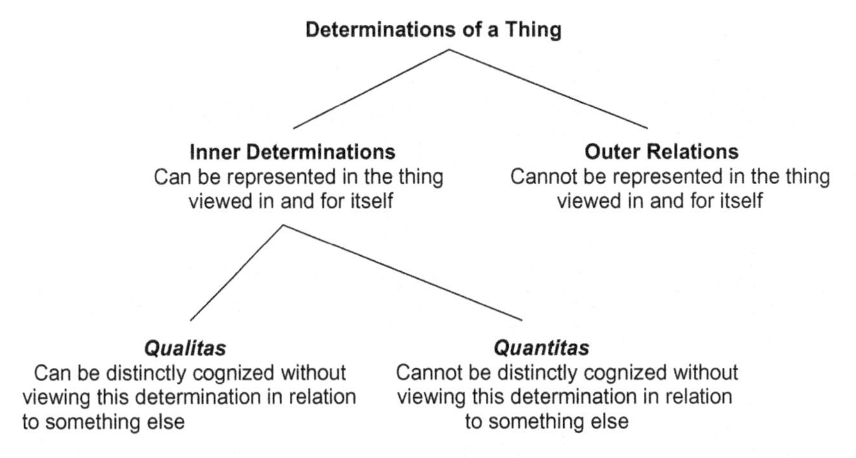

Figure 8.1 Baumgarten's metaphysics of quantity.

on their own cannot be distinctly cognized, it follows that when an object is viewed on its own, the quantitative properties are represented sensibly. In contrast, qualitative properties are capable of being intellectually represented when an object is viewed on its own. This explains Wolff's claim that similarity is the agreement of those properties that are distinguished from one another through the intellect.[22]

In summary, Leibniz, Wolff, and Baumgarten defined similarity as identity of quality and equality as identity of quantity; they in turn distinguish between quality and quantity by appeal to whether or not a property can be grasped or known or distinctly cognized in an object regarded on its own. This way of drawing the quality/quantity distinction is relative to what one takes as one's object; what might be a quantitative property of parts of an object may be a qualitative property of the object as a whole. Their account of the quantity/quality distinction has the consequence that the metaphysical definitions of similarity and equality subsume Euclid's geometrical notion of equality and his geometrical definition of similarity as special cases, fulfilling their objective of finding a place for these mathematical relations in an overarching metaphysics. But Wolff and Baumgarten characterize quantity in epistemological terms in a different way from Leibniz, and Baumgarten in particular does so in a way that allows him to hold that quantitative properties are inner determinations of a thing, while also allowing that a meaningful articulation and distinct cognition of the quantity of a thing requires reference to something outside it. We are now in a position to consider how Kant reacts to and revises this view.

[22] Wolff (1973, p. 118).

8.4 Kant's New Understanding of Quality in Relation to *Quanta* and *Quantitas*

As already noted in previous chapters, Kant used Baumgarten's *Metaphysica* in his lectures on metaphysics for many years, and his notes in his copy of Baumgarten provide a great resource for understanding his reaction to Leibnizian-Wolffian metaphysics. One must use these sources carefully, however; one can be most confident that they reflect Kant's views when he consistently disagrees with the Leibnizian-Wolffian metaphysics. In his early career, apparent agreement might simply reflect his teaching from those texts, while as Kant enters the critical period, one can be increasingly confident that a correspondence between his lectures and Baumgarten's *Metaphysica* indicate actual agreement. This is especially true when what else Kant says, including his departures from Baumgarten, presuppose some common ground.

With these precautions in mind, it is clear that Kant adopts and agrees with many features of the metaphysics of quantity found in Leibniz, Wolff, and Baumgarten. In lectures on metaphysics present during 1782–3, for example, Kant takes issue with various features of Baumgarten's account, but also asserts, without objection or qualification, the traditional philosophical definitions of equality as identity of quantity and similarity as identity of quality (*Metaphysik Mongrovius*, 1782–3, 29:838). And in lectures given in 1783–4, Kant again presents the traditional definitions without any critical comment (*Metaphysik Volkmann*, 1784–5, 28:414). Kant also states in the *Critique* that the principle that equals added to equals or subtracted from equals are equal is analytic, because it rests on the consciousness of the identity of two productions of magnitude (A164/B204f). I examine this passage and the connection between equality and magnitude production more closely elsewhere.[23] For now, it suffices to note that Kant ties equality to the representation of magnitudes, and that there is an analytic dependence of equality on identity. Several passages in the Amphiboly of the *Critique* provide direct evidence that Kant accepts the traditional metaphysical definitions of similarity and equality, where he indicates that identity of the qualitative and quantitative inner determinations of two objects entails their similarity and equality, respectively (A263f/B319, A272/B328).[24] Kant reaffirms this connection in a first draft of *What Real Progress Has Metaphysics Made in Germany since the Time of Leibniz and Wolff?* (20:282). Significantly, as we shall see, Kant also agrees with Baumgarten that quantitative and qualitative properties of a thing can be distinguished by whether the properties can be clearly cognized in the thing on its own.

[23] See Sutherland (2005a).

[24] I return to this passage in Section 9.8.

While Kant preserves this framework of the metaphysics of quantity, however, he also radically departs from Leibniz, Wolff, and Baumgarten. As we saw in the previous chapter, Kant held that qualities are properties of an object that are specifically different, that is, heterogeneous, from each other:

> quality differs from quantity in that, and to the extent that, the [former][25] indicates something in the same object which is inhomogeneous [*ungleichartiges*] with regard to other determinations found in it. Therefore, quality is that determination of a thing according to which whatever is specifically different finds itself under the same genus, and can be distinguished from them [i.e., other specifically different determinations under the same genus]. This is heterogeneous [*heterogen*] in contrast to that which is not specifically different, or to the homogeneous [*homogen*]. (*Metaphysik Vigilantius*, 1794–5, 29:992)

Qualities are specific differences and are heterogeneous; quantities lack specific difference and are for that reason homogeneous.[26] More carefully, it is *quanta* that are homogeneous:

> Homogeneity is specific identity with numerical diversity [*numerischen Diversitaet*], and a *quantum* consists of homogeneous parts [*partibus homogeneis*] . . . (*Metaphysik von Schön*, late 1780s, 28:504)

Against the backdrop of Leibniz, Wolff, and Baumgarten's metaphysics of quantity, we can now see how innovative his position is. These authors have nothing remotely like Kant's notion of *quantum* and its dependence on the representation of strict homogeneity.

As noted already in Chapter 1, philosophical terminology concerning quantity, and in particular German philosophical terminology, was not well fixed in Kant's time, so that quantity (*Quantität*) and magnitude (*Größe*) do not have precise and stable meanings and are used differently by different authors. In particular, *Größe* was used ambiguously by Kant's contemporaries, who sometimes used "magnitude" to refer to an object insofar as it is measured or quantitatively determined, and sometimes used it to refer to the measurement or quantitative determination itself. Kant uses the terms "*quanta*" and "*quantitas*" for the first and second senses, respectively. On the other hand, Kant uses the term "magnitude" (*Größe*) for both *quanta* and *quantitas*. Sometimes, when it is particularly important, he disambiguates, but more often, he does not disambiguate his use of "magnitude," and we must determine which sense

[25] I follow the editors and translators of Kant's lectures on metaphysics, Ameriks and Naragon, in thinking that "former" was mistakenly replaced by "latter." See Kant (1997, p. 460, note b).

[26] As discussed in the previous chapter, Kant states in other lectures that difference or diversity without specific difference is numerical difference (28:561, 1790–1?, 28:504, late 1780s), linking quantity to bare numerical difference.

of magnitude he has in mind from the context. He also uses *Quantität* ambiguously in mathematical contexts, for example, when drawing the quality/quantity distinction, though in mathematical contexts it usually means *quantitas*.[27]

What is particularly important for our present topic is that when Kant characterizes a magnitude as a homogeneous manifold in intuition, he is explicit that he means *quanta*, and, as the last quoted passage makes clear, it is *quanta* to which Kant refers when characterizing quantity as specific identity with numerical diversity and contrasting that with quality as that which represents specific differences through concepts. I will call this notion of quality "specific-difference quality." This way of characterizing quality in contrast to *quanta* is important, since in Kant's view concepts represent specific differences and hence qualities; as we saw in the previous chapter, concepts without the aid of intuition cannot represent specific identity with numerical diversity, and hence cannot represent *quanta*.

We can also now better appreciate Kant's notion of *quantitas*, which aligns much more closely with the metaphysical notion of *quantitas* discussed by Leibniz, Wolff, and Baumgarten. As discussed in Chapter 5, Kant distinguishes between a concrete and a more abstract sense of magnitude: *quanta* and *quantitas*. *Quanta* are magnitudes "as such" (A163/B204), which he defines as a homogeneous manifold in intuition in general (B202f). In contrast, *quantitas* answers the question of how big something is, that is, what its size is (A163/B204). This presupposes a standard of measure, or at a minimum, a comparison of two objects with respect to whether one is larger than, equal to, or smaller than another. Kant's distinction between *quantum* and *quantitas* makes room for a notion of quantity more concrete and more basic than the determination of size, in addition to a notion of quantity as the determination of size, particularly a determination of size with respect to some measure. The former, being a *quantum*, is an inner determination of a thing; the latter, *quantitas*, is something that can be distinctly cognized only relative to some unit.

We saw that Kant draws a contrast between *quantum* and quality, but he also draws a contrast between *quantitas* and quality, and in this case, the quantity/quality distinction takes on a different character. The distinction no longer turns on what is a specific difference, but on whether something can be distinctly cognized in the object taken on its own or only in comparison to some other object. Thus, Kant's *quantitas*/*qualitas* distinction introduces a different sense of quality, one that aligns with Baumgarten; a quality of a

[27] Kant also uses *Quantität* in logical contexts, such as the logical functions of judgment and of course as the general heading for the first group of categories. I can detect no important distinction between Kant's use of "quality" (*Qualität*) and of "*qualitas*," and I use them interchangeably.

Inner Determinations of a Thing

Quantum	vs.	**Quality (*sensu* specific difference)**
Strictly homogeneous manifold. Bare numerical difference. Only representable with the aid of intuition.		A heterogeneous difference. Specific difference. Can be represented by means of concepts without dependence on intuition for its content
⤊ Determination		↑ differing notions of quality ↓
Quantitas	vs.	**Quality (*sensu* distinctly cognizable)**
Determination of the size of something, and hence a determination insofar as it is itself a *quantum*, e.g., a line being four times longer than another line, or an area being 5 square inches.		Determination of a thing that can be distinctly cognized in the thing itself. Includes some determinations of a thing insofar as it is a*quantum*, but not determinations with respect to size; e.g., includes a line as straight or curved, a triangle as isosceles or equilateral, but not that a line is four times longer than another, or an area being 5 square inches.

Figure 8.2 Kant's metaphysics of quantity and quality

thing in this sense is a property that can be distinctly cognized in the thing taken on its own. I will call this notion of quality "isolated-cognition quality." The distinct cognition of the *quantitas* of something, that is, a determination with respect to size, requires comparison to some other thing – another object, say, or a unit of measure. And as for Leibniz, Wolff, and Baumgarten, Kant's view entails that whether a property is a quality in this sense or a *quantitas* depends on what one takes as the object; as described above, the *quantitas*, or "how-muchness" of the relative sizes of the sides of a triangle is a quantitative property of those sides, but it is a qualitative property of the triangle as a whole. To summarize, we have two distinctions in Kant concerning quantity that count as inner determinations of a thing, as shown in Figure 8.2.

The first sense of quality is of a property that is a specific difference and has particular importance when considering the roles of concepts and intuitions in cognition. The second sense of quality is a property that can be distinctly cognized on its own, and corresponds to the quantity/quality distinction inherited from Leibniz, adopted by Wolff, and modified by Baumgarten.

There are two points to note about the relation between Kant's two versions of the quality/quantity distinction. First, the two distinctions are connected by the relation between *quantum* and *quantitas*. *Quantitas* is a quantitative determination of a *quantum*; that is, *quantitas* is a determination with respect to size of the object insofar as that object is a *quantum* and hence is a strictly homogeneous manifold. There are, however, other determinations of a thing insofar as it is a *quantum* that do not concern size and do not count as *quantitas*, such as the straightness or curvature of a line; these are qualities in the second sense (a point to which I will return shortly).

Second, the relation between the two distinctions is not straightforward; it is not that one distinction is simply a further specification falling under one of the concepts of the other distinction in a genus–species hierarchy, as was the case in Baumgarten's distinctions above. The two distinctions are partially orthogonal, and this can be confusing. To begin with, something being a *quantum* can be cognized in the thing taken on its own, and hence the property of a being a *quantum* counts as a quality in the second sense. If one has the first sense of quality in mind, which contrasts *quantum* and quality, this is confounding, and in fact contradictory. It does not help that some of the texts we have also make claims about closely related issues, such as the nature of number, which adds to the complexity and are difficult to disentangle. But Kant is explicit that the property of being a *quantum* is something that can be cognized in a thing on its own and not just relative to or in relation to something else, contrasting it to *quantitas* on this point.

Quite significantly, Kant distinguishes between *quantum* and *quantitas* in notes he wrote in the margin of his copy of Baumgarten's *Metaphysica*. As we saw, Baumgarten distinguishes between quality (*qualitas*) and quantity (*quantitas*) in §69, and defines equality and similarity in §70. Between §69 and §70, Kant pens "*quantum* und *quantitas*," introducing a distinction that does not appear in Baumgarten. At the beginning of §69 Kant inserts the following comment:

> The distinction of one thing from another through the plurality [*Vielheit*] of the homogeneous, which is contained in both, is magnitude [*Größe*]. The distinction of the non-homogeneous is quality [*Qualität*]. (Reflexion 3539, 1776–89, 17:42)

As we have seen, the notion of magnitude Kant is describing here is *quantum*, while the notion of quality is the first sense discussed above. Kant is introducing a quite different notion of magnitude from Baumgarten. Finally, next to §69, in which Baumgarten characterizes *quantitas* as something that can be distinctly cognized only by assuming or relating it to something else, Kant writes:

> Better *quanta*; that which is a *quantum* can be cognized absolutely, but how big (*quantitas*) only relative. (Reflexion 3541, probably after 1790, 17:42)

Kant is explicitly correcting Baumgarten by introducing the notion of a *quantum* in contrast to a *quantitas*, and asserts that a *quantum* can be cognized "absolutely," that is, not relative to something else. Kant expresses the same view more clearly in later lectures on metaphysics:

> Through the comparison of a thing with itself and its parts one can clearly cognize that there is a *quantum*, but one can never determine, without comparison of a thing with other things, what it would actually have for a magnitude or – how large it is. (*Metaphysik Vigilantius* 1794–5, 29:992)

Furthermore, Kant follows this by expressly classifying the property of being a *quantum* as a quality:

> that something is a *quantum* merely expresses something that determines the quality of the thing . . . (*Metaphysik Vigilantius*, 1794–5, 29:993)

Thus, being a *quantum* – that is, being a homogeneous manifold in intuition – is a qualitative property of a thing in the second sense of quality, because it can be clearly cognized in the thing through "comparison with itself," that is, without comparison to any other thing.

Furthermore, as noted above, there are some properties that are determinations of a thing insofar as it is a *quantum* that are not determinations of size and count as quality rather than *quantitas*. For example, a line being straight or curved is a quality in the second sense, because this property can be cognized in the line itself, without comparison to other things. We will look more closely at such qualities in Section 8.5 below.

Although Kant's notion of a *quantum* is novel, Kant's notion of *quantitas* corresponds to Wolff's and Baumgarten's understanding of quantity. Significantly, Kant follows Baumgarten's departure from Leibniz in allowing that quantity understood as *quantitas* can be an inner determination of a thing despite requiring comparison to other objects for its complete determination, and that *quantitas* cannot be distinctly cognized in a thing without comparison to other things. In the Amphiboly of the *Critique* and in a draft of *What Progress*, Kant refers to differences of quality and quantity as inner differences and to *qualitas* and *quantitas* as inner determinations (A263f/B319f, A272/B328, 20:282). Just as in Wolff and Baumgarten, the relation to something outside the object does not make the *quantitas* a mere outer relation; it is still an inner determination.

Crucially, however, and unlike Wolff and Baumgarten, Kant gives a further account of *quantitas* that does not simply classify it as an inner determination merely because it can be represented in the object. It is an inner determination of a thing because it is a determination of the size of a thing insofar as it a *quantum*. The *quantum* – the homogeneous manifold in the object – is the concrete feature of the object to which the size belongs; the *quantitas* is a determination of the *quantum*. Although the passage is fraught with interpretive challenges, this relation between *quanta* and *quantitas* is at least suggested in a passage of the Schematism of the *Critique* that we have already examined in Section 3.6:

> The pure image of all magnitudes (*quantorum*) for outer sense is space; for all objects of the senses in general, it is time. The pure **schema of magnitude** (*quantitatis*), however, as a concept of the understanding, is **number**, which is a representation that summarizes the successive addition of one to one (homogeneous). (A142f/B182)

While a difficult passage, one that will require a close look at Kant's conception of number to untangle, it suggests that *quantitas* is a determination of the size of a *quantum* expressed by the number of units contained in the *quantum*, in which the image of a *quantum* is spatial or temporal.

Let us return to the sensible nature of our cognition of quantity. Kant's introduction of the notion of *quantum* ties our cognition of magnitudes directly to intuition, and hence to the sensible. We also saw above that Wolff's and Baumgarten's distinction between intellectual and sensible representation turns on the logical distinction between distinct and confused representations; in their account, a quantity is sensible insofar as it is a confused representation, and a quantity is a confused representation when an object is regarded alone and not in relation to other objects. Thus, like Kant, Wolff and Baumgarten tie our cognition of quantity to the sensible. Kant often views Leibniz through the lens of Wolff and Baumgarten, and as he makes clear in the Amphiboly, Kant thought that Leibniz also distinguished intellectual and sensible intuitions on the basis of whether they are confused. So at least as Kant understands him, Leibniz agrees with Kant in tying our cognition of quantity to sensible representations.

Despite this point of similarity between Leibniz, Wolff, and Baumgarten, on the one hand, and Kant, on the other, the relation to sensibility in Kant's account couldn't be more different. Kant holds that cognition of a *quantum* depends on sensibility because it can be represented only with the aid of an intuitive form of representation that stands in contrast to conceptual representation; moreover, a *quantum* can be distinctly cognized in intuition. As I have noted, Kant began to focus attention on the special role of intuition in our mathematical cognition, and hence in our cognition of quantity and magnitude, in about 1763; by the time of his mature critical philosophy, intuitions are a quite different sort of representation from concepts, and belong to their own faculty, distinct from the faculty of understanding. Intuitions are far from simply being confused intellections; they are their own form of representation entirely different from concepts. Kant's insertion of intuition into the metaphysics of quantity by means of the notion of a *quantum* thoroughly transforms the metaphysics of quantity and how quantity is cognized.

8.5 The *Qualitas* and *Quantitas* of *Quanta*

I noted in the previous section that the contrast between quantity and quality (*qua* independently cognizable) can be drawn within mathematics. There is one particularly interesting text in which Kant contrasts the quality and quantity of a geometrical figure that is worth considering in detail. Kant argues in the *Prolegomena* that some propositions of geometry are synthetic and hence require intuition. As an example, he cites the proposition that the shortest distance between two points is a straight line, and claims that it cannot be analytic. To argue the point, he states that the distance between two points is a quantity, while a line being straight is a quality, which he thinks makes clear that no analysis of one could establish the other:

> That the straight line between two points is the shortest is a synthetic proposition. For my concept of the straight contains nothing of magnitude [*Größe*], but only a quality [*Qualität*]. The concept of the shortest is therefore wholly an addition and cannot be extracted by any analysis from the concept of straight line. (4: 269)

Kant's classifying straight as a quality presumably entails that being curved is also a quality. Counting them as qualities conforms to the distinction between properties that can be cognized in the object on its own and those that cannot; one can cognize that a line is straight or curved without considering any other object. But the claim that a straight line between two points is the shortest is implicitly a comparison to every other line that could be drawn between the two points.

We have seen above that some qualities of an object can be reduced to quantitative properties of the parts of those objects. In particular, the similarity of two rectilinear figures is an identity of quality, but can be reduced to the equality of the corresponding angles and the identity of the ratios of the corresponding sides, which is a quantitative relation among these parts. Kant apparently holds that straightness is not such a quality reducible to quantitative features of its parts, for if it were, then his argument would have to address the relation between the quantitative property of the length of a line and whatever quantitative properties determine the straightness of a line in order for his argument to hold. Yet the straightness or curvedness does seem to depend on the relative spatial positions of the parts of the line. This suggests that he thinks that such relative positions are qualitative.

Unfortunately, Kant does not say a great deal more about the quality/quantity distinction within geometry, even in unpublished notes. It is possible that we have met the limit of Kant's explicit theorizing about the relation between his metaphysics of quantity and quality, on the one hand, and mathematics and mathematical cognition, on the other; he may simply have not had time to pursue the issue further. One can nevertheless attempt to work out a position consistent with Kant's commitments.

Since Kant holds that being straight or curved is a quality, he is committed to the possibility of these qualities being distinctly cognized in the line itself, without comparison to other figures. However, Kant does not mean that perfect straightness or any extremely small curvature can be distinctly cognized by examining a line. Perfect straightness plays an important role in pure geometry, but it is always either a stipulated property or it is inferred from other properties; pure Euclidean geometry never relies on a judgment based on examination or perception that a line is perfectly straight. In this respect, straightness is like equality.[28] On the other hand, there are clear cases in which we can perceive that a line is curved, just as there are clear cases in which we can perceive that two lines are unequal in length. Kant's point is that straightness and curvature are the kinds of property that can in general and in principle be distinctly cognized in a thing itself. And that does seem to be the case.

In support of this point, Kant may have in mind something like Euclid's definition of straight line, which I introduced in Section 6.3:

I.D4 A straight line is one which lies evenly with itself.

While the meaning of "lying evenly with itself" is not transparent, the idea behind the definition might be expressed this way: if one were to look along a line segment or a ray from an end point, the points of the line would all "line up," and you would "see" only a point. This is not the case for a curved line.[29] And in either case, one can cognize that it is straight or curved without reference to any other object.

There is, however, another definition of straightness that Kant most likely had in mind. While Kant nowhere refers to it, the definition was commonly known, and in fact Wolff cites it in his *Mathematisches Lexicon*: a straight line is a line that is similar to all its parts.[30] As noted earlier, similarity entails that two things, each examined on its own, and hence abstracting from its size, cannot be distinguished.[31] Thus, two isosceles triangles with the same ratio of side to base will be indistinguishable, even if one were double the size of the other. Another way to regard this point: if two figures are similar and their sizes are allowed to vary, one could change their relative size and then

[28] See Sutherland (2005a) for an account of perfect equality as a stipulated or inferred property.

[29] One could perhaps state the property using projective geometry under suitable constraints; for a straight line, there is a plane onto which the line can be projected that yields just a point; this is not so for a curved line. But the idea is intuitive enough as it stands.

[30] Wolff (1965, pp. 806–7).

[31] At least in Euclidean space, which is the space Euclid and Kant were considering.

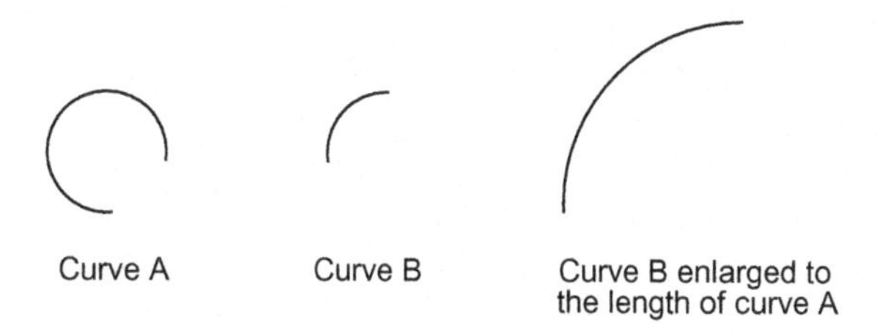

Figure 8.3 Similarity and curvature

superimpose them.[32] This is the case for a straight line and any of its parts, but it is not the case for the relation between a curved line and some of its parts. To illustrate with a simple case, consider a line of constant curvature. Every same-sized part will be similar. But now imagine a line segment of constant curvature constituting roughly three-fourths of a circle, and compare its shape with a third part of it. They are clearly not similar; they have a different shape (see Figure 8.3, curves A and B). We can make the lack of similarity vivid: recalling that the relation of similarity is size invariant (in Euclidean geometry), suppose the part were enlarged to be equal in length to the original curve; it is clear that the two curves are not similar even if they are the same length (see Figure 8.3, curves A and C). Thus, a line is similar to all its parts if and only if it is straight.

This definition of straightness in terms of similarity fits with the claim that straightness or curvedness of a line is a quality of a line, since it can be cognized without reference to any other figure than the line. Moreover, since similarity is defined as identity of quality, it does so in a way that does not ultimately appeal to quantity.

This is not say, however, that we can cognize *how much* a line is curved by considering the curve on its own (again, think of size invariance). The amount of a constant curvature, for example, is measured by the distance from the line to the center of the circle on which the curve can be superimposed, that is, the radius of curvature. For a line of smoothly varying curvature, the curvature at a point of the line is measured by the radius of curvature of an arbitrarily small segment of the curve including that point. In either case, the measure of curvature – the question of "how much?" – is determined by the radius of curvature, that is, the line segment which is the radius of the circle that measures it; because it requires a comparison to something outside the curve,

[32] At least for plane geometry, if you allow motion in three-space. See Sutherland (2005a) for a discussion of superposition and its relation to similarity and equality.

it counts as quantitative. But whether a line is simply straight or curved can be clearly cognized in the line itself and hence is a quality of the line.

The property of being straight or curved is a function of the relative spatial orientation of the parts of the line. Hence, to make Kant's approach plausible, the relative spatial orientation of parts of a homogeneous manifold would have to count as a qualitative relation among the parts, in contrast to the quantitative relations among parts, such as the ratios of the sides of a triangle.[33]

8.6 Conclusions

We have seen that Kant reworks the metaphysics of quantity found in Leibniz, Wolff, and Baumgarten by means of his distinction between *quanta* and *quantitas*. Kant's account of *quanta* ties the representation of magnitudes to intuition. This radically reforms the nature of magnitudes, our cognition of them, and the explanation of what is unique to mathematical method. But in Kant's characteristic way, his radical reformation preserves a great deal of the metaphysics of quantity. He does so primarily by capturing the traditional properties of quantity in his notion of *quantitas*. In this way, Kant can agree with Baumgarten's version of the Leibnizian and Wolffian distinction between quality and quantity, which appeals to whether or not a property can be clearly cognized in a thing on its own. Kant can also agree with the venerable philosophical definitions of equality and similarity as identity of quantity and quality already found in Aristotle, while also maintaining agreement with the Euclidean definition of similarity.

Stepping back further, we can see that Kant both radically revises the metaphysics of quantity while preserving many of its features and at the same time radically reworks the Eudoxian conception of magnitudes while preserving the Eudoxian theory of proportions. The two reforms dovetail in a comprehensive restructuring of the understanding of mathematics and mathematical cognition in his time, a reformation that retains what Kant considers to be correct and valuable in previous theorizing about quantity and magnitude.

Kant's fundamental altering of both the metaphysics of quantity and the Eudoxian account of magnitudes helped to spur and also shaped his critical philosophy. What motivated Kant to reformulate the quality/quality distinction and tie it to intuition? I believe there were several factors that convinced Kant that fundamental reform was necessary. The first was his growing conviction that any attempt to account for all truths solely by means of

[33] If this is an accurate reconstruction or filling-out of Kant's views, it suggests interesting connections to Kant's understanding of incongruent counterparts, though I won't be able to pursue this idea here.

concepts and their relations, and definitions in particular, would fail. The second was his reflection on the nature of mathematical cognition and what made it certain. As I have indicated at previous points in this work, the emergence of Kant's views on these issues can be traced to 1763–4, and mark the beginning of Kant's turn to the critical philosophy. Those included his reflections on the logical/real ground distinction under the influence of Crusius. Kant wrote *Attempt to Introduce the Concept of Negative Magnitudes into Philosophy* during this period, in which he argued that the opposition of magnitudes could not be accounted for in terms of logical opposition, a serious challenge to Leibniz's and Wolff's metaphysical systems.

These developments were instigated or reinforced by a contest organized by the Berlin Academy of Sciences. The contest concerned the relative certainty attainable in philosophy and mathematics, for which Kant wrote and submitted *Inquiry Concerning the Distinctness of the Principles of Natural Theology and Morality*, receiving second prize. It prompted Kant to think more deeply about what is required for mathematical cognition, and he took issue with Leibniz and Wolff. They shared an ambitious project to depart from Euclid and provide a completely new foundation for geometry; at the same time, they aimed to subsume geometry under a unified metaphysical system of concepts by means of the philosophical definitions of equality and similarity. Kant's confidence in Euclid and his belief that Leibniz and Wolff failed in their project provided a strong impetus for him to reform their metaphysics of quantity and to reconsider the notion of magnitude underlying the Eudoxian theory of proportions.[34] Kant's reaction to Leibniz and Wolff on this issue was tremendously important for the development of his critical philosophy.

Kant's refashioning of both the Euclidean theory of magnitudes and the metaphysics of quantity laid the groundwork for his understanding of mathematical cognition. Nevertheless, our analysis of Kant's theory of magnitudes is not yet complete, for we have not fully considered the extent to which Kant's account of magnitudes and their composition succeeds in explaining mathematical cognition and the mathematical character of experience. Is that account really sufficient to explain mathematical cognition and the mathematical features of the world? I take up this question in the next chapter.

[34] See Sutherland (2010) for an argument that Wolff's unsuccessful attempt to reform Euclid's geometry contributed to Kant's conviction.

9

From Mereology to Mathematics

9.1 The Gap between Kantian and Euclidean Magnitudes

The previous chapters have argued that Kant aims to provide an account of mathematical cognition historically rooted in the Euclidean theory of magnitudes employed in the Eudoxian theory of proportions. Making allowances for changes to that theory over the centuries, one might reasonably presume that when Kant discusses magnitudes, he means magnitude in the sense found in that theory. In fact, as I mentioned in Chapter 1, Kant's eighteenth-century audience would have understood Kant to be employing a conception of magnitude in wide currency, one inherited from the Greek mathematical tradition, and in Section 6.8 we saw that philosophers and mathematicians as diverse as Wolff, Kästner, and Euler described mathematics as the science of magnitudes and their measurement. While magnitude is not explicitly defined in Euclid's *Elements*, it is implicitly characterized by its role in the theory of proportions. And as we saw in Sections 6.4 to 6.6, that theory provides a body of propositions that allows for a sophisticated mathematical treatment of magnitudes, one that had a tremendous influence on the history of mathematics. The notion implicitly defined by that theory is a rich mathematical notion of magnitude.

Kant himself seems to indicate that he has in mind the rich mathematical notion of magnitude inherited from Eudoxus and Euclid. Kant calls the principles of the Axioms of Intuition and the Anticipations of Perception, which are principles concerning magnitudes, "mathematical," because they explain the possibility of mathematics, and Chapters 2 and 3 provided further evidence that the Axioms of Intuition concern pure mathematics as well as the applicability of mathematics to appearances, insofar as appearances are extensive magnitudes whose real qualities have intensive magnitudes. In fact, Kant asserts that the principle of the Axioms "is that alone which makes pure mathematics in its complete precision applicable to objects of experience" (A165/B206), lending support to the view that Kant is using the Euclidean notion of magnitude.[1]

[1] As argued in Friedman (1992). Friedman is, as far as I know, the first to describe the connection between Kant's discussions of magnitude and the Eudoxian theory of proportion.

On the other hand, we have seen that Kant is employing a refashioned notion of magnitude. We saw in chapter 7 that although Euclid provides no explicit definitions of magnitudes or their homogeneity, Kant has his own characterization of these notions that reflects a deeper analysis of them and their connections to human cognition. Kant's explication of homogeneity and magnitude align with the Eudoxian conception of magnitude in dividing the kinds of homogeneous spatial magnitudes into lines, areas, and planes. Kant also gives an account of the synthesis of composition that captures something essential to the Euclidean notion of homogeneous magnitudes: the fact that they allow composition into larger magnitudes that are homogeneous with what one started. As emphasized in the chapter 7, Kant seems to have his finger on something essential to the mathematical character of magnitudes in pointing to the special properties of the synthesis of such a manifold, for it yields more of exactly the same kind of magnitude, which is purely a difference in quantity without any change in quality. Moreover, because a composed magnitude is more than any of its parts, it is in that sense greater or larger, which imposes an ordering on the part–whole relations of magnitudes with respect to size.

Unfortunately, the notions of magnitude and composition Kant employs are still a far cry from the rich mathematical Euclidean conception of magnitude. The primary function of the categories of quantity is to represent the part–whole relations of magnitudes; their role is mereological. The definitions of extensive and intensive magnitudes, which turn on the relation of the representation of the parts to the whole, are also mereological. We saw that in Kant's view, intuition plays a role in allowing us to represent the part–whole relations of extensive magnitudes in a way that makes their part–whole relations cognitively accessible, so that intuition also serves a mereological role.

Even Kant's notion of composition is only mereological, since it is defined as the synthesis of a strictly homogeneous manifold, that is, the synthesis of a continuous or discrete manifold into a totality. Moreover, even if Kant's account of composition indicates something essential to the mathematical character of magnitudes, the ordering with respect to relative size is severely limited in two ways. First, it will order magnitudes only in relation to fully contained parts; it establishes that a magnitude is smaller than a second magnitude of which it is a fully contained part, that the second magnitude is smaller than a third of which it is a fully contained part, and so on. It does not, however, order the relative sizes of magnitudes that are disjoint or only partially overlapping; the ordering based on part–whole composition is a partial ordering restricted to full part–whole containment, rather than a total ordering across all magnitudes that are homogeneous with each other.

Second, even within the partial ordering of parts and subparts, all that is established is that one magnitude is larger than another; the ordering does not determine *how much* greater the whole is than any of its parts. It does not, for example, determine whether one magnitude is twice as large as one of its parts or whether the part and the whole stand in some other ratio.[2] In sum, because Kant's account of homogeneity, magnitude, and composition is only mereological, there is a gap – a quite significant gap – between Kant's mereological notion of magnitude and the Euclid's mathematical conception of magnitude.

There is a parallel point about applied mathematics. Kant's mereological account of magnitudes and their composition fulfills an important task for Kant in establishing the employment of the categories of quantity in our cognition of appearances and the employment of the categories of quality to sensations and the real corresponding to them. It establishes that appearances, sensations, and the real corresponding to sensations are a strict homogeneity that can be represented as composed, a necessary condition of applying mathematics to them. It is nevertheless not sufficient for the applicability of mathematics. Thus, Kant does not appear to be entitled to his claim at the end of the Axioms that its principle "is that alone which makes pure mathematics in its complete precision applicable to objects of experience" (A165/B206).

In summary, Kant's account of magnitudes falls short of explaining the mathematical character of Euclidean magnitudes handed down through the Euclidean geometrical tradition. In order to probe Kant's views more deeply, we must first return to Euclid for a closer look at the *Elements*, which is the topic of Sections 9.2–9.5. We will see that the Eudoxian theory of proportions, and hence the Euclidean conception of magnitude implicitly defined by it, rests on an unarticulated theory of measurement. A consideration of the fate of this theory in the Euclidean tradition and in the views of Kant's immediate predecessors in Section 9.6 then paves the way for a closer examination of Kant's own views in Sections 9.7 and 9.8.

9.2 Euclidean Presuppositions: Aliquot Measurement

We begin by looking more closely at the way in which magnitudes are first introduced in Book V of Euclid's *Elements*. The first two definitions of Book V state:

V.D1 A magnitude is part of a magnitude, the less of the greater, when it measures the greater.

V.D2 A magnitude is a multiple of the less when it is measured by the less.

[2] In the language of the theory of measurement, the part–whole ordering relation allows at best the creation of an ordinal scale and allows ordinal but not extensive measurement. See Krantz (1971, chapters 1 and 3) for an overview. Nevertheless, to repeat the previous point, in the present case even the ordinal scale is only a partial ordering, restricted to magnitudes that are fully contained within another, and hence even the sense of ordinal measurement in this case is severely restricted.

Euclid is defining the notions of part and multiple, while presupposing the notion of magnitude, relations of greater and less, and the notion of measuring. He is defining when a magnitude counts as a part or multiple of another by appealing to when one measures the other, but both measure and part have idiosyncratic meanings here. As noted in Section 6.4, the notion of part is not the same as the notion employed in the common notions when Euclid states that the whole is greater than the (proper) part. In one way, it is narrower and in another it is much broader. First, it is a narrower notion of part. In the sense of the common notions, a part could be any (proper) part of a whole whatsoever. In contrast, V.D1 defines parthood in terms of measuring, and "measure" here has a quite specific meaning: one magnitude measures another if and only if an integral number of the former is equal to the latter; otherwise, the former does not measure the latter. This is also reflected in the definition of multiple in V.D2: one magnitude is a multiple of another if and only if it is equal to an integral number of the latter.

The notion of part employed here is what is termed "aliquot" part, and to distinguish the notion of measure on which it is based from our ordinary, more inclusive sense of measure, I will sometimes call it "aliquot measure." To illustrate with a concrete example, a five-inch section of a twenty-inch rope aliquot measures, and hence is an aliquot part of, the twenty-inch rope, while a four-and-a-half-inch section of the twenty-inch rope does not aliquot measure, and hence is not an aliquot part of, the twenty-inch rope.

Although the notion of part is restricted to an aliquot part and hence is narrower than the notion of part found in the Common Notions, it is at the same time much broader. Common Notion CN5, which claims that a whole is greater than its (proper) part, applies only to parts that are entirely contained in the whole. This is in keeping with the mereological notion of magnitude found in Kant, which I discussed above. In contrast, one magnitude can count as an aliquot part of another even if the former is not completely contained in the latter; the aliquot part taken multiple times need only be *equal* to another magnitude to measure it. Reverting to our example, a five-inch rope counts as an aliquot part of an entirely distinct twenty-inch rope. Thus, any magnitude could potentially measure any magnitude that is homogeneous with and larger than it. This is made possible by the relation of equality ranging over all magnitudes homogeneous with each other. Thus, the notions of aliquot measure and aliquot part presuppose that homogeneous magnitudes can stand in the relation of equality.

The notion of aliquot measure is not restricted to continuous spatial magnitudes. Euclid also applies it to numbers understood as collections of units. We saw in Chapter 6 that Euclid provides a separate treatment of continuous spatial magnitudes in Book V from the treatment of number in Book VII, despite the many parallels between them. The use of the notion of aliquot measure is a further, even deeper and profoundly significant parallel.

As noted in Section 6.6, Book VII begins with the definition of a unit and the definition of a number as a multitude composed of units, which are then followed by:

VII.D3 A number is part of a number, the less of the greater, when it measures the greater.
VII.D4 But parts when it does not measure it.
VII.D5 The greater (number) is a multiple of the less when it is measured by the less.

In this case as well, "part" means "aliquot part," and one number measures another only if there is an integral multiple of the former in the latter. Note the idiosyncratic notion of "parts" (in the plural) to mean that a lesser magnitude does not measure the greater an integral number of times.[3]

There is thus a sense of measuring that applies to collections and is common to continuous magnitudes and collections. The notion of measuring the sizes of collections is found in Aristotle. As discussed in Section 6.6, Aristotle held that numbers are multiplicities and hence that one is not a number. But one is the principle of number, and in discussing the relation between one and number, he appeals to measure:

> Each number is said to be many because it consists of ones and because each number is measurable by one.. . . And one and many in numbers are in a sense opposed, not as contrary, but as we have said some relative terms are opposed; for inasmuch as one is measure and the other measurable, they are opposed . . . plurality is number and one is measure. (Met. I, 1056b–1057a)

> For measure is that by which quantity is known; and quantity *qua* quantity is known either by a "one" or a number, and all number is known by a "one." (Met. I, 1052b, 20–3)

> Being one in the strict sense, if we define it according to the meaning of the word, is a measure . . . (Met. I, 1053b, 4)

On this view, the counting of distinct individuals can be thought of as a kind of measuring. As in measuring continuous magnitudes, we specify a unit by which to count and then identify the units one by one. For example, we might specify that our unit is matching pairs of shoes and then count off the pairs of shoes in a room. As Julia Annas puts it in an elaboration of Aristotle's

[3] The idea is that if a collection is less than another but does not aliquot measure it, then it must be at least some multiple of the unit part, and the lesser collection is called "parts" in virtue of being a collection of parts of this unit part. For example, a collection of two is an aliquot part of a collection of eight, while a collection of three is parts of the collection of eight, because the unit is a part of eight and a collection of three is three of those units, and hence a collection of parts of those units.

suggestion: "Whereas we regard it as natural to think of measurement as applied arithmetic, Aristotle regards counting as a kind of pure measuring."[4]

In the case of the aliquot measure of collections, what plays the role of equality is equimultiplicity. That is, two collections are equal if and only if they are equimultiple. For example, the collection of pairs of shoes in my closet might be equimultiple with the fingers on my hand. The possibility of any two collections standing in the relation of equimultiplicity means that aliquot measure of collections ranges over all collections. Note that in contrast to continuous spatial magnitudes, there is no need to restrict the range of equimultiplicity to collections whose units are homogeneous with the units of the other collection. As we saw in Section 6.6, to regard a collection as a number is to regard it simply insofar as it is composed of units, and all discrete things, so regarded, are homogeneous, so no explicit mention of homogeneity is required. In summary, aliquot measure plays a foundational role in Euclid's treatment of both the spatial continuous magnitudes of geometry and in his arithmetic; it is a common presumption of both.[5]

Let us return now to V.D1 and V.D2, and consider in more detail what Euclid's notion of aliquot measure of continuous magnitudes presupposes. As noted above, they define aliquot part and multiple in terms of the unexplained notions of magnitude, greater and less, and aliquot measure. Aliquot measure in turn presupposes that magnitudes can be composed to make a larger magnitude. It also presupposes that taking a magnitude an integral number of times can be equal to the possibly entirely distinct magnitude it measures; that is, it assumes that distinct magnitudes can stand in the relation of equality. Finally, it also assumes something more specific about composition; it assumes that a magnitude can be multiplied, that is, that a magnitude can be composed with itself some number of times, or, to express it more carefully, that it can be composed with magnitudes distinct from yet equal to itself to make a larger magnitude. There are thus two appeals to the relation of equality: in the multiplication of a magnitude with itself and in the relation between a composed magnitude and the magnitude that it aliquot measures. And the same point applies to what the aliquot measure of collections presupposes in VII.D3–D5; there are two appearances of equimultiplicity, first in the multiplication of the measuring collection and second between the collection resulting from multiplication and the measured collection (see Figure 9.1).

Euclid clearly holds that distinct homogeneous spatial magnitudes can be equal to each other and that magnitudes can stand in greater than and less than relations even when one magnitude is not completely contained within another. This is more than the bare mereological notion of greater and less

[4] Julia Annas (1975, p. 99).

[5] This is a point I owe to Ian Mueller (1981); see especially pp. 61ff. My understanding of the theory of measurement presupposed by Euclid is indebted to his careful analysis.

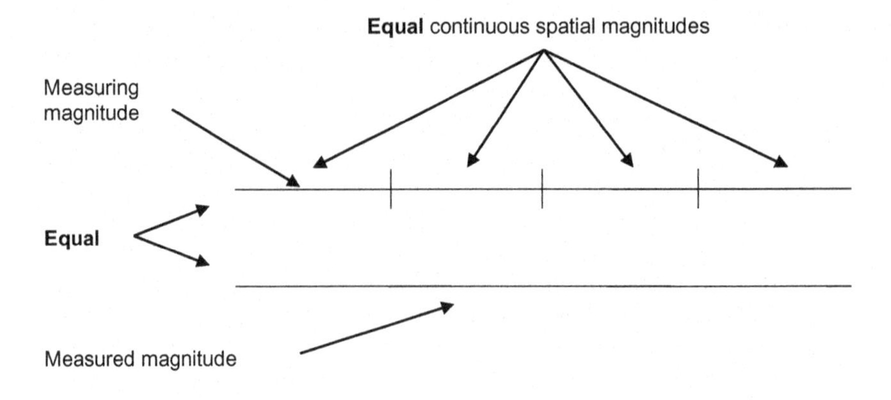

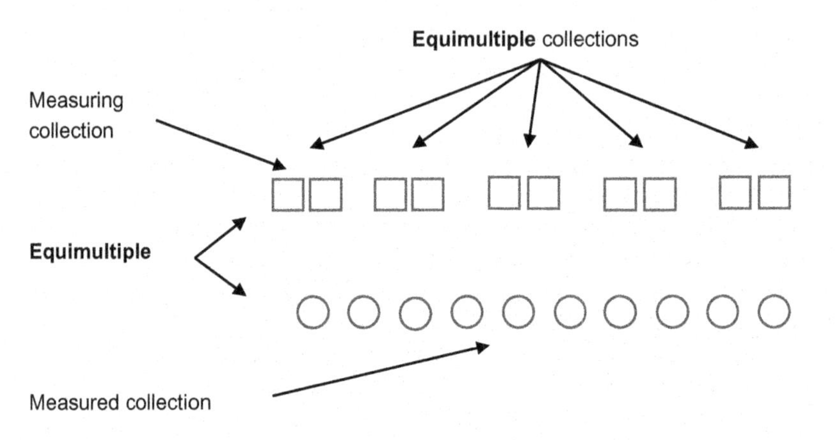

Figure 9.1 Two employments of equality and two of equimultiplicity

discussed in relation to Kant's mereological conception of magnitude in the previous section, in which the greater is defined as that which completely contains another and more besides. That, as we saw, yields only a partial ordering of magnitudes completely contained within another magnitude.

We can now highlight some of the key presuppositions underlying Euclid's appeal to aliquot measure. I will sometimes refer to these presuppositions as "assumptions," but I do not thereby mean explicit starting points of Euclid's theory taken on board to see what follows from them. Rather, I mean "assumptions" in the sense of presuppositions, that is, tacit commitments underlying that theory, commitments I am attempting to articulate on Euclid's behalf. I will not draw up a completely exhaustive list of all of the properties that Euclid implicitly assumes relations have – for example, that greater than and less than are transitive and antisymmetric, and so on. Instead, I focus on those principles that bring out features of the Eudoxian theory of proportions for continuous magnitudes and collections of units that are most important for our inquiry.[6] Euclid's appeal to aliquot measure presupposes the following:

A1. **Size and Ordering Relations of Homogeneous Spatial Magnitudes**

A1.1 **Comparative Size Relations of Homogeneous Spatial Magnitudes**

Spatial magnitudes that are homogeneous stand in relations of greater, less and equal.

A1.2 **Total Ordering of Homogeneous Spatial Magnitudes**

These relations impose a total ordering on spatial magnitudes that are homogeneous.

As noted, aliquot measure presupposes that magnitudes can be composed to form a larger magnitude, and more specifically that magnitudes can be multiplied, that is, can be composed with other magnitudes distinct from yet equal to it. This requires that a larger magnitude can be represented as composed out of smaller magnitudes.

The role of construction in Euclid's geometry suggests something stronger: that given a spatial magnitude, a larger spatial magnitude can be constructed using the construction postulates. In that case, it would require the ability to construct spatial magnitudes equal to a particular magnitude, and to construct them contiguously. Thus, Euclid assumes the following:

A2. **Aliquot Measure of Continuous Homogeneous Spatial Magnitudes**

A2.1 **Continuous Spatial Magnitude Composition**

Spatial magnitudes can be composed to make larger magnitudes of the same kind.

A2.2 **Continuous Spatial Magnitude Multiplication**

A spatial magnitude can be composed with magnitudes distinct from but equal to it to make a continuous larger magnitude of the same kind.

A2.3 **Continuous Spatial Magnitude Construction**

Composition and multiplication can be carried out using the construction postulates.

A1 and A2 underlie Euclid's of the notion of aliquot measure in Book V.

Let us turn to Euclid's arithmetical books and his account of number. As we saw, Euclid also appeals to the notion of aliquot measure in Book VII.

[6] Mueller (1981) gives a much more fine-grained presentation of the assumptions underlying Euclid's *Elements*, and does not isolate the principles I am articulating here.

Thus, parallel assumptions to those found under A1 are made concerning collections of individuals:

A3. **Size and Ordering Relations of Collections**

A3.1 **Comparative Size Relations of Collections**

Collections of individuals regarded as units stand in relations of greater than, less than, and equimultiple.

A3.2 **Total Ordering of Collections**

These relations impose a total ordering on collections.

There are also assumptions parallel to those found under A2:

A4. **Collection Composition and Equimultiplicity**

A4.1 **Collection Composition**

Individuals and collections can compose larger collections.

A4.2 **Collection Multiplication**

Collections equimultiple with a given collection can compose a larger collection.

As noted earlier, Euclid appeals to a notion of aliquot measure common to spatial magnitudes and collections, which is apparent in the parallels between A1 and A3 and between A2 and A4. Nevertheless, the nature of the composition for spatial magnitudes and for collections are fundamentally different. The first composition is continuous, such as that found in the generation of lines in accordance with the construction postulates P1–P3, that is, drawing a line between two points, extending a line segment, and drawing a circle around a point. The second is a composition of discrete individuals in a collection.

There is also a further important difference between the two forms of composition. As noted in Section 6.6, Euclid treats collections of units as if they are simply found rather than constructed; composition appears to be a relation in which collections stand, rather than the result of an explicit construction. To mark this difference, I will use the phrase "can compose" to mean that it is possible for collections to stand in a composition relation, rather than the passive "can be composed," which might more strongly suggest something that is brought about. Note as well that there is no need to state a homogeneity requirement on the composition relations among individuals or collections of individuals. As noted above and in Section 6.6, abstracting from all properties of discrete things other than their unity renders all individuals homogeneous.

Articulating A1–A4 immediately brings out a few important points that have been just below the surface of our discussion. The first concerns the relation between Euclid's theory of proportions for continuous magnitudes and his theory of number, and hence the relation between Books V and VII. The assumptions underlying the application of the notion of measure to continuous spatial magnitudes expressed in A2.2 and A2.3 presuppose that continuous spatial magnitudes can be regarded as discrete units standing in composition relations to collections of units that make up a larger continuous spatial magnitude. That is, the composition mentioned in A2.2 and A2.3 presupposes the composition mentioned in A4. Viewed from the other side, the second form of composition expressed in A4 includes the composition of a collection of units in which each unit is a homogeneous spatial magnitude. More particularly, it includes the composition of discrete homogeneous spatial magnitudes each of which is itself continuous and is also contiguous and continuous with the other individuals once composed. Even more particularly, it includes the composition of discrete individuals each of which is contiguous, continuous, and equal. For example, the four continuous and contiguous line segments that compose the upper line segment in Figure 9.1 stand in the second form of composition relation to the whole line they compose. Thus, Euclid's theory of continuous magnitudes presupposes the notion of aliquot measure, which in turn presupposes that a continuous magnitude can be multiplied, and that in turn presupposes that parts of a continuously generated magnitude can be regarded as individual units that stand in a composition relation to the whole magnitude.

These observations draw out and articulate the way in which Euclid's theory of continuous magnitudes depends on his concept of number as a collection of units and how much of the theory of number is presupposed: just enough to support the aliquot measure of continuous magnitudes. This dependence is obscured by several factors. As we have seen, Euclid simply assumes the notion of aliquot measure for both continuous magnitudes and collections; he thereby buries the difference between the two sorts of composition required for the aliquot measure of continuous magnitudes. As discussed in Section 6.3, Euclid likely thought that the notions of unit and collection and their relations were sufficiently obvious on their own to make use of them, and articulates them further in Book VII only when he needs to develop the theory of number further for the later books of the *Elements*. Even in Book VII, however, he assumes the notion of aliquot measure applied to collections and does not articulate the assumptions on which it rests.

The dependence of the theory of continuous magnitude on the concept of number and on the minimal theory of number required for aliquot measure is further obscured by the overlap of the two forms of composition. That is, the geometer both generates a continuous spatial magnitude in accordance with the construction postulates, and, when employing the notion of aliquot measure, regards parts of that continuous magnitude as units composing a collection. The two kinds of composition are intertwined in the aliquot measure of continuous magnitudes. I believe that this intertwining complicated attempts to sort out the relation between continuous magnitudes and number into the eighteenth century, and is found in Kant as well.

The analysis I've provided also brings a second important point into sharp relief. The theory of proportions and the full theory of number depends on aliquot measure, which rests in turn on the notion of equality, in two forms: equality between continuous spatial magnitudes and equimultiplicity between collections. I will follow up on this point, but first we need to complete the analysis of the role of aliquot measure in the Eudoxian theory of proportions by returning for a closer look at the crucial definition of sameness of ratio in V.D5.

9.3 Sameness of Ratio Revisited

Euclid's definition of sameness of ratio, discussed in some detail in Section 6.5, is as follows:

> Magnitudes are said to be in the same ratio, first to second and third to fourth, when equal multiples of the first and the third at the same time exceed or at the same time are equal to or at the same fall short of equal multiples of the second and the fourth when compared to one another, each to each, whatever multiples are taken.

We will now consider what the definition presupposes. Although it can be misleading in ways that we will address below, our modern anachronistic way of expressing the definition will once again make our task easier:

For any four magnitudes a, b, c, and d and any two positive integers m and n

$$
\begin{aligned}
&a{:}b = c{:}d \text{ iff for all } m, n{:}\\
&ma > nb \rightarrow mc > nd, \text{ and}\\
&ma = nb \rightarrow mc = nd, \text{ and}\\
&ma < nb \rightarrow mc < nd.
\end{aligned}
$$

Note first that ma consists of a collection, where the collection happens to consist of continuous and contiguous spatial magnitudes equal to a. Thus, A3 and A4 governing collections are applied to the special case when the individuals are continuous, contiguous, and equal spatial magnitudes.

There is, however, also a new notion relied on that has not yet been covered in A1–A4; it is assumed that we can multiply a magnitude an equal number of times as we multiply another magnitude – more specifically, we can multiply c an equal number of times as we multiply a, and multiply d an equal number of times as we multiply b. This is distinct from the equality assumed in the comparative relations among continuous homogeneous spatial manifolds – it is an equality of the multiplicity of the collections, regardless of the relative sizes of a and c (and b and d). In fact, as noted in Section 6.5, a and b need not be homogeneous with c and d; a and b may stand for lines, for example, while c and d stand for areas.

What V.D5 assumes, then, is that continuous spatial magnitudes can be multiplied equimultiply. (See Figure 9.2.) It is important not to confuse multiplication with equimultiplicity. In V.D5, multiplication is an operation

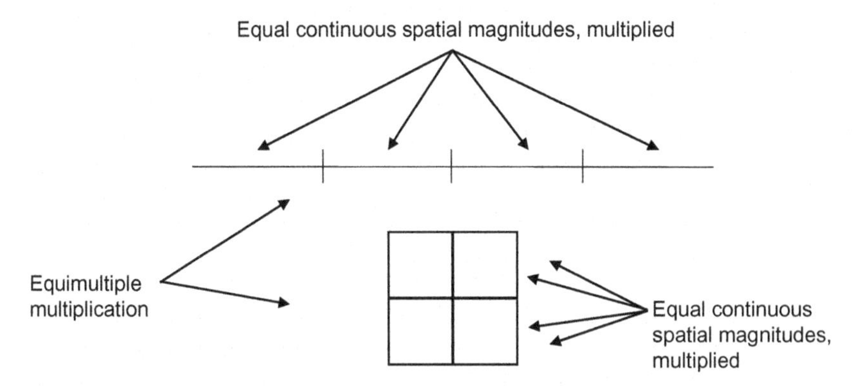

Figure 9.2 Equimultiple collections of equal continuous spatial magnitudes

on continuous spatial magnitudes and is a kind of composition, namely, a composition of continuous spatial magnitudes in which the parts that are composed are all equal to some magnitude, and in which the composition relations among those parts are regarded as units of a collection. In contrast, equimultiplicity is a relation between collections. Equimultiplicity is a quite general relation: the individuals can be discrete and spatially separated units or parts of some continuous spatial magnitude; the individuals of one collection can be inhomogeneous with the individuals of the other; and the individuals can be equal or unequal to each other.

When Euclid refers to the construction of equal multiples in V.D5, it is a special case of equimultiple collections, namely, when the individuals of each of two collections are equal and contiguous parts of a continuous spatial magnitude, although the individuals of one collection need not be homogeneous with the other. An example is helpful here. (See Figure 9.3.) In this example, although $a > b$ and $c > d$, $a{:}b$ is not the same ratio as $c{:}d$, which is shown by the fact that for $m = 4$ and $n = 8$, $ma < nb$ while $mc = nd$. Note that we have two appearances of equality and one of equimultiplicity: equal-sized parts making up a multiplied magnitude, equimultiplicity of the parts of pairs of collections, and equality of whole continuous magnitudes (in this example, equality of the whole areas.)

In summary, the Eudoxian theory of proportions includes additional presuppositions not already covered by the assumptions of aliquot measure given in A1–A4. I express those assumptions in the following two principles. The second is really a corollary of the first, but bears explicit statement due to its importance to the Eudoxian theory of proportions:

A5. **Equimultiple Collection Multiplication**

A5.1 **Equimultiple Collection Multiplication**

It is possible to multiply an individual or collection equimultiple times as another collection.

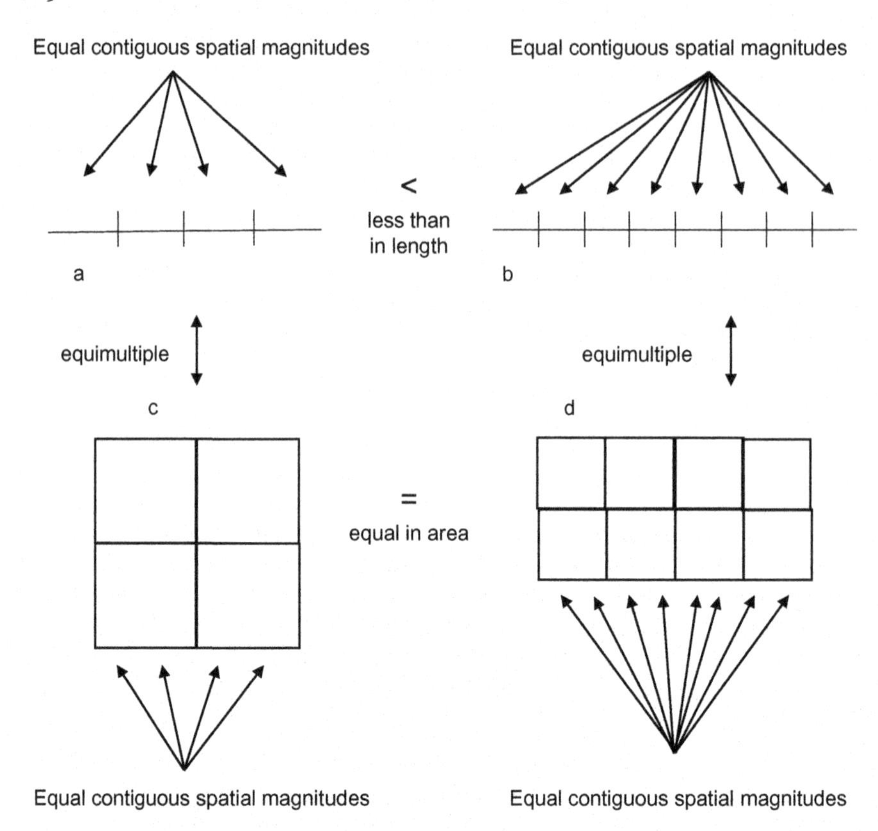

Figure 9.3 Example of a Eudoxian comparative size relation

A5.2 **Equimultiple Collection Multiplication of Continuous Spatial Magnitudes**
It is possible to multiply one spatial magnitude equimultiple times as one multiplies another spatial magnitude, regardless of whether the two magnitudes are homogeneous with each other.

A1–A4 give support to Euclid's notion of aliquot measure and A5 extends its application, allowing it to be to be exploited in the Eudoxian theory of proportions. Because they are so closely related to aliquot measure and are necessary for the employment of aliquot measure in the Eudoxian theory of proportions, I include the assumptions articulated in A5 under the title "Euclid's theory of measurement."

We are now in a position to clarify and more exactly express Eudoxus' remarkable achievement in his theory of proportions. It was not to find a way of defining sameness of ratio without appeal to numbers, for numbers are collections of units, which play an essential role. Eudoxus' contribution was more nuanced. The Pythagoreans had assumed that all ratios between homogeneous magnitudes could be directly expressed as the ratio between two numbers, and hence, that there was always a part that could serve as a lowest common denominator. Stated in terms we've explicated, they assumed that among any two homogeneous magnitudes, there is some part that aliquot measures both. A proof was then discovered that many magnitudes, such as the side of a square and its diagonal, fail this condition. In contrast, Eudoxus starts by assuming that aliquot measure is possible for some continuous spatial magnitudes. More specifically, a continuous spatial magnitude aliquot measures any continuous magnitude generated from it by multiplication. He then leverages aliquot measure to define when any two pairs of homogeneous magnitudes stand in the same ratio – even when that ratio cannot be directly expressed as a ratio of two collections of parts.

We have been investigating the underpinnings of Euclid's mathematical notion of magnitude implicitly defined by the Eudoxian theory of proportions. We found that it rests on the unexplained notion of aliquot measure, which in turn rests on tacit assumptions that constitute what I have called Euclid's theory of measurement. The principles A1–A5 articulate the assumptions of that theory. We are now in a position to highlight some important features of Euclid's theory of measurement.

9.4 Euclid's General Theory of Pure Concrete Measurement

I said in the previous section that although it is helpful to employ modern algebraic notation to explain the linchpin of the Eudoxian theory – the definition of sameness of ratio in V.D5 – it is also potentially misleading. Explaining how will bring out ways in which Euclid's theory of measurement differs from our modern understanding.

We expressed the antecedents of the conditionals given in the definition of sameness of ratio by saying that a and c are multiplied by the same number m, and similarly for the consequents. More carefully, the algebraic formulation refers to an arbitrary specific number m that belongs to each collection ma and mc, and an arbitrary specific number n that belongs to nb and nd. And according to a common articulation of the modern view, there is just a single abstract number m, such that two collections are multiplied by the same particular number. Furthermore, the algebraic formulation assumes that a definition has been given of specific numbers such as 7 or 8 over which m and n range. Finally, it suggests a modern understanding of measurement, according to which an independent grounding has been given for an abstract system of numbers and their arithmetic, which are then applied to magnitudes in measuring them. If we read Euclid in this way, it seems as if his theory of proportions simply assumes a fully and independently grounded theory of natural numbers and arithmetic.[7] Thus, from a modern perspective, is seems that even by the lights of Greek mathematics, Euclid's Book V assumes the arithmetical books VII–IX.

This fundamentally misconstrues Euclid, however. Recall once again the way in which Euclid states the defining condition of V.D5:

> when equal multiples of the first and the third at the same time exceed or at the same time are equal to or at the same fall short of equal multiples of the second and the fourth when compared to one another, each to each, whatever multiples are taken.

We can see in this passage two closely related differences between the view suggested by the modern algebraic formulation and Euclid's account. To begin with, Euclid does not appeal to an abstract number m or n. He refers instead to "multiples," which are themselves continuous spatial magnitudes represented as distinct collections of equal parts. Although he does not here refer to the multitudes as numbers (*arithmoi*), they count as numbers as Euclid conceives of them, but this is number only in the sense that they are collections of units. As emphasized in Section 6.6, this notion of number is concrete in that particular units compose distinct collections. Moreover, because Euclid does not have abstract numbers in mind, he also does not think of abstract numbers as applied to magnitudes in measuring them. Instead, concrete collections measure each other. Euclid's theory of measurement is in this sense *concrete*.

This difference is directly tied to a further one. Euclid does not appeal to the sameness of an abstract number shared by more than one collection. The identity of abstract numbers presupposes that each number has its specific properties that distinguish it from all others. Euclid instead only employs equimultiplicity. Equimultiplicity is a more general relation in that it applies to collections of individuals; it assumes that collections are determinate with respect to standing or failing to stand in the relation of equimultiplicity, but it does not depend on the specific size of a collection. Of course, one might think that two collections are equimultiple in virtue of the fact that the same specific number applies to both, but that is not Euclid's view. The sense in which equimultiplicity is more general can be brought out by a standard epistemological observation: I can know that there are as many knives as forks on the

[7] See Mueller (1981, pp. 62–4). Mueller argues that Euclid takes for granted not only the relation of measurement, but the arithmetical relation of measuring "a certain number of times," and uses "m" to symbolize this particular number of times. Thus, Mueller renders Euclid's expression as "k measures l according to the units in m" into the claim "k measures l m times." Despite how natural this move is and its necessity for expressing the relation in modern symbols, I think it can mislead us into thinking that Euclid simply assumes all the natural numbers over which m ranges.

banquet table without knowing how many knives or how many forks there are. Knowing the latter requires knowledge of something more specific. Because Euclid relies only on the relation of equimultiplicity between collections, his theory of measurement is in this sense *general.*

This is of course not to deny that ancient Greek mathematicians employed a more abstract conception of specific numbers such as 5; indeed, it is hard to imagine that they didn't have them in mind. The point is that the theory of measurement on which Books V and VII rest does not rely on them. In fact, even the arithmetical books do not provide definitions of specific numbers; as noted in Section 6.6, Book VII defines number in general and then divides numbers into different kinds, such as odd and even and prime and composite.[8] There is nothing, recursive or otherwise, that corresponds to the definition of specific numbers in modern foundations of arithmetic.

An additional feature of Euclid's theory of measurement is that it concerns the spatial constructions of Euclidean geometry, and not any empirically given spatial unit of measure. Thus, in the same sense that his geometry is pure, Euclid's theory of measurement is *pure.* The theory of measurement has implications for empirical measurement insofar as it rests on or appeals to the pure theory of measurement, but the latter is nevertheless pure. In summary, the theory of measurement we have articulated in A1–A5 constitutes a general theory of pure concrete measurement.

Contrasting Euclid's theory to that suggested by the algebraic formulation of V.D5 helps explain the relation between the theory of proportions in Book V and the arithmetical Books VII–IX. Euclid does not assume the arithmetical books in Book V. Instead, he presupposes aliquot measure employed for both continuous spatial magnitudes and collections. This in turn presupposes the Euclidean notion of number as a collection of units and the general theory of pure concrete measurement. And, to emphasize once again, they rest on the equality and equimultiplicity relations among continuous magnitudes and collections of units. This is what underlies the mathematical treatment of both spatial magnitudes in Book V and collections in Books VII–IX.

This result leads us back to the sense in which Euclid provides a foundation for geometry. As emphasized in Section 6.3, Euclid had a looser notion of what foundations of geometry require than later geometers. He does not aim to lay bare every assumption prior to invoking it, and appears content to delve more deeply into number – what it is and what the properties of number are – when a more immediate need arises; in particular, when the theory of proportions of

[8] As discussed in Section 6.6, Euclid sometimes proves a result for the case of a collection and the case of a unit separately. At other times, however, he treats a unit as a number, and the Euclidean tradition eventually considered a unit as a number one. If we take a unit to be a number, then this is an exception to the claim that Euclid does not employ specific numbers.

number is needed for Books X–XIII of the *Elements*. With his more permissive conception of providing a foundation, it is not surprising that Euclid might have presupposed a theory of measurement he did not explicitly articulate, and that he simply regarded the notion of collections of units and the notion of aliquot measure as sufficiently clear for his purposes.

It is also not surprising, however, that the Greek mathematical tradition following him sought to improve the *Elements*; we saw in Chapter 6 that subsequent generations of philosophers and mathematicians thought that Euclid's geometry was flawed in its foundations and that they added axioms to supply justifications for inferences that are required to meet higher standards of rigor. Those included axioms concerning number and measurement. It is also not surprising that some later mathematicians attempted to improve on Euclid by rearranging the *Elements* and placing the arithmetical books earlier.

To summarize, Euclid's more permissive understanding of the requirements of providing a foundation for geometry allows the general theory of pure concrete measurement to be obscured in three ways. First, some of the most important presuppositions of the general theory of measurement that are relied on in Book V, such as the comparative size relations among homogeneous continuous spatial magnitudes and the composition and construction of them, are already presupposed or explicitly assumed at the very beginning of the *Elements*. Moreover, some important further requirements of aliquot measure that presuppose equality are established early in Book I. (I will return to these in the next section.) Second, yet further assumptions concerning collections required for the Eudoxian theory of proportions in Book V are simply presupposed, namely, the comparative size relations among collections of individuals, and in particular, the relation of equimultiplicity, that is, the possibility that collections of individuals can be equimultiple with each other. It is also assumed that these constructions and relations hold for collections consisting of equal, continuous, contiguous spatial magnitudes. Finally, these assumptions concerning collections of individuals concern numbers in the Euclidean sense, but number is first explicitly defined in Book VII, and that part of the general theory of measurement concerning collections comes into full force only in developing the theory of proportions for number elaborated in Books VII–IX.

Let us take stock. Our aim was to uncover what is required to bridge the gap between a merely mereological conception of magnitude and the rich mathematical conception of magnitude implicitly employed by the Eudoxian theory of proportions. We found that it rests on the unexplained notion of aliquot measure and tacit assumptions about aliquot measure that we have articulated in assumptions A1–A5. These principles constitute Euclid's general theory of pure concrete measurement. It is possible, however, to take the analysis one level deeper to shed light on the relation between a bare mereology of magnitudes and collections, on the one hand, and their mathematical treatment, on the other.

9.5 From Mereology to Measurement to Mathematics: Equality

The assumptions given in A1 and A3 make plain the dependence of Euclid's theory of measurement on the comparative size relations of greater than, less than, and equality and the total ordering they impose on continuous spatial magnitudes and the same comparative size relations they impose on collections. The further assumptions found in A2, A4, and A5 build on these relations, with a particularly prominent and important role for equality and equimultiplicity, as noted in the previous section and illustrated in Figures 9.1–9.3.

The relations of equality and equimultiplicity are akin. In fact, equimultiplicity can be regarded as a species of a more general notion of equality covering both continuous spatial magnitudes and collections. As discussed in Section 6.3, Euclid gives common notions governing equality, and as common notions they can be viewed, and certainly were viewed by Proclus and later geometers, as also governing the relation of equimultiplicity between collections.[9] If we consider equimultiplicity as a species of equality, then we can see that equality in this broader sense plays a pivotal role in Euclid's theory of measurement.

The role of equality extends even more deeply, however, for we can give a definition of greater than and less than in terms of bare mereological part–whole relations and equality. We begin with Euclid's last common notion, CN5, "the whole is greater than the part." By "part," he means "proper part," and to say that the whole is greater than the part is to say that the whole contains the part and something more, something distinct from the part. The less-than relation is simply the reciprocal; to say that the part is less than the whole is just to say that the part and something more are contained in the whole. Since greater than and less than are defined simply in terms of the part–whole relation, we can think of them as "mereological greater than and less than." Now recall two points made in the discussion of Kant's mereological account of magnitude in Section 9.1. First, the ordering imposed by the part–whole relation is only partial because it only orders magnitudes that are completely contained in another magnitude; a whole is greater than its part, which is in turn greater than its part, and so on, but the greater-than relation

[9] This is certainly true of CN1–3, although CN4, "And things which coincide with each other are equal to one another," appears to be an exception, since coincidence is a specifically spatial notion in Euclid. The notion of *equality* invoked in CN4 can nevertheless be regarded as itself a general notion of equality. On CN4, see Section 6.3, especially note 5; on equimultiplicity as a species of equality, see Section 6.6, especially note 17.

does not apply to two magnitudes unless one is fully contained in the other. Second, the ordering does not establish how much greater a whole is than its part, only that it contains the part and yet more. That means that no sense can be given to, say, the whole being three times larger than its part, or one part of the whole being smaller than another disjoint or overlapping part. These features of greater than and less than also hold for collections of units and their part–whole relations.

In contrast, the equality relation can hold between homogeneous magnitudes that are disjoint or partially overlapping; any two homogeneous magnitudes are either equal or unequal. This allows us to define greater than in a broader sense using part–whole relations and equality as follows: one homogenous magnitude is greater than another if the second is equal to a part of the first. The assumption that any two homogeneous magnitudes are either equal or not equal, and the additional assumption that magnitudes have no smallest part, transforms greater than and less than into a total ordering.[10] It is now possible to say of any two homogeneous magnitudes that one is equal to, greater than, or less than another. These enhanced comparative size relations can be thought of as "mathematical greater than and less than" in light of the role they play in a mathematical treatment of magnitudes. And here, too, the greater than and less than relations also hold for collections. As a result, the theory of measurement articulated in A1–A5 can be grounded in the part–whole relations of continuous spatial relations and of collections supplemented by the relation of equality.

This analysis shows that the relation of equality, in the broad sense that covers both continuous magnitudes and collections, is what bridges the gap between the simple mereological properties of magnitudes and the general theory of pure concrete measurement, which in turn grounds the rich mathematical conception of magnitudes found in Euclid's theory of proportions.

The pivotal role of equality in moving from mereology to measurement and on to a full mathematical treatment of magnitudes naturally raises a question: How does Euclid introduce and appeal to equality? Recall first that Euclid's construction postulates correspond only to an employment of a straight edge, not a ruler, so that one cannot use it to mark off or measure out equal distances, and it corresponds only to employment of a collapsible compass, that is, one that collapses when lifted from the page, so that it cannot be used

[10] Strictly speaking, the greater-than and less-than relations would not be a total ordering on the set of magnitudes, since some magnitudes will be equal to one another and hence neither greater nor less than each other. This is merely a technicality, however; we can define the relation greater-than-or-equal, which will not be a strict ordering but will be a total ordering. Alternatively, one could take a modern approach, and define equivalence classes of magnitudes of equal size and then order these equivalence classes using the greater-than or less-than relations.

to mark off or measure out an equal span at another location. How, then, does equality enter into Euclid's geometry?

As discussed in Section 6.3, Euclid does not define equality. In fact, as the example definitions listed in Section 6.3 make clear, he simply assumes equality in the Book I definitions of other notions, including right angle, square, and circle. The circle, in particular, is defined as a plane figure with equal radii.[11] On the other hand, Euclid does provide three common notions governing equality that can be taken to implicitly define it, which I repeat here:

CN1: Things equal to the same thing are also equal to one another.
CN2: And if equals are added to equals the wholes are equal.
CN3: And if equals are subtracted from equals the remainders are equal.

Equality then plays a role in demonstrating the very first proposition of Book I, which shows how to construct an equilateral triangle. That construction rests on the third construction postulate, which is to describe a circle with any center and distance; the key point is that the demonstration exploits the fact that the radii of a circle are all equal. The second proposition of Book I then demonstrates how to "place at a given point a straight line equal to a given straight line." Again, Euclid's demonstration exploits the fact that the radii of a circle are all equal. The power of this proposition is immediately apparent. It allows one to construct equal line segments anywhere, circumventing the restriction corresponding to a collapsible compass and straight edge. Thus, the equality of rectilineal line segments required for the theory of measurement is introduced by and rests on the definition of a circle and its construction.

Equality is next extended to rectilineal angles and areas. Early in Book I, Euclid proves various equality, greater than, and less than relations that hold among angles of intersecting straight lines and angles of different sorts of triangles, and the equality of areas of rectilineal figures. Proposition I.4 begins the treatment of angles by proving the side-angle-side theorem, that is, that two triangles with two sides equal and the angle enclosed by them equal will have the remaining sides and angles equal and will also have equal areas. (The demonstration appeals to coincidence, that is, the possible superposition, of figures. It thereby implicitly appeals to the likely interpolated CN4.) In I.23, Euclid demonstrates how to construct a rectilineal angle at a point on a line equal to another given angle. Equality is also involved in constructing equal geometrical figures, such as a square equal in area to a given rectilinear figure. Using equality to find such relations among different figures, including solids,

[11] More precisely, Euclid defines a circle as "a plane figure contained by one line [which is called its circumference], such that all the straight lines falling upon it [the circumference] from one point of those lying inside the figure are equal to one another." The next definition defines that point as the center of the circle; Greek geometers did not have a word corresponding to radius. See Mueller (1981, pp. 317–18).

is not a trivial matter. Nor is it always even possible in Euclid's geometry; one need only think of the attempts to square the circle. Nevertheless, it is a central concern of, and appears to be an underlying motivation for, a great number of the propositions of the *Elements*. This quick overview makes clear how Euclid introduces and employs the crucial relation of equality. We can now turn to the way in which the presuppositions of the general theory of measurement were treated in the subsequent geometrical tradition.

9.6 The General Theory of Measurement in the Euclidean Tradition

As already noted, an important part of the geometrical tradition that followed Euclid introduced a more exacting standard for foundations; it sought to identify crucial assumptions in Euclid's proofs and add axioms that would cover those assumptions. A paradigm example is the claim that two straight lines do not enclose a space, which is required in the proof of Book I, 4. Other axioms sought to make explicit the properties of relations such as equality. The history of this tradition is of course long and complex, with some Greek versions of the *Elements* lost and then rediscovered, others preserved and developed in the Arabic tradition, and most editions of Euclid undergoing transformations. To speak in general terms, the late medieval and Renaissance periods saw both efforts to recover as faithfully as possible Euclid's original *Elements* and increasingly aggressive criticisms and amendments to Euclid. As noted in Section 6.7, Vincenzo De Risi identifies 318 additional axioms articulated by various geometers from the early twelfth century and into the eighteenth. They include the foundations of the theory of magnitudes; the mereological and compositional properties of magnitudes; the theory of ratios and proportions; arithmetic and number theory; spatial properties such as intersection and incidence, continuity, congruence, equality; measurement; and the parallel postulate. I reviewed these axioms in Chapter 6, but I now return to the Euclidean tradition with a focus on the presuppositions of the general theory of measurement. For ease of reference, I list the assumptions of that theory again in one place:

A1. **Size and Ordering Relations of Homogeneous Spatial Magnitudes**

A1.1 **Comparative Size Relations of Homogeneous Spatial Magnitudes**
Spatial magnitudes that are homogeneous stand in relations of greater, less and equal.

A1.2 **Total Ordering of Homogeneous Spatial Magnitudes**
These relations impose a total ordering on spatial magnitudes that are homogeneous.

A2. **Aliquot Measure of Continuous Homogeneous Spatial Magnitudes**

A2.1 **Continuous Spatial Magnitude Composition**
Spatial magnitudes can be composed to make larger magnitudes of the same kind.

A2.2 **Continuous Spatial Magnitude Multiplication**
A spatial magnitude can be composed with magnitudes distinct from but equal to it to make a continuous larger magnitude of the same kind.

A2.3 **Continuous Spatial Magnitude Construction**
Composition and multiplication can be carried out using the construction postulates.

A3. **Size and Ordering Relations of Collections**

A3.1 **Comparative Size Relations of Collections**
Collections of individuals regarded as units stand in relations of greater than, less than, and equimultiple.

A3.2 **Total Ordering of Collections**
These relations impose a total ordering on collections.

A4. **Collection Composition and Equimultiplicity**

A4.1 **Collection Composition**
Individuals and collections can compose larger collections.

A4.2 **Collection Multiplication**
Collections equimultiple with a given collection can compose a larger collection.

A5. **Equimultiple Collection Multiplication**

A5.1 **Equimultiple Collection Multiplication**
It is possible to multiply an individual or collection equimultiple times as another collection.

A5.2 **Equimultiple Collection Multiplication of Continuous Spatial Magnitudes** It is in possible to multiply one spatial magnitude equimultiple times as one multiplies another spatial magnitude, regardless of whether the two magnitudes are homogeneous with each other.

What did the geometers following Euclid have to say about the general theory of measurement?[12] Relatively few axioms specifically addressed aliquot measure. Herigone (1634) included perhaps the most basic such principle: "The measure is not greater than the measured thing." Commandino (1572) lists three, including, "Any magnitude that measures another magnitude also measures any magnitude that is measured by it," while Borelli (1658) states, "If a magnitude measures two other magnitudes, it also measures their sum and their difference." De Risi (2016) lists another four axioms specifically concerning number, that is, collections of units. Candale (1566), for instance, included: "if a number measures a part [i.e., an aliquot part], it also measures the whole." There are only eight axioms that explicitly concern measure. Nevertheless, there are other axioms that refer to (aliquot) part or one magnitude dividing

[12] See De Risi (2016). I'm indebted to this article and multiple conversations with De Risi in what follows.

another. As we saw in Section 9.2, Euclid defined aliquot part in terms of measure, so these axioms are closely related to measure.[13] Commandino, for example, added: "Those numbers which are the same part (or the same parts) of the same number (or of equal numbers) are equal to one another."

We are not, however, solely interested in principles that mention measure or govern notions defined in terms of measure; we are looking for axioms that support the general theory of measurement, and hence axioms that correspond to those listed in A1–A5, so we should look more broadly.

I first consider the assumptions of A1 and A3 regarding the ordering relations of greater than, less than, and equality for continuous magnitudes and for number, respectively. Many axioms concerned properties of relations; for example, Herigone (1634) included an axiom of transitivity for the greater-than relation: "If a magnitude is greater than another magnitude, and the latter is greater than a third magnitude, the first magnitude is also greater than the third"; and Schott (1661) added "a magnitude which is not greater or lesser than another magnitude is equal to it," a claim that takes a step toward trichotomy, which leads to a total ordering on magnitudes; trichotomy was stated in three axioms by Hill (1726). These axioms support the assumptions we identified under A1.

If collections are understood as discrete magnitudes and hence as a species of magnitude, then these order axioms also cover collections, and support the assumptions listed under A3. As discussed in Section 6.7, there were attempts shortly after Euclid to broaden the notion of magnitude to include nonspatial continuous magnitudes such as time and motion, and discrete magnitudes, that is, collections of units. And as I have emphasized, obvious similarities between the propositions of Books V and VII spurred hopes of a unified theory of proportions that subsumed both continuous spatial magnitudes and discrete units, and these hopes persisted in the later Euclidean tradition.

It is not always clear if an axiom concerning magnitudes was interpreted as applying just to spatial magnitudes, to other continuous magnitudes, or also to discrete magnitudes, though sometimes there are indications. Gottignies (1669), for example, included an axiom specifically referring to continuous magnitudes;[14] by singling out continuous magnitudes, he seems to be implicitly contrasting them with discrete magnitudes. Some authors refer to quantity rather than magnitude, introducing a further interpretive complication. More significantly, other authors state common notions governing "things," and thus hold a very general version of the common notion, but do not, at least in their expression of their axioms, commit themselves to a general notion of magnitude. To cite just one example, Clavius (1589) states

[13] This is a point made by De Risi (2016).

[14] See De Risi (2016, p. 34).

that "If a thing is greater, or lesser, than another thing, it is also greater, or lesser, than a thing equal to the latter." Whether the term "magnitude" or "quantity" or "thing" is used, almost all of the basic axioms concerning order relations and equality in particular either were or seem to be taken to apply to collections of units as well. Thus, many if not all of the axioms supporting the order relations of continuous magnitudes listed under A1 also support those for collections of units listed under A3. Understanding the axioms in this way requires taking equimultiplicity to be a species of equality, for which we have seen a long precedent.

Let us turn now to A2, that is, assumptions about the composition and multiplication of continuous magnitudes. The most basic such principle is one that Proclus attributes to Pappus, which, if his attribution is accurate, is as old as the fourth century. It states: "Magnitudes are susceptible of the unlimited both by way of addition and by way of successive diminution, but in both cases potentially only." Clavius (1574) states a closely related principle: "Given any magnitude, it is possible to take a magnitude greater or lesser than it."[15] Both of these lend support to A2.1, that continuous (spatial) magnitudes can be composed to make larger magnitudes of the same kind, while also adding that they can be decomposed, and that both operations can be carried out *ad indefinitum*. A2.2 states: "A spatial magnitude can be composed with magnitudes distinct from but equal to it to make a continuous larger magnitude of the same kind." We do not find a principle exactly corresponding to A2.2, which introduces equality into the composition of continuous magnitudes to ground their multiplication. Nevertheless, De Risi identifies roughly fifty principles proposed by various geometers concerning magnitude: about twelve principles concerning the foundations of a theory of magnitude and thirty-seven principles specifically concerning the composition of magnitudes and multiples of magnitudes.[16] They show that geometers in the Euclidean tradition gave a fair amount of attention to principles governing comparative size relations, and in particular equality, including operation of composition of equal magnitudes. All of these principles assume the relation of equality and that equal continuous magnitudes can be composed.[17]

A4 comprises principles of the composition and multiplication of collections, which mirror the principles of A2 for continuous magnitudes. We have just seen that the principles of A1 sometimes refer to magnitudes in a way that

[15] For both principles, De Risi (2016, p. 29).

[16] See De Risi (2016, pp. 29–32).

[17] Note that De Risi titles the latter set "Principles on Mereological Composition and Multiples of Magnitudes," and all those concerning composition presuppose equality, such as multiplication. I have been careful to distinguish between bare mereological composition of magnitudes, on the one hand, and composition that presupposes equality, on the other.

could be construed to cover discrete magnitudes, or refer to "things" in a way that can include collections of units. The same holds for the principles of A2. As De Risi points out, the mereological, order, and composition axioms concerning magnitudes and number coincide with and were often treated as axioms of a *mathesis universalis* that applied to both continuous and discrete quantities; moreover, they were employed in the theory of proportions, which was itself understood as a kind of universal mathematics.[18]

In addition, however, there were some who gave a separate treatment of number and articulated principles specifically for them. As noted in Section 6.7, Proclus's commentary on Euclid indicates that by the fifth century, geometers were willing to consider axioms and construction postulates for numbers understood as collections of units. In the late twelfth century, Tynemouth translated An-Nayrizi's commentary on Euclid, which included principles of arithmetic, and in the first half of the thirteenth century, Jordanus de Nemore wrote a treatise on arithmetic not directly related to the Euclidean tradition. The widely influential Campano (1255–1259) drew from both sources in adding axioms concerning number to his version of Euclid's Book VII. Recall that in Euclid, only Book I includes axioms; Books V and VII, in particular, do not include any.[19] Axioms of the sort that Campano introduces support the assumptions listed under A4.

Campano states a particularly significant example concerning number that is a version of a construction postulate for numbers found in De Nemore: "To take any quantity of numbers equal to, or multiples of, any given number." There are three features of this particular principle worth noting. First, the ability to take any quantity of numbers equal to any given number still reflects the Euclidean notion of number as collections of units that can be distinct but equal. Second, Campano uses the term "equal" for equimultiple, indicating that he understands equimultiplicity as a species of equality. In fact, most of the principles introduced in the Euclidean tradition that explicitly concern number refer to equimultiplicity as equality. Finally, Campano, following De Nemore, departs from Euclid in introducing construction postulates for collections. We saw in Section 6.6 that in Euclid, the emphasis on construction of geometrical objects is lacking in the case of arithmetic, in which collections of units are introduced as if they are simply found. By this point in the tradition, at least some mathematicians viewed this as an oversight and sought to remedy it; in their view, collections also need construction postulates. This puts continuous and discrete magnitudes on a par with respect to construction and thereby brings them closer. In conclusion, whether we consider general

[18] See De Risi (2016, p. 16).

[19] Thus, even if Euclid himself did not introduce axioms for arithmetic, the idea that it is possible to provide axioms for arithmetic had a long history before nineteenth-century axiomatizations, as De Risi (2016) emphasizes.

principles that concerned magnitudes in a broad sense, or principles particular to numbers, there were principles that supported the assumptions under A4.

There were also axioms that specifically concerned equimultiple collections of individuals, where those individuals are magnitudes standing in the relation of equality. For example, De Nemore included axioms concerning equimultiplicity of equals, such as: "The equimultiples of the same thing, or of equal things, are equal to one another." These axioms hold for collections of contiguous and equal spatial magnitudes, and thus undergird the assumptions articulated in A5.

In summary, the great drive in the Euclidean tradition to articulate all the hidden assumptions on which Euclid relies hit upon a great many variations of different kinds of axioms, and among them are many that concern magnitudes and many that lend support to the general theory of pure concrete measurement described in Sections 9.2–9.4.

There are, however, two important points to make about the wide variety of axioms found in these authors. First, I am aware of no one who carved out a set of axioms and identified them as constituting a general theory of measurement on which both Euclid's geometry and his arithmetic implicitly relies. Second, there was no distinction drawn between bare mereological axioms concerning magnitudes and collections and those axioms that presuppose equality for aliquot measure, nor between a bare mereological notion of composition and one that assumed and appealed to the notion of equality. Although axioms abound concerning both magnitudes and equality, no geometer highlights the role of equality as the notion bridging mereology and mathematics. This is noteworthy in light of the contrast we've been investigating between, on the one hand, Kant's bare mereological conceptions of magnitude and composition, and on the other, the richer mathematical conception of magnitudes implicitly defined by Euclid's theory of proportions.

Let us briefly turn to Kant's immediate predecessors and contemporaries. We saw in Section 6.8 that Wolff, Kästner, and Euler, three of Kant's influential immediate predecessors, all regard mathematics as the science of magnitudes, and that Wolff and Euler explicitly describe mathematics as the science of magnitudes and their measurement. Although they had access to the works of only some of the geometers we have been considering, we can now see that their emphasis on magnitudes and measurement appears to be a natural outgrowth of the Euclidean tradition; moreover, it points to an understanding of measurement that underlies pure as well as applied mathematics, and arithmetic as well as geometry. Despite its broad foundational role in their understanding of mathematics, however, they say relatively little about it; we do not find these authors probing for principles that underlie a fleshed-out theory of measurement. It follows that like the long Euclidean tradition before them, they also do not distinguish between a merely mereological account of magnitudes and a mathematical understanding of magnitudes, or the role of equality in bridging the gap between them. Kästner both describes all mathematics as about magnitudes and states that we can think of magnitudes in mereological terms (see Section 6.8. Recall that at the beginning of his work on mathematical foundations, Kästner states:

> One can regard magnitudes merely as a collection of parts, as a **whole** (*totum*); or one can at the same time view the connection or order of these parts that constitute a certain **composed thing** (*compositum*). (Kästner (1800, p. 3))[20]

Nevertheless, Kästner does not go further to explain how one gets from part–whole relations of magnitudes to the mathematics of magnitudes.

9.7 Kant on Equality and the General Theory of Measurement

Like the geometers of the Euclidean tradition and Kant's immediate predecessors, Kant does not identify and articulate the general theory of pure concrete measurement underlying the Eudoxian theory of proportions. On the other hand, as we saw in Chapters 7 and 8 and as I emphasized in Section 9.1 above, Kant does give an account of magnitudes and their composition, and that account is mereological. Perhaps Kant was partly inspired or influenced by Kästner's claim, quoted in the previous section, that a mathematical treatment of magnitudes can regard them as a collection of parts that constitute a *compositum*. Regardless, his mereological account goes well beyond both Euclid and Kästner in emphasizing the mereological properties of magnitude and, furthermore, explaining how we cognize them. But Kant also does not directly address the relationship between his mereological account of magnitudes and Euclid's mathematical magnitudes, nor does he emphasize the role of equality in making that transition possible. There is a gap in Kant's explanation corresponding to the gap between a mereological account of magnitudes and Euclid's mathematical magnitudes.

I do not think, however, that Kant simply overlooked the difference between his mereological account of magnitudes and the mathematical properties of magnitudes implicitly defined by the Eudoxian theory of proportions. I believe he thought his account of magnitudes was not complete, but rather an account of the most important features of magnitudes sufficient for the task of the *Critique*, which was to provide not a complete system of metaphysics but only a critical examination of the most basic foundations of human cognition and its limits. The role of the categories of quantity and intuition in our cognition

[20] Kant owned this work, though his was an earlier edition than the one I was able to obtain and have quoted here. While it is likely that Kant read this passage, I have not verified that it appears in Kant's edition.

of magnitudes, together with the Schematism and the principles of the Axioms and Anticipations, largely fulfill that task.

Furthermore, the possibility of saying more about the cognition of magnitudes and its connection to mathematical cognition is not foreclosed by Kant's mereological account of magnitudes in the *Critique*. Recall from Section 3.3 that Kant's characterization of magnitude in the Axioms does not count as a definition in the strict sense; it is, rather, an exposition of a concept, and hence an identification of essential marks of being a magnitude that may not be exhaustive. It is therefore consistent with Kant's characterization of magnitude as a homogenous manifold in intuition that there are further important properties of magnitudes not included in that characterization, properties necessary for a full account.

It should be noted, moreover, that Kant made significant further clarifications concerning magnitudes in the B-edition of the *Critique*, indicating that he thought more could and should be said than he had said in the A-edition. Overall, one is left with the impression that Kant had more to say than he had space or time to include.

As I noted in Section 1.5, Kant refers to the topic of metaphysical foundations of mathematics in several lecture notes and reflections, indicating that it would concern the cognition of magnitudes. For example, in one of his reflections, Kant refers to the "Metaphysics of the doctrine of magnitude or the metaphysical foundations of mathematics" (Reflexionen 14:195–6, 1764–1804; see also 29:755, 28:636). While Kant may be referring to the quite incomplete account of mathematical cognition in the *Critique*, these passages suggest a systematic work parallel to what Kant's *Metaphysical Foundations of Natural Science* does for Newton's natural philosophy.[21]

This is not to imply that Kant had already worked out the content of a text falling under that title. A survey of Kant's extant writings, including his notes and marginalia, suggests that we have reached the limits of his explicit theorizing about the relation between our mereological cognition of magnitudes and mathematical cognition – not the limits of his abilities, but the limits of the time he had to further develop and articulate his views. Kant goes beyond Euclid in characterizing what he takes to be the most fundamental properties of magnitudes and explaining our cognition of them; he nevertheless follows Euclid in making tacit assumptions required to fully account for the mathematical character of magnitudes. While we cannot determine how far Kant's thoughts on this topic reached, we can investigate what his views

[21] Recall the caveat noted in Section 1.5; such a text would include not a justification of *a priori* principles of mathematics, which the *Metaphysical Foundations of Natural Science* was to provide, but an account of the fundamental concepts on which mathematical cognition depends.

might be by looking at what Kant has to say about the assumptions of the general theory of pure concrete measurement given in A1–A5.

9.7.1 Kant on Comparative Size Relations of Continuous Magnitudes

The first set of assumptions articulated in A1 concern the comparative size relations in which magnitudes stand: greater than, less than, and equal to (see the list of assumptions on pages 260–61). We saw in Section 9.5 that equality is the linchpin of the entire theory of measurement. On the assumption that any two homogeneous magnitudes are either equal or unequal and the further assumption that magnitudes have no smallest part, it can be used to define greater than and less than and extends a mere partial ordering to a total ordering over magnitudes that are homogeneous with one another. And by so doing, it provides a crucial element of aliquot measure and the general theory of pure concrete measurement, which in turn supports the Euclidean theory of proportions that implicitly defines Euclid's mathematical conception of magnitude.

What is Kant's own understanding of these fundamental relations among magnitudes? How do they relate to Kant's mereological understanding of magnitude? The best way to bring out Kant's views is to first review Kant's discussion of the part–whole relations among concepts. In Kant's view, both things and concepts fall under a concept, and from a purely logical point of view, which considers the relation between concepts in abstraction from their relation to objects, the extensions of concepts consist of the concepts that fall under it. More precisely, from a logical point of a view, the extension of a concept consists of the possible concepts that fall under it (see Section 5.3). Kant explains various logical forms of judgment in terms of the relations between the extensions of concepts and uses Euler's circles (a precursor to Venn diagrams) to illustrate them (9:109). For example, "All Clydesdales are horses" would be illustrated by a circle or square representing the concept <Clydesdale> completely contained within a larger circle or square representing the concept <horse>. Kant thus uses the part–whole relations of space to explain the relations among the extensions of concepts.

Kant also draws explicit further analogies between the part–whole relations of concept extensions and the part–whole relations of space. First, just as space is infinitely divisible into further spaces, concept extensions are infinitely (logically) divisible into extensions distinguished by further concepts (A655–6/B683–4). This is another way of stating his doctrine that in logic there are no *infimae species* of concepts (again, see Section 5.3). Second, points are to lines as individuals are to the extension or "horizon" of concepts; that is, a point is a limit of the extension of a line just as an individual is the limit of the extension of successive concepts in a hierarchy of concepts (A658–9/B686–7).

Since Kant gives a solely mereological account of magnitudes and draws strong analogies between the part–whole relations among concepts and those of space, one might be tempted by the thought that the part–whole relations between concepts could somehow be exploited to represent the part–whole relations on which Kant's mathematics is based. In other words, perhaps by Kant's own lights, a purely conceptual and hence logical foundation would do just as well, obviating Kant's appeal to intuition.

This purely logical approach will not work, however. As we saw in Chapter 7, the extensions of concepts consist of other concepts that represent specific differences and hence heterogeneous manifolds, and in Kant's view such manifolds cannot be combined in a manner that represents mathematical composition.

Even if this problem were set aside, however, we are now in a position to appreciate an important further point. In Kant's view, concept extensions cannot stand in the relation of equality. (Kant does state that distinct concepts can have identical extensions (9:98), but he nowhere allows that extensions could be distinct yet equal.) As a consequence, equality cannot be used to supplement the comparative size relations between extensions to form what we would call a total ordering. In fact, Kant explicitly mentions this limitation of the part–whole relations between concept extensions:

> One concept is not broader than another because it contains more under itself – for one cannot know that – but rather insofar as it contains under itself the other concept and besides this still more. (9:98)

Keep in mind that in Kant's view, the extension of a concept comprises the possible concepts that can fall under it, which are limitless, since the division of the sphere of a concept can continue *ad infinitum*. There is no sense in which we could count the number of concepts falling under two concepts in order to reach the judgment that one contains more under it. Hence, the only sense in which the extension of one concept can be more than another is if the former contains the latter and yet more.

What is most important for our present purposes is what the passage reveals about his understanding of the part–whole relations of concepts. Despite his use of Euler's circles to illustrate logical relations, Kant is well aware, and even goes out of his way to state, that the comparative size between concept extensions is what we would call a merely partial relation, restricted to full part–whole containment. Because Kant does not allow the relation of equality between extensions, there is no sense that can be given to the comparative size relations between concepts with only partially overlapping or disjoint extensions. Moreover, even when a concept is fully contained in another, there is no way to introduce a measure of size; one can know only that the latter concept contains the former "and besides this still more." In other words, the ordering is not only partial; even this partial ordering is at best ordinal.

Kant's treatment of magnitudes sharply contrasts with this. For Kant explicitly appeals to equality to define a size relation between distinct magnitudes that are not related by part–whole composition. In an extended lecture note on the part–whole relations of magnitudes, Kant defines larger and smaller for magnitudes using the relation of equality as follows:

> A > than B if a part of A = B; in contrast A < B, if A is equal to a part of B. (28:506, late 1780s)[22]

Thus, one magnitude will be larger than another as long as a part of it is equal in size to the other. Kant thereby gives a sense to one magnitude being greater or less than another even when the magnitudes are only partially overlapping or do not overlap at all, and this extension of the part–whole relation orders any two homogeneous magnitudes according to their size. Moreover, we saw in Section 9.5 that greater than and less than can be defined in terms of equality and part–whole relations, making equality alone the crucial element that bridges mereology with mathematics.

In summary, Kant is aware of and sensitive to the limitations of mereology on its own, and explicitly draws attention to its limitations in his account of the part–whole relations among the extensions of concepts. He is also aware that equality allows us to move beyond a solely mereological account of magnitudes and ground comparative size relations across magnitudes. Finally, in defining greater than and less than in terms of part–whole relations and equality, Kant isolates equality as the key element in moving beyond a bare mereology of magnitudes. It is Kant's comments of this sort that bring me to believe that Kant had more to say about mathematics and mathematical cognition than we find in the published works, and that he would have further developed his views had he written a work devoted to the foundations of mathematics.

9.7.2 Kant on Composition and Multiplication of Continuous Magnitudes

We can now consider what Kant's views would be with regard to the assumptions listed under A2, that is, those concerning composition and multiplication of continuous magnitudes. As I have repeatedly noted, Kant strongly emphasizes the synthesis of composition underlying mathematical cognition, a kind of synthesis particular to homogeneous magnitudes (see Sections 3.2, 4.1, and 7.8). As pointed out in Section 9.1, however, Kant's understanding of composition is only mereological. That is, it is a synthesis of a homogeneous manifold,

[22] Kant makes the same point in his discussion of magnitudes at 28:424, 1784–5; 28:561, 1790–1; and 28:637, 1792–3.

without further specification of other properties of the manifold, much less the size of elements of the manifold; in fact, for the synthesis of a continuous manifold, there are no distinguished elements; no determinate parts are represented (see Section 4.6 and Chapter 5). Kant's account of composition includes the claim that a composed magnitude is homogeneous with any of its parts and yet greater than them, but that is all that composition on its own entails. Nevertheless, Kant's understanding of composition is sufficient to support A2.1.

On the other hand, A2.2 articulates a further requirement of the Eudoxian theory – what is required is not just composition, but multiplication, that is, the composition of magnitudes all equal to a given magnitude. We have just seen, however, that Kant introduces the relation of equality into his account of homogeneous magnitudes in the definition of greater than and less than. Moreover, he clearly assumes that equal continuous homogeneous magnitudes can be composed into a larger continuous whole. Kant thus presupposes what is required for multiplication.

9.7.3 Kant on Comparative Size Relations among Collections

The assumptions listed under A3 cover the total ordering of collections with respect to less than, greater than, and equality, where the last is understood as equimultiplicity. Kant's understanding of number is nuanced and does not exactly correspond to the Euclidean, but it incorporates the notion of number as a collection of units. And as noted in Sections 3.4 and 4.6, Kant both treats and sometimes refers to collections of units as discrete magnitudes. Kant was influenced by developments in the theory of magnitudes, including the drive for a broadened theory of magnitudes that would include spatial magnitudes, other sorts of continuous magnitudes, and collections of individuals as well. In Kant's understanding of that unified theory, the Euclidean common notions concerning equality – the transitivity of equality, equals added to equals are equal, and equals taken away from equals are equal – receive a broader interpretation; they cover both equality of continuous magnitudes and equality of discrete magnitudes. His assimilation of number into a broader framework of thinking of mathematics as a science of magnitudes follows an important trend I have highlighted in the Euclidean tradition that is also found in his immediate predecessors. These claims concerning Kant's understanding of number require further clarification and argument that will have to be postponed for a more in-depth look at the foundations of Kant's philosophy of arithmetic.[23] For now, however, it suffices to note that Kant uses the term "equal" (*gleich*) in numerical

[23] See Sutherland (2017) for a defense of the claim that Kant's conception of number includes the notion of a collection of units as well as their ordering.

contexts, and regards numbers as governed by the common notions governing equality, understood broadly to include equimultiplicity.[24] Kant's account therefore supports the assumptions listed under A3.

9.7.4 *Kant on Composition and Multiplication of Collections*

The A4 assumptions concern the composition and multiplication relations in which collections can stand. We saw in Sections 3.4 and 4.6 that Kant appears to have two sorts of composition in mind, that of a continuous magnitude and of a discrete magnitude, that is, a collection of units. The latter sort of composition appears in the Schematism, which I quote one more time. Kant states:

> The pure image of all magnitudes (*quantorum*) for outer sense is space; that of all objects of the senses in general is time. The pure **schema of magnitude** however (*quantitatis*), as a concept of the understanding, is **number**, a representation which collects [*zusammenbefaßt*] the successive addition of one to one (homogeneous). Number is thus nothing other than the unity of the synthesis of the manifold of a homogeneous intuition in general, brought about through my producing time itself in the apprehension of the intuition. (A142–3/B182)

The fact that Kant both identifies the unity of the successive synthesis with number and describes that successive synthesis as of "one to one (homogeneous)" suggests a kind of synthesis of composition that takes units as its elements, in contrast to the continuous synthesis of a homogeneous manifold in the representation of a continuous magnitude, such as the drawing of a line.

The idea that there are two sorts of synthesis, continuous and discrete, finds further support in Kant's defense of his claim that all appearances are continuous magnitudes. Kant argues that a collection of coins is not an appearance properly described as a discrete magnitude, because it isn't an appearance, but an aggregate, in which each coin is a continuous appearance.[25] In the course of his argument, he states:

> If the synthesis of the manifold of appearance is interrupted, then it is an aggregate of many appearances, and not really appearance as a *quantum*,

[24] Recall as well from Chapters 6 and 7 that Kant defines equality as identity of quantity. He does not have a separate analysis of equimultiplicity; rather, equimultiplicity is treated as identity of quantity for collections.

[25] While Kant's wording suggests that all *quanta* are continuous and hence that there are no discrete *quanta*, that is not Kant's view. Even elsewhere in the *Critique*, Kant refers to discrete *quanta*. But to corroborate that Kant allows that *quanta* can be discrete, one needs to look beyond this passage and the *Critique* to all that Kant says in his work bearing on discrete magnitudes. See Section 3.5, especially note 39. I will have to postpone that task for a later work more narrowly focused on Kant's conception of number and philosophy of arithmetic.

> which is not generated through the mere continuation of productive synthesis of a certain kind, but through the repetition of an ever-ceasing synthesis. (A170/B212)

Kant here contrasts the continuation of a productive synthesis with a repeated "ever-ceasing" synthesis, which suggests a synthesis of a collection into an aggregate. This is reinforced by Kant's claim that the representation of "so many coins" is the representation of "an aggregate, i.e., a number of coins." Nevertheless, to complete his argument that all appearances are continuous, Kant insists that this sort of synthesis applied in the cognition of an aggregate of appearances presupposes that the unit (*Einheit*) grounding number is an appearance as unity that is itself "a *quantum*, and is as such always a continuum."[26]

Together, these texts suggest that Kant thought there were two sorts of synthesis of composition, a synthesis of continuous magnitudes and a synthesis of collections. In that case, Kant's account of composition certainly supports A4.1, that individuals and collections can compose larger collections. Kant does not directly address A4.2, the multiplication of collections. Nevertheless, because he holds that collections can stand in the relation of equality, his account has all that is required to support A4.2, and he appears to presuppose it. Similarly, Kant also does not directly address A5, the final set of assumptions concerning equimultiple multiplication and its application to continuous magnitudes. But what Kant says about equality, comparative size relations, and composition strongly suggests that he supports these final presuppositions required for the Euclidean theory of proportions for both continuous magnitudes and collections.

I have already noted that the synthesis of composition of a continuous manifold and of a discrete manifold are closely connected in the presuppositions of the general theory of measurement underlying Euclid's *Elements*. In Section 9.2, we saw that A2.2 and A2.3 presuppose the concept of collections and their aliquot measure, because they presuppose the possibility of representing a continuous magnitude as composed of a collection of continuous, contiguous, and equal magnitudes. And we also saw that this has two important implications. First, Euclid's appeal to aliquot measure for continuous magnitudes in Book V already presupposes the concept of number as a collection and aliquot measure of collections that is also at the foundation of his arithmetic in Book VII. Second, the notion of composition presupposed in A2.2 and A2.3 incorporates two sorts of composition, so that in such cases there is both a generation of a continuous magnitude and a

[26] There are two sorts of unity in the representation of a number, i.e., a collection: the unity of the units that are synthesized, and the unity of the aggregate. The sort of unity Kant means in this last quote is that of the unit, which is how I have translated *Einheit*.

representation of that magnitude as composed of a collection of discrete equal units. The doubling and intertwining of the two sorts of composition in the presuppositions of Euclid's geometry not only complicated the understanding of mathematics in the Euclidean tradition. That same doubling and intertwining also contribute to the obscurity of understanding Kant's account of number and its relation to our representation of continuous and discrete magnitudes.

I would like to emphasize again that Kant did not identify or articulate the general theory of pure concrete measurement on which the Euclidean mathematical tradition rests. I argued at the beginning of this section that it would have been out of place for him to do so in the *Critique*, and that he also did not outline that theory in any unpublished texts that we have. Kant, like Euclid, tacitly assumes the general theory of measurement whose assumptions I have listed in A1–A5. Nevertheless, this section has brought out two points. First, Kant's understanding of magnitudes, composition, and the role of equality all support the tacitly assumed theory of pure concrete measurement. Second, Kant sees that equality is crucial to moving beyond his mereological account of magnitudes and bridging the gap to their mathematical treatment. Had Kant had the opportunity to fully flesh out his understanding of the foundations of mathematical cognition and its role in our experience of the world, I believe that this would have been his starting point.

9.8 The Place of Equality in Kant's Account of Human Cognition

The pivotal role of equality in Euclid's tacit theory of measurement and hence in his mathematical treatment of magnitudes raised questions about the place of equality in Euclid's geometry, that is, how Euclid introduces and employs equality into the *Elements*, which I addressed in Section 9.5. In a parallel fashion, the importance of equality in bridging Kant's mereological account of magnitudes, on the one hand, and both mathematical cognition and the mathematical character of experience, on the other, raises questions about the place of equality in Kant's account of human cognition. The importance of the concepts of part and whole is reflected in the fact that we cognize part–whole relations through the categories of quantity – unity, plurality, and allness (totality). But how does the concept of equality fit into Kant's account of human cognition? Given the evidence for his awareness of its importance, one might even have thought it would find a place somewhere in the table of categories or in the predicables that follow from them. If not, however, where does Kant find a place for the concept of equality?

Kant does not directly address this question. Here, too, I believe we have reached the limits of Kant's explicit theorizing about the theory of magnitudes and their relation to mathematical cognition. Nevertheless, it may be that Kant had more to say on the issue than he had time to develop, and I believe he may have thought that his doctrine of the concepts of reflection found in the Amphiboly provides the resources to account for the concept of equality. A full account of the concepts of reflection would take us too far afield, so I will be brief; moreover, my suggestions for how this might work should be considered speculative.

Kant holds that there are four pairs of concepts we use to make comparisons: identity and diversity, agreement and opposition, inner and outer, and form and matter. Kant was in agreement with other early modern philosophers as diverse as Hume and Baumgarten on the need for concepts of comparison.[27] In Kant's view, the concepts of reflection are distinguished from the categories because they are not employed to cognize a corresponding object; they are instead used to compare representations with each other (A269/B325).[28] There are two very different ways in which representations can be compared; the first uses them in the comparison of concepts alone, while the second uses them for the comparison of the things to which the concepts refer (A262/B318, A269/B325).[29] Kant maintains that cognition of a thing represented by a concept requires intuition, so that a thing can be cognized only if it is an object of sensibility. As a consequence, the comparison that results in cognition must take into account the sensible conditions of space and time. The application of the concept of equality in the comparison of magnitudes will be a comparison of this latter type.

Chapter 8 explained how Kant both adopted and reworked the traditional metaphysics of quantity found in Leibniz and the Leibnizian-Wolffian philosophy. Since Kant embraced the traditional view of similarity as identity of quality and equality as identity of quantity, it is possible that the concepts of similarity and equality could be grounded in the first pair of concepts of reflection, identity and diversity. As mentioned in Section 8.4, the Axioms of Intuition includes a discussion of Common Notions 2 and 3, the propositions

[27] See Hume (1978, p. 14) and Baumgarten in Kant (17:83f). Identity was thought to be fundamental. Hume counts identity as one of the seven philosophical relations which arise from the comparison of objects, while Baumgarten devotes the first section on the fundamental relations of things to identity and diversity.

[28] Béatrice Longuenesse, in particular, has emphasized the importance of the concepts of reflection in Kant's theory of cognition; see Longuenesse (1998, especially chapter 6). A full account of these matters would need to carefully address the relation between fundamental geometrical relations and the general role Kant assigns to the concepts of reflection in human cognition (A262/B318, A269/B325).

[29] The point of the Amphiboly is to diagnose where the Leibnizian theory of cognition (and implicitly the Wolffian theory of cognition) has gone wrong, and Kant locates Leibniz's error in a failure to distinguish these two ways of employing the concepts of reflection. As we saw in Section 7.5, Kant's criticism of Leibniz in the Amphiboly revealed that intuition can represent numerical difference without qualitative difference, and hence a homogeneous manifold. I set aside Kant's criticisms of Leibniz here to focus on Kant's understanding of equality.

"Equals added to equals are equal" and "Equals subtracted from equals are equal" (A163–4/B204). Kant states that both propositions are analytic and that they are grounded in consciousness of the identity of the production of magnitudes. Kant's appeal to identity here lends further support to the idea that the concept of equality is based on the concepts of reflection.

Objective judgments of similarity and equality are judgments that objects are identical with respect to *qualitas* and *quantitas*, respectively. Since *quantitas* is a determination of the "how much" of a *quantum*, equality would be an identity of the quantitative determinations of *quanta*.

If Kant accounts for the concepts of similarity and equality by defining them as identity of *qualitas* and *quantitas*, it would be natural to ask after the source of the concepts of *qualitas* and *quantitas* themselves. As we also saw in Section 8.4, Kant thinks that *qualitas* (in the second sense I identified) is a determination that can be cognized through the thing itself, while *quantitas* is a determination that can be determinately cognized only through comparison of the thing with other things. Thus, *qualitas* and *quantitas* are distinguished by whether the grounds of their determinate cognition are inner or outer relative to the object cognized. The concepts of inner and outer are the third pair of concepts of reflection. Perhaps Kant would appeal to them in defining *qualitas* and *quantitas*. I will not pursue these suggestions further here; my present purpose is merely to show that Kant seems to have the resources to find a place for the concept of equality in his theory of cognition.

9.9 Conclusion

We have seen that Kant's mereological account of homogeneity, magnitudes, and composition falls short of the Euclidean mathematical notion of magnitude implicitly defined by the theory of proportions. That led us to consider what is required to fill that gap, which revealed that Euclid tacitly assumes a general theory of pure concrete measurement, a theory that supports the crucial notion of aliquot measure lying at the foundation of both Books V and VII. Articulating the assumptions of that theory allowed us to see what it in turn requires, which brought into focus the way in which the relation of equality plays a pivotal role in accounting for Euclid's mathematical treatment of magnitudes. The long geometrical tradition that followed proposed a great number and variety of additional axioms to buttress the foundations of Euclid's *Elements*; many of them concerned magnitudes, and some specifically concerned aliquot measure. No one, however, articulated and isolated the general theory of pure concrete measurement that lies at the foundation of Books V and VII. One can nevertheless see why Wolff, Kästner, and Euler all came to characterize mathematics as a science of magnitudes, and why Wolff and Euler more specifically characterize mathematics as a science of magnitudes and their measurement.

Although Kant makes the same unstated presuppositions as Euclid concerning the general theory of pure concrete measurement, there is still a crucial difference between them. Because Euclid does not define magnitude or homogeneity, magnitudes and their properties are implicitly defined by their role in Euclid's theory of proportions. That leads to a rich mathematical conception of spatial magnitudes; lines, areas, volumes, and angles are those sorts of things that can stand in ratios and are subject to the powerful set of propositions established in Book V. Kant, in contrast, defines magnitude and composition in a way that draws on only mereological properties. Thus, it does not follow from his characterizations of them that magnitudes are subject to the Eudoxian theory of proportions. On the one hand, Kant's restricted mereological notions of homogeneity, magnitude, and composition, and in particular the principles of the Axioms of Intuition and the Anticipations of Perception, help to explain the employment of the categories of quantity in relation to intuitions and to appearances in regard to their intuition. They also establish an important necessary condition for the applicability of mathematics to objects of experience – that appearances, as regards their intuition, are extensive magnitudes. On the other hand, they are not sufficient to explain the mathematical character of experience, nor is the principle of the Axioms on its own sufficient to establish the applicability of mathematics to appearances. What is needed are two additional claims. First, magnitudes characterized solely in mereological terms can stand in relations of equality (where that encompasses both the equality of continuous magnitudes and equimultiplicity of collections of discrete magnitudes), which supports the tacit assumptions articulated in A1 and A3. Second, equal magnitudes can be composed in accordance with the assumptions articulated in A2, A4, and A5.

As already noted in Section 9.1 above, Kant states that the principle of the Axioms is "that alone which that makes pure mathematics in its complete precision applicable to objects of experience" (A165/B206). This certainly sounds like the assertion of a sufficient condition of the applicability of mathematics, and warrants a clarifying digression. Interpreting it in that way both misunderstands Kant's primary aim in the Axioms of Intuition and misconstrues Kant's claim about the applicability of mathematics.

First, Kant's most important aim in the Axioms is to explain why and how the categories of quantity, under the conditions of sensibility, are employed in the cognition of all appearances. Since the categories of quantity – unity, plurality, and allness – are mereological, it makes sense that the principle corresponding to these categories concerns basic mereological properties alone, and that appearances having these basic mereological properties are necessary to account for the employment of the categories of quantity, but not that they are sufficient for pure or applied mathematics.

This is an instance of an important general point about the *Critique*, which is an unfolding account of necessary conditions of experience, and an uncovering of more conditions, or more details concerning the same conditions, as it progresses; Kant's claims about the conditions of experience are rarely, if ever, about sufficient conditions. This feature of the *Critique* is reflected in Kant's Transcendental Aesthetic claims about the applicability of mathematics (see Section 2.7). He argues that time and space are two sources of cognition from which synthetic cognitions can be drawn. He also asserts there that they are valid only for objects considered as appearances, which is alone the field of their validity. That is, the fact that space and time are pure forms of sensibility accounts for both the limitation of and the validity of mathematics for objects appearing in space and time (A39/B56). Thus, a necessary condition of the applicability of mathematics to objects of experience is that space and time are pure *a priori* forms of intuition, and this partly explains the applicability of mathematics to appearances. Yet this is hardly a sufficient condition for the applicability of mathematics to experience. The Axioms reveals a further necessary condition of the applicability of mathematics to experience, which is that the cognition of objects of appearance be subject to the same synthesis of composition as the synthesis underlying mathematical cognition. Since the apprehension of appearances relies on that synthesis, this necessary condition of applicability is met and partly explains the applicability of mathematics, while still not being a sufficient condition of the applicability of mathematics to appearances. In particular, the latter also requires that homogeneous magnitudes stand in the relation of equality, so that they are subject to aliquot measure and the general theory of measurement.

Despite the unfolding nature of Kant's description of the necessary conditions for cognition, however, it is still quite natural to read Kant's claim that the Axioms principle is "*that alone* that makes pure mathematics in its complete precision applicable to objects of experience" (my emphasis) as an articulation of a sufficient condition of all that would be needed for the full applicability of mathematics. But the import of the phrase "is that alone" is quite different; Kant is claiming that without this principle in particular – no other will do – the applicability of mathematics would not be possible, and that is an articulation of a necessary condition. For mathematical cognition, and the content and justification of mathematical propositions, depends not just on the fact that space and time are pure *a priori* forms of intuition, but on the synthesis of composition of a homogeneous manifold in intuition, and is the very same synthesis required for the apprehension of an appearance. It is this fact alone – and no other – that fuses our cognition of mathematics and objects of experience in the required way.

Furthermore, the reference to complete precision is a reference not to all applications of mathematics but to a quite specific feature of its application. In elaborating his point, Kant mentions geometrical constructions that establish that lines and angles are infinitely divisible and that objects of experience have to be in agreement with this result; they too are infinitely divisible. Kant's argument here is a rebuttal of Wolff and his followers who, for metaphysical reasons, wished to claim that real substances in real space are not infinitely divisible. They therefore claimed that geometry is an idealization and hence only approximately represents the properties of reality. That is, they claimed that geometry does not apply in all its precision. The Axioms principle rules out this view, since the synthesis on which the geometrical demonstrations of infinite divisibility rests is the very same synthesis required for the apprehension of objects of experience. In conclusion, it is wrong to take Kant's claim about the applicability of mathematics at the close of the Axioms of Intuition as evidence that Kant thinks the argument and principle of the Axioms of Intuition supply a sufficient condition for the applicability of mathematics.

Let us return to the theme of this chapter – the gap between the mereological notions of magnitude and composition and the additional tacit assumptions required for the general theory of pure concrete measurement. There is further evidence that Kant appreciated the role of a general theory of measurement in an account of mathematics and the mathematical character of the world. When Kant presents the Table of Categories in the *Prolegomena* (4:303), he follows each of the categories of quantity with a term in parentheses:

Transcendental table of concepts of the understanding

I.
According to Quantity

Unity (measure [*das Maß*])
Plurality (magnitude [die *Größe*])
Allness (whole [*Das Ganze*])

While I cannot give a full account here, Kant's parenthetical additions anticipate the role the categories of quantity play in measurement.[30] This is an indication of the central role that measurement plays in an understanding of mathematical cognition and the mathematical character of the world, expressed in terms of a theory of magnitudes.

Remarkably, we have seen that Kant is aware of the limitations of mereology, which is revealed in his discussion of the relations among concepts. Moreover, he explicitly defines greater than and less than for magnitudes in terms of part–whole relations and equality, which establishes a total ordering of magnitudes with respect to size. Thus, even if Kant does not articulate the pure theory of concrete measurement, in a certain sense he sees even more deeply into the issue. He understands that equality is what is crucial to move

[30] I will return to consider this passage more closely in future work, when considering Kant's conception of number and philosophy of arithmetic in more detail.

from his mereological account of magnitude to Euclid's mathematical notion of magnitude.

There is additional evidence that Kant appreciated the distinction between a bare mereology of magnitudes and the role of equality in bridging the gap from mereology to mathematics and a mathematical account of the world. First, Kant's *Metaphysical Foundations* begins with phoronomy, the pure doctrine of motion, and Kant's treatment includes a proof that the composition of motions, understood as intensive magnitudes, can be constructed. While Kant refers to adding speeds that are equal to each other and hence the doubling of a speed, he does so only to provide a vivid example; a close reading of the argument shows that it relies only on the mereological composition of motions.[31] This reflects Kant's sensitivity to the distinction between a bare mereology of magnitudes and one supplemented with the relation of equality, and a desire to establish this necessary, though not sufficient, condition of the mathematical treatment of motion on a minimal mereological basis.

Furthermore, as Michael Friedman has emphasized, one of Kant's aims in the *Metaphysical Foundations* is to explain the possibility of a universal mathematical concept of quantity of matter (mass), one that allows it to be measured. Kant holds that the quantity of matter can be estimated only by the quantity of motion at a given velocity, by which he means momentum. But momentum can in turn be measured only through the communication of motion between bodies, that is, through their mechanical interactions.[32] Underlying all of Kant's theorizing is a focus on the conditions for measurement; it is not sufficient that these physical features of the world are magnitudes; one requires an ability to introduce, even if indirectly, a relation of equality for that kind of magnitude.

[31] See Sutherland (2014) for a detailed defense of this claim.

[32] Kant's account is far more complex than this suggests; see Friedman (2013) for a comprehensive treatment. For the features I emphasize here, see, for example, pp. 381 and 293ff.

10

Concluding Remarks

As I noted in Chapter 1, a resurgence of interest in Kant's of mathematics in the last half-century has led to a great deal of high-quality scholarship that has significantly improved our understanding of Kant's philosophy of mathematics, despite the fact that most of that work does not rely on an explicit engagement with Kant's theory of magnitudes. My aim has been to fundamentally transform our current understanding of both Kant's philosophy of mathematics and his account of our cognition of the world by broadening and deepening our understanding of that theory. Doing so required recovering a very different way of thinking about mathematics as a science of magnitudes and their measurement prior to the arithmetization of mathematics in the late nineteenth century, a way of thinking that in many ways had more in common with Euclid and the Euclidean tradition than with our modern views.

I began my book with the famous quote from Galileo because he so eloquently expressed the idea that the world is mathematical – in his words, the world is written in the language of mathematics whose characters are triangles, circles, and other geometrical figures. As I understand him, Galileo meant not simply that mathematics as a language is a useful a tool to describe the world, but that the world was itself the locus of mathematics, a sentiment with which Kant would agree. But I also chose this quote because it was Galileo and his followers in the early seventeenth century who focused particular attention on developing the Eudoxian theory of proportions as a method for understanding the mathematical character of the magnitudes that constitute that world.

Of course, mathematics was transformed by tremendous advances in the time between Galileo and Kant. Algebra overtook the theory of proportions as the preferred method of solving problems, including geometrical problems. Nevertheless, algebra was still often thought to be a tool for representing the relations among magnitudes expressed in the theory of proportions. Mathematics was also fundamentally transformed by the development of finite and infinite analysis, while subsequent advances in analysis promoted an increasingly abstract understanding of mathematics. An oft-told story about the foundations of the calculus emphasizes this.[1] Influenced by his teacher Barrow, Newton was inspired by and appealed to Euclid's geometry to secure the meaning of mathematical terms and the certainty of results, anchoring mathematics in the representation of spatial magnitudes. Newton in turn influenced British mathematicians following him. On the other hand, mathematicians on the Continent, who looked to Leibniz rather than Newton, were driven by successful mathematical breakthroughs that used equations to describe the world and were on the whole less concerned with foundational questions. Mathematics continued to advance at a remarkable rate into the eighteenth century, which was accompanied by a shift in attention from what mathematical symbols in mathematical equations stand for to the properties of equations themselves. Mathematics was outstripping its foundations. Some mathematicians had always been less concerned about foundations than others, a difference in temperament that widened after the development of the calculus. Nevertheless, among those who were concerned with foundations, it was still thought that mathematics was grounded in magnitudes. Whether a mathematician was driven to say something substantive about the foundations of mathematics or mathematical cognition, or was merely prompted or felt compelled to say something about foundations as preamble to a mathematics text, it was still common to regard and describe mathematics as ultimately founded in some way or other on magnitudes. We have seen this in the work of Wolff, Kästner, and Euler, for example.

This was the backdrop for Kant's project of explaining the foundations of mathematical cognition. Kant could count on his readers to understand his references to the notion of magnitude inherited from the Euclidean tradition. Kant's views are strongly influenced by that tradition, but we also have seen that Kant attempted both to clarify the fundamental notions of magnitude, homogeneity, and composition and to explain the fundamental elements of human cognition – the categories as well as the pure forms of intuition – that make our cognition of magnitudes possible. Moreover, his early reflections on mathematical cognition eventually convinced him that intuition plays a role not just in mathematics but in all human cognition. His subsequent critical account of the role of intuition in representing magnitudes, as well as the categories of quantity in making the cognition of magnitudes possible, informed Kant's account of human experience generally. We have seen that Kant does not separate mathematical cognition from the cognition involved in everyday experience. Mathematical cognition consists in the cognition of magnitudes, while our cognition of the world, which consists of extensive

[1] This story is inaccurate in ways that matter greatly for a full understanding of this history and Kant's place in it; nevertheless, it accurately reflects the persisting connection between foundations of mathematics and magnitudes despite advances. For a more nuanced account of this history and its relation to Kant, see Sutherland (2020b).

and intensive magnitudes, rests on that very same cognition. As a result, Kant's account of mathematics is not an abstract and independently grounded body of knowledge applied to the world; it is immanent in the world as we experience it. The *Critique* is not simply providing an account of our cognition of everyday experience, with a nod to the applicability of mathematics to those objects. Kant's theory of magnitudes fuses our cognition of experience with all mathematical cognition, pure and applied, and includes our counting of coins on a table as well as the most advanced eighteenth-century developments in pure mathematics and mathematical physics.

An understanding of Kant's theory of magnitudes thus provides insight not just into his philosophy of mathematics but into Kant's critical philosophy, and above all his account of theoretical cognition in the *Critique*. It clarifies the relationship between the claims Kant makes about space, time, and mathematics in the Transcendental Aesthetic and elsewhere. It also improves our understanding of the contributions of intuition to human cognition in several ways. Perhaps most importantly, it allows us to see a previously unacknowledged role for intuition in representing a strictly homogeneous manifold. I have also argued that understanding Kant's theory of magnitudes reveals the nature of the singularity of intuition and clarifies the sense in which intuitive representation is concrete. The theory of magnitudes explains why Kant calls the categories of quantity "mathematical categories," how they are employed in making mathematical cognition possible, allows us to understand the argument of the Axioms of Intuition, and clarifies Kant's distinction between extensive and intensive magnitudes and Kant's references to measurement. It also sheds light on Kant's account of the mathematically sublime in the *Critique* of Judgment and the relation between the aesthetic estimation of magnitude and the cognition of determinate spaces and times described in the *Critique of Pure Reason*. Finally, the interpretation of Kant's theory of magnitude provided in these pages indicates an important line of inquiry into how Kant's critical philosophy developed. Kant's thoughts and writings concerning mathematical cognition during the *Prize Essay* period of 1762–64, in particular his essay introducing negative magnitudes into philosophy and the *Inquiry* comparison of philosophical and mathematical cognition, played a crucial role in his appreciation of the limitations of the Leibnizian-Wolffian philosophy; it both spurred and informed Kant's commitment to a role for intuition in all theoretical cognition, and what that role looked like.

I have focused on a broad reconstruction and analysis of the foundations of Kant's account of mathematical cognition and its role in our cognition of the world. It is impossible in one book to do justice to the long, rich history of mathematical thought from Euclid to the eighteenth century, but I have done my best to bring out a few features of Euclid's geometry, his theory of magnitudes, and their legacy in order to explain Kant's views. Perhaps most importantly for understanding Kant's place in the history of mathematics preceding him, our investigation of Euclid's *Elements* revealed a tacit general theory of pure concrete measurement, one that Kant also tacitly assumes. Uncovering that theory helps to clarify the relationship between, on the one hand, Euclid's notion of magnitude and the theory of proportions, and on the other, Euclid's notion of number and this theory of arithmetic. That clarification is, I hope, valuable for understanding Euclid as well as Kant.

There are more specific implications of this reconstruction for Kant's philosophy of geometry, arithmetic, algebra, and the calculus that I have not had the space to describe here. The complexity of the relation of magnitude and number in the Euclidean tradition emerges in Kant's account as well. Subsequent work will make it possible to further untangle Kant's views of *quanta*, *quantitas*, and number, and to provide a more complete interpretation of Kant's philosophy of arithmetic and the role of intuition in it. The key role played by equality in Kant's account of mathematical cognition and his reformation of the metaphysics of quantity sets the stage for a thorough investigation of Kant's understanding of equality, similarity, and congruence in his philosophy of geometry. An appreciation of the influence of the Euclidean tradition on Kant's thought helps to explain Kant's views on algebra, in particular, the role of symbolic construction and its relation to the theory of proportions. An appreciation of the centrality of the mereological properties of magnitude offers insights into Kant's understanding of motion as an intensive magnitude and his argument for the composition of motions in the Phoronomy of the *Metaphysical Foundations of Natural Science*. Finally, Kant's theory of magnitudes also makes his understanding of the representation of infinitesimals and limits in analysis intelligible, as well as the role of intuition in those representations. I look forward to further clarifying Kant's views in future work.

I have attempted to pursue Kant's theory of magnitudes to the limits of his reflections on the topic. While I have necessarily interpolated from disparate texts and have reconstructed the presuppositions underlying his theory of magnitudes, I have attempted to hew as closely as possible to Kant's stated views and to avoid anachronistic distortions. I have explicitly set aside an evaluation of the foundations of Kant's philosophy of mathematics from a modern perspective in this work, believing it best to first ascertain his actual views. The deep differences between Kant's understanding of mathematics and our own are all the more reason to bring out those differences rather than dismiss or read past them; an attempt to skip over Kant's references to magnitudes risks missing what is distinctive about Kant's approach and the historical perspective it provides for thinking about our own. This is true, for example, of what the task of foundations of mathematics should be. I suspect that Kant's very different perspective might prompt us to reflect on how to integrate our contemporary approaches to foundations with our best and most detailed accounts of the foundations of mathematical cognition.

Kant had a tremendous influence on the history of mathematics that followed him, and I hope the work I have done here will also help readers to understand and evaluate the reactions to Kant's views – both those that rejected his philosophy of mathematics and those that saw him as in some way an ally. An accurate history of the foundations of mathematics and analytic philosophy more generally should be based on a correct understanding of Kant's views, which I've done my best to provide.

Despite my strong conviction in the overall interpretation given here and my best efforts to avoid mistakes, I do not doubt that there are flaws, in no small part because of my attempt to place Kant's views in the context of the very rich and long Euclidean mathematical tradition. I hope that whatever faults may come to light, the reader will have benefited from this reorientation in thinking about Kant's philosophy of mathematics and its relation to his theory of experience, and will be convinced that understanding Kant's views of the nature of both mathematics and the world of experience requires taking his theory of magnitudes seriously.

BIBLIOGRAPHY

Adams, R. 1994. *Leibniz: Determinist, Theist, Idealist*. Oxford: Oxford University Press.

Allais, L. 2015. *Manifest Reality*. Oxford: Oxford University Press.

2016. "Conceptualism and Nonconceptualism in Kant: Survey of the Recent Debate." In *Stanford Encyclopedia of Philosophy*, accessed fall 2020.

2017. "Synthesis and Binding." In *Kantian Nonconceptualism*, Dennis Schulting, ed. London: Palgrave Macmillan.

Allison, H. 2004. *Kant's Transcendental Idealism: An Interpretation and Defense*. New Haven, CT: Yale University Press.

Anderson, L. 2005. "The Wolffian Paradigm and Its Discontents." *Archiv für Geschichte der Philosophie* 87 (1): 22–74.

Anderson, R. L. 2015. *The Poverty of Conceptual Truth*. Oxford: Oxford University Press.

Annas, J. 1975. "Aristotle, Number and Time." *The Philosophical Quarterly* 25: 97–112.

Aristotle. 1984. *The Complete Works of Aristotle*. Jonathan Barnes, ed. Princeton, NJ: Princeton University Press.

1990. *Aristotle's Categories and De Interpretatione*. J. Ackrill, trans. Oxford: Oxford University Press.

1993. *Metaphysics: Books Γ, Δ, and E*. 2nd ed. Christopher Kirwan, trans. and notes. Oxford: Clarendon Press

Baumgarten, A. 2014. *Metaphysics*. Courtney Fugate and John Hymers, eds. and trans. London: Bloomsbury.

Bennett, J. 1966. *Kant's Analytic*. London: Cambridge University Press.

Beth, E. W. 1956–7. "Über Lockes 'Allgemeines Dreieck.'" *Kant-Studien* 49: 361–80.

Boniface, J. 2007. "The Concept of Number from Gauss to Kronecker." In Goldstein et al. (2007), pp. 315–42.

Brittan, G. 1978. *Kant's Theory of Science*. Princeton, NJ: Princeton University Press.

Campbell, N. R. 1928. *An Account of the Principles of Measurement and Calculation*. London: Longmans Green.

Carson, E. 1997. "Kant on Intuition in Geometry." *Canadian Journal of Philosophy* 27 (4): 489–512.

1999. "Kant on the Method of Mathematics." *Journal of the History of Philosophy* 37 (4): 629–52.

2006. "Locke and Kant on Mathematical Knowledge." In *Intuition and the Axiomatic Method*, E. Carson and R. Huber, eds. Kluwer Academic Publishers, pp. 3–21.

2012. "Pure Intuition and Kant's Synthetic *a priori*." In *Debates in Modern Philosophy*, Antonia LoLordo and Stewart Duncan, eds. Routledge.

Forthcoming. "Arithmetic and the possibility of experience." To appear in a volume on Kant's philosophy of mathematics edited by Carl Posy and Ofra Rechter, Cambridge University Press.

(unpublished manuscript) "Number, the Category of Quantity, and Non-conceptual Content in Kant."

Cassirer, E. 1954. *Kant's First Critique: An Appraisal of the Permanent Significance of Kant's Critique of Pure Reason*. New York: Macmillan.

Clagett, Marshall. 1959. *The Science of Mechanics in the Middle Ages*. Madison: University of Wisconsin Press.

Cohen, M., and Ernst Nagel. 1968. *An Introduction to Logic and Scientific Method*. Mumbai: Allied Publishers.

Couturat, L. 1901. *La logique de Leibniz*. Paris: Ancienne Librairie Germer Bailliére et Cie.

Dedekind, R. 1872. "Stetigkeit und irrationale Zahlen." Braunschweig: Vieweg. Repr. in *Gesammelte mathematische Werke*, R. Fricke, E. Noether, and O. Ore, eds., vol. 3, pp. 315–34. Braunschweig: Vieweg, 1932. English trans. in Ewald (1996), pp. 765–79.

De Risi, V. 2016. "The Development of Euclidean Axiomatics." *Archive for the History of the Exact Sciences* 70 (6): 591–676.

Dunlop, Katherine. 2009. ""The Unity of Time's Measure: Kant's Reply to Locke." *Philosophical Imprint* 9: 1–31.

2012. "Kant and Strawson on the Content of Geometrical Concepts." *Nous* 46: 86–126.

Einstein, A. 1922. "Geometry and Experience." In *Sidelights on Relativity: Ether and Relativity II: Geometry and Experience*. London: Methuen.

Ellis, B. 1966. *Basic Concepts of Measurement*. Cambridge: Cambridge University Press.

Euclid. 1956. *Euclid The Thirteen Books of The Elements*. 2nd ed. Sir T. Heath, trans. and commentator. New York: Dover.

Euler, L. 1738. *Einleitung zur Rechen-Kunst zum Gebrauch des Gymnasii bei der Kaiserlichen Academie der Wissenschaft in St. Petersburg*. Also reprinted in Euler Opera Omnia, ser. 3 v. 2, pp. 1–303. Berlin: Springer Verlag.

1802. *Vollständige Einleitung zur Algebra*. Vol. 1. St. Petersburg: Kaiserlichen Academie der Wissenschaften.

1911–2015. *Opera Omnia. Schweizerische Naturforschende Gesellschaft.* Berlin: B. G. Teubneri.

1984. *Elements of Algebra.* Reprint of 1890 of John Hewlett, trans. of Vollständige Einleitung zur Algebra. Berlin: Springer Verlag.

Ewald, W. 1996. *From Kant to Hilbert: A Source Book in the Foundations of Mathematics*, vol. 2. Oxford: Oxford University Press.

Friedman, M. 1992. *Kant and the Exact Sciences.* Cambridge, MA: Harvard University Press.

2000. "Geometry, Construction and Intuition in Kant and His Successors." In *Between Logic and Intuition: Essays in Honor of Charles Parsons*, G. Sher and R. Tieszen, eds. New York: Cambridge University Press.

2012. "Kant on Geometry and Spatial Intuition." *Synthese* 186: 231–55.

2013. *Kant's Construction of Nature: A Reading of Kant's Metaphysical Foundations of Natural Science.* Cambridge: Cambridge University Press.

2020. "Space and Geometry in the B-Deduction." In Posy and Rechter (2020).

Gardner, Sebastian. 1999. *Kant and the Critique of Pure Reason.* New York: Routledge.

Goldstein, C. et al., eds. 2007. *The Shaping of Arithmetic after C. F. Gauss's Disquisitiones arithmeticae.* Berlin: Springer Verlag.

Golob, S. 2011. "Kant on Intentionality, Magnitude, and the Unity of Perception." *European Journal of Philosophy* 22 (4): 505–28.

Grattan-Guiness, I. 1996. "Numbers, Magnitudes, Ratios and Proportions in Euclid's *Elements*: How Did He Handle Them?" *Historia Mathematica* 23: 355–75.

Guyer, P. 1987. *Kant and the Claims of Knowledge.* New York: Cambridge University Press.

1992. *The Cambridge Companion to Kant.* Cambridge: Cambridge University Press.

Hahn, H. 1956. "The Crisis of Intuition." Translated in *The World of Mathematics*, James R. Newman, ed. New York: Simon and Schuster. (From lectures given in 1933.)

Heis, J. 2014. "Kant (versus Leibniz, Wolff, and Lambert) on Real Definitions in Geometry." *Canadian Journal of Philosophy* 44 (5–6): 605–30.

Forthcoming. "Kant on Parallel Lines." In *Kant's Philosophy of Mathematics: Modern Essays.* Vol. 1: *The Critical Philosophy and Its Background*, Ofra Rechter and Carl Posy, eds.

Hintikka, J. 1969. "On Kant's Notion of Intuition (Anschauung)." In *The First Critique: Reflections on Kant's Critique of Pure Reason*, J. J. MacIntosh and T. Penelhum, eds. Belmont, CA: Wadsworth Publishing Company.

1972. "Kantian Intuitions." *Inquiry* 15: 341–5.

1974a. "Kant's 'New Method of Thought' and His Theory of Mathematics." In *Knowledge and the Known.* Dordrecht: D. Reidel.

1974b. "Kant on the Mathematical Method." In *Knowledge and the Known.* Dordrecht: D. Reidel.

Hogan, D. 2009. "Three Kinds of Rationalism and the Non-spatiality of Things in Themselves." *Journal of the History of Philosophy* 47 (3): 355–82.

Hume, D. 1978. *A Treatise of Human Nature*. 2nd ed. L. A. Selby-Bigge and P. H. Nidditch, eds. Oxford: Clarendon Press.

Ishiguro, H. 1990. *Leibniz's Philosophy of Logic and Language*. Cambridge: Cambridge University Press.

Jolley, N., ed. 1995. *The Cambridge Companion to Leibniz*. Cambridge: Cambridge University Press

Kant, I. 1902–. *Kant's Gesammelte Schriften*. 29 vols. Berlin: G. Reimer, subsequently Walter de Gruyter & Co.

1926. *Kritik der reinen Vernunft*. Raymund Schmidt, ed. Leipzig: F. Meiner.

1965. *The Critique of Pure Reason*. Norman Kemp Smith, trans. New York: St. Martin's Press.

1992. *Theoretical Philosophy, 1755–1770*. D. Walford, trans. and ed. The Cambridge Edition of the Works of Immanuel Kant. New York: Cambridge University Press.

1996. *The Metaphysics of Morals. M. Gregor, trans. and ed. The Cambridge Edition of the Works of Immanuel Kant*. New York: Cambridge University Press.

1997. Kant's Lectures on Metaphyics. *K. Ameriks and S. Naragon, trans. and ed.* The Cambridge Edition of the Works of Kant. New York: Cambridge University Press

1998. *Critique of Pure Reason. P. Guyer and A. W. Wood, trans. The Cambridge Edition of the Works of Immanuel Kant*. New York: Cambridge University Press.

Kästner, A. 1796. *Geschichte der Mathematik*. Göttingen: Johan Georg Rosenbusch.

1800. *Mathematische Anfangsgründe Teil I: Anfangsgründe der Arithmetik: Geometrie, ebenen und sphärischen Trigonometrie, und Perspektiv*. Göttingen: Johan Georg Rosenbusch.

Kitcher, P. 1975. "Kant's Foundations of Mathematics." *Philosophical Review* 84: 23–50.

1982. "How Kant Almost Wrote 'Two Dogmas of Empiricism.'" In *Essays on Kant's Critique of Pure Reason*, J. N. Mohanty and R. W. Shahan, eds. Norman: University of Oklahoma Press.

Klein, J. 1992. *Greek Mathematical Thought and the Origin of Algebra*. New York: Dover.

Krantz, D. Lee, P. Suppes, and A. Tversky. 1971. *Foundations of Measurement*, vol. 1. New York: Academic Press.

Land, T. 2013. "Intuition and Judgment: How Not to Think about the Singularity of Intuition (and the Generality of Concepts) in Kant." In *Kant and Philosophy in a Cosmopolitan Sense*, Claudio La Rocca et al., eds. Berlin: De Gruyter, 221–31.

2014. "Spatial Representation, Magnitude, and the Two Stems of Cognition." *Canadian Journal of Philosophy* 44: 524–50.

2016. "Moderate Conceptualism and Spatial Representation." In Schulting (2016).

Leibniz, G. W. 1956. *Philosophical Papers and Letters*, vols. I and II. L. Loemker, trans. and ed. Boston: Kluwer Publishing.

1961. *Opuscules et fragments inédits de Leibniz: Extraits des manuscrits de la Bibliothèque royale de Hanovre*. L. Couturat, ed. Hildesheim: G. Olms.

1971. *Mathematische Schriften*. Vols. 1–7. C. I. Gerhardt, ed. Hildesheim: Georg Olms.

1989. *Philosophical Papers and Letters*. Vols. 1 and 2. L. Loemker, trans. and ed. Boston: Kluwer Publishing.

1995a. *La caractéristique géométrique*. J. Echeverría and M. Parmentier, eds. and trans. Paris: J. Vrin.

1995b. *Philosophical Writings*. G. H. R. Parkinson, trans. and ed. Rutland, VT: Rowman and Littlefield.

Liang, Chen. 2020. "Form of Intuition and Formal Intuition." PhD dissertation, University of Illinois at Chicago.

Longuenesse, B. 1998. *Kant and the Capacity to Judge: Sensibility and Discursivity in the Transcendental Analytic of the Critique of Pure Reason*. Princeton, NJ: Princeton University Press.

2009. *Kant on the Human Standpoint*. Cambridge: Cambridge University Press.

Mates, B. 1989. *The Philosophy of Leibniz: Metaphysics and Language*. Oxford: Oxford University Press.

McLear, Colin. 2014. "The Kantian (Non)-conceptualism Debate." *Philosophical Compass* 9 (11): 769–90.

2015. "Two Kinds of Unity in the Critique of Pure Reason." *Journal of the History of Philosophy* 53 (1): 79–110.

2020. "Kantian Conceptualism/Nonconceptualism." In *Stanford Encyclopedia of Philosophy*. Accessed September 2020.

Melnick, Arthur. 1973. *Kant's Analogies of Experience*. Chicago: University of Chicago Press.

Michell, J. 2006. "Psychophysics, Intensive Magnitudes, and the Psychometrician's Fallacy." *Studies in History and Philosophy of Biological and Biomedical Sciences* 17: 414–32.Mueller, I. 1970a. "Aristotle on Geometrical Objects." *Archiv für Geschichte der Philosophie* 52: 156–71.

1970b "Homogeneity in Eudoxus' Theory of Proportion." *Archive for History of Exact Sciences* 7: 1–6.

1981. *Philosophy of Mathematics and Deductive Structure in Euclid's Elements*. New York: Dover.

Parsons, C. 1983. *Mathematics in Philosophy*. Ithaca, NY: Cornell University Press.

1983a. "Infinity and Kant's Conception of the "Possibility of Experience." In *Mathematics in Philosophy*.

1983b. "Kant's Philosophy of Arithmetic." In *Mathematics in Philosophy*.

1984. "Arithmetic and the Categories." *Topoi* (3): 109–21.

1992. "The Transcendental Aesthetic." In Guyer (1992).

2012. *From Kant to Husserl: Selected Essays*. Cambridge, MA: Harvard University Press.

Paton, H. J. 1965. *Kant's Metaphysic of Experience: A Commentary on the First Half of the Kritik der reinen Vernunft*. 2 vols. London: Allen & Unwin.

Penelhum, T., and J. MacIntosh, eds. 1969. *The First Critique*. Belmont, CA: Wadsworth.

Petri, B., and N. Schappacher. 2007. "On Arithmetization." *In Goldstein et al.* (2007), pp. 243–374.

Place, U. T. 1956. "Is Consciousness a Brain Process?" *British Journal of Psychology* 47 (1): 44–50.

Popkin, R. 1966. *The Philosophy of the Sixteenth and Seventeenth Centuries*. New York: Free Press.

Posy, C., and Ofra Rechter, eds. 2020. *Kant's Philosophy of Mathematics, vol. 1: The Critical Philosophy and Its Roots*. Cambridge: Cambridge University Press.

Proclus. 1970. *A Commentary on the First Book of Euclid's Elements*. G. Morrow, trans. Princeton, NJ: Princeton University Press.

Rechter, O. 2006. "The View From 1763: Kant on the Arithmetical Method before Intuition." In *Intuition and the Axiomatic Method*, Emily Carson and Renate Huber, eds. Springer, pp. 21–46.

Rosen, G. 2017. "Abstract Objects." In *Stanford Encyclopedia of Philosophy*. Accessed online November 2019.

Rusnock, P., and Rolf George. 1995. "A Last Shot at Kant on Incongruent Counterparts." *Kant-Studien* 86 (3): 257–77.

Russell, B. 1903. *The Principles of Mathematics*. Cambridge: Cambridge University Press.

Rutherford, D. 1995. "Philosophy and Language in Leibniz." In *The Cambridge Companion to Leibniz*, N. Jolley, ed. Cambridge: Cambridge University Press, pp. 224–69.

Ryle, G. 1949. *The Concept of Mind*. Chicago, IL: University of Chicago Press.

Schepers, H. 1966. "Leibniz's Arbeiten zu einer Reformation der Kategorien." *Zeitschrift für philosophische Forschung* 20 (vols. 3 and 4): 539–67.

1969. "Begriffsanalyse und Kategorialsynthese Zur Verflechtung von Logik and Metaphysik bei Leibniz." In *Studia Leibnitiana Supplementa*, Vol. III, Akten des Internationalen Leibniz-Kongresses Hannover, 14–19 November 1966. Wiesbaden: Franz Steiner Verlag: 34–9.

Schulting, D. 2016. *Kantian Nonconceptualism*. London: Palgrave Macmillan.

Sellars, W. 1968. *Science and Metaphysics: Variations on Kantian Themes*. London: Routledge and Kegan Paul.

Shabel, L. 1998. "Kant on the 'Symbolic Construction' of Mathematical Concepts." *Studies in the History and Philosophy of Science* 29: 589–621.

Smit, H. 2000. "Kant on Marks and the Immediacy of Intuition." *The Philosophical Review* 109 (2): 235–66.

Smith, N. K. 1979. *A Commentary to Kant's Critique of Pure Reason*. 2nd ed. London: Macmillan.

Smyth, D. 2014. "Infinity and Givenness: Kant on the Intuitive Origin of Spatial Representation." *Canadian Journal of Philosophy* 44 (5–6): 551–7.

Stein, H. 1990. "Eudoxus and Dedekind: On the Ancient Greek Theory of Ratios and Its Relation to Modern Mathematics." *Synthese* 84: 163–211.

Stevens, S. S. 1946. "On the Theory of Scales and Measurement." *Science* 103: 677–80.

1951. "Mathematics, Measurement, and Psychophysics." In *Handbook of Experimental Psychology*, S. S. Stevens, ed. New York: Wiley, pp. 1–49.

Strawson, P. F. 1989. *The Bounds of Sense: An Essay on Kant's Critique of Pure Reason*. London: Routledge.

Suppes, P.A. 1951. "A Set of Independent Axioms for Extensive Quantities." *Portugaliae Mathematica* 10: 163–72.

Sutherland, D. 2004a. "The Role of Magnitude in Kant's Critical Philosophy." *Canadian Journal of Philosophy* 34(4): 411–42.

2004b. "Kant's Philosophy of Mathematics and the Greek Mathematical Tradition." *Philosophical Review* 113 (2): 157–201.

2005a. "Kant on Fundamental Geometrical Relations." *Archiv für Geschichte der Philosophie* 87 (2): 117–58.

2005b. "The Point of Kant's Axioms of Intuition." *Pacific Philosophical Quarterly* 86: 135–59.

2006. "Kant on Arithmetic, Algebra, and the Theory of Proportions." *Journal of the History of Philosophy* 44 (4): 533–58.

2008. "From Kant to Frege: Numbers, Pure Units, and the Limits of Conceptual Representation." *Royal Institute of Philosophy Supplement* 63.

2010. "Philosophy and Geometrical Practice in Leibniz, Wolff, and the Early Kant." In *Discourse on a New Method: Reinvigorating the Marriage of History and Philosophy of Science*, Michael Dickson and Mary Domski, eds. Chicago: Open Court.

2014. "Kant on Construction and Composition of Motion in the Phoronomy." Special supplement to *Canadian Journal of Philosophy* 44 (5–6): 686–718.

2017. "Kant's Conception of Number." *Philosophical Review* 126 (2).

2020a. "Kant's Philosophy of Arithmetic: A New Approach." *In Posy (2020).*

2020b. "Continuity and Intuition in 18th Century Analysis and in Kant." In *The History of Continua: Philosophical and Mathematical Perspectives*. Oxford: Oxford University Press.

Tait, W. 2005. "Frege versus Cantor and Dedekind: On the Concept of Number." In *The Provenance of Pure Reason*. Oxford: Oxford University Press.

Thompson, Manley 1972. "Singular Terms and Intuitions in Kant's Epistemology." *Review of Metaphysics* 26(2): 314–43.

Vaihinger, H. 1900. "Siebzig textkritische Randglossen zur Analytik." *Kant-Studien* 4: 452–63.

1922. *Kommentar zu Kant's Kritik der reinen Vernunft*. 2 vols. Aalen: Scientia Verlag.

Walsh, W. H. 1975. *Kant's Criticism of Metaphysics*. Edinburgh: Edinburgh University Press.

Warren, D. 2001. *Reality and Impenetrability in Kant's Philosophy of Nature*. New York: Routledge.

Wilson, J. 2017. "Determinables and Determinates." In *Stanford Encyclopedia of Philosophy*, first published February 7, 2017.

Wilson, K. D. 1975. "Kant on Intuition." *Philosophical Quarterly* 25: 247–65.

Wolff, C. 1736. *Philosophia Prima Sive Ontologia*. 3. Auflage. Verona: Dionisio Ramanzini.

1962–2009. *Christian Wolff: Gesammelte Werke*. Hildesheim: G. Olms.

1962. *Philosophia Prima Sive Ontologia*. 1736. In Christian Wolff, *Gesammelte Werke*. II. Abteilung Lateinische Schriften, Band 3.

1965. *Mathematisches Lexikon*. 1716. In Christian Wolff, *Gesammelte Werke*. I. Abteilung Deutsche Schriften, Band 11.

1968. *Elementa Matheseos Universae. 1742. In Christian Wolff, Gesammelte Werke. II. Abteilung Lateinische Schriften, Band 29.*

1973. *Anfangsgründe aller Mathematischen Wissenschaften. Erstausgabe 1710–17. (7. Auflage 1750–7). In Christian Wolff, Gesammelte Werke, I. Abteilung Deutsche Schriften, Band 12.*

2009. *Auszug aus den Anfangs-Gründen aller Mathematischen Wissenschaften, in multiple editions. Auflage 1728. In Christian Wolff,* Gesammelte Werke, *1. Abteilung, Deutsche Schriften, Band 25.*

Wolff, R. P. 1963. *Kant's Theory of Mental Activity*. Cambridge, MA: Harvard University Press.

Wood, A. 1979. *Kant's Rational Theology*. Ithaca, NY: Cornell University Press.

INDEX

Printed by Printforce, United Kingdom